Advances in Thermal-Fluids Engineering (ATFE 2021)

Held online due to COVID-19

IOP Conference Series: Materials Science and Engineering
Volume 1146

Gandhinagar, India
25 - 26 March 2021

ISBN: 978-1-7138-3416-8
ISSN: 1757-8981

Printed from e-media with permission by:

Curran Associates, Inc.
57 Morehouse Lane
Red Hook, NY 12571

Some format issues inherent in the e-media version may also appear in this print version.

Additional copies of this publication are available from:

Curran Associates, Inc.
57 Morehouse Lane
Red Hook, NY 12571 USA
Phone: 845-758-0400
Fax: 845-758-2633
Email: curran@proceedings.com
Web: www.proceedings.com

TABLE OF CONTENTS

Preface

Advances in Thermal-Fluids Engineering (ATFE 2021)

25-26 March 2021, Gandhinagar, India

About the Conference

A two-day conference was organized by Department of Mechanical Engineering, Pandit Deendayal Petroleum University (PDPU), Gandhinagar with the aim to bring together leading academic scientists, researchers and research scholars across the country to share their experiences and research findings in all aspects of Advances in Thermal-Fluids Engineering. It also provides a premier interdisciplinary platform for researchers, academicians and industry personal to present and discuss the most recent innovations, trends, and challenges encountered and the solutions adopted in the field of Thermal-Fluid Engineering.

Motive:

The main aim of organizing this conference was to introduce the participants to new advancements in the field of thermo-fluid engineering and to provide an interdisciplinary platform to put forward their research explorations and findings during the conference in the field of Thermal-Fluid Engineering.

Conference topics include but not limited to:

- Fuels and Combustion
- Computational Methods in Thermal-fluid Systems
- Energy and Sustainability
- Energy-Water-Food Nexus
- Engineering Equipment and Environmental Systems
- Experimental Methods/Tools and Instrumentation in Fluid Mechanics and Heat/mass Transfer
- Flow Instability
- Fluid Mechanics and Rheology of Nonlinear Materials and Complex Fluids
- Multiphase Flow and
- Turbulent Flows
- Advanced Energy Systems
- CFD and Numerical Heat/Mass Transfer
- Energy Conservation and Storage Techniques
- Heat/Mass Transfer Enhancement Techniques
- Heat/Mass Transfer In Renewable and Clean Energy Systems
- Heat/Mass Transfer In Pollutant Emission Control Systems

- Micro/Nano Heat/Mass Transfer
- Air Conditioning and Refrigeration
- Water Desalination and Waste Water Treatment

Event Description:

The preparation of the event was started in the month of September 2020 and the event was proposed by Department of Mechanical Engineering, PDPU that was supported by the Management. Gujarat Council on Science and Technology (GUJCOST) also funded the event. The conference was held on 25th – 26th March, 2021. Around 80 papers were received of which 56 were shortlisted for presentation based on the review of experts from industry and academia. These papers were presented under **Eight Tracks** namely:

1. Experimental and numerical heat transfer
2. Energy and sustainability
3. Fluid flow and CFD
4. Thermal system analysis and optimization
5. IOT and Industry 4.0
6. Fuels and Combustion
7. Heat and fluid flow
8. Water Desalination and Waste Water Treatment

The event started with Saraswati Vandana followed by Welcome address by Conference Convener Dr. Rajesh Patel. This was followed by Speech from Dr. Vishvesh Badheka HoD, Mechanical Engineering Department, PDPU who gave a brief overview of the Department and research activities and facilities. This was followed by Director's address by Prof. Sunil Khanna and DG's Address by Prof. S.S. Manoharan, honorable Director General, PDPU. Message from Guest of Honour Dr. Narottam Sahoo, Advisor & Member Secretary, Gujarat Council on Science and Technology, Govt. of Gujarat was delivered. This was followed by the address of the Chief Guest Mr. Prashant Pathak, President: ISHRAE, Ahmedabad. The inauguration ceremony was concluded with Vote of thanks from the conference Co-convener Dr. Vivek Patel.

There were four Keynote Speeches. The first keynote speech was delivered by Prof. PMV Subbarao, Professor, Mechanical Engineering Department, IIT Delhi on topic- "Analysis of Fluid Flow in Micro-scale Channels". This was followed by two parallel sessions on Track 1 and Track 2. The second half on Day 1 started with Keynote Speech by Prof. MV Rane Professor, Mechanical Engineering Department, IIT Bombay. His topic was "Rotating Contacting Disks Enabling Efficiency Improvements: In Cooling Towers, Evaporative Coolers

& Condensers, Waste Water Vaporizers, Biogas Scrubbers, etc." This was followed by two parallel sessions on Track 3 and Track 4.

Day 2 began with the Keynote address by Prof. Santanu Bandyopadhya, Professor, Department of Energy Science and Engineering, IIT Bombay, on topic "Pinch Analysis for Heat Integrated Water Allocation Networks." This was followed by two parallel sessions on Track 5 and Track 6. The second half on Day 2 started with Keynote address by Prof. Philip Davis, Professor, University of Birmingham, Edgbaston, Birmingham, UK on "Batch Reverse Osmosis for Energy Efficiency in High Recovery Desalination."

Event Overview:

This two-day conference was a big success with lot of information shared through this platform. The participants thanked the organizers for knowledge transfer that took place during the two days of the event. The event finally concluded by Vote of Thanks by Conference Convener, Dr. Rajesh Patel.

- **Reasons why the meeting was changed to a virtual format:**
  The conference Advances in Thermal-Fluids Engineering (ATFE 2021) was announced in the month of September 2020. In the beginning it was very difficult for anyone to predict the severity of Covid-19 Corona virus infection and from the day one we were open for all the possible format for organizing ATFE 2021 i.e. offline, hybrid (online and offline) and online. However, In January 2021, the Covid-19 pandemic situation was not favourable to conduct ATFE 2021 meeting offline as spreading of corona virus was very fast and number of active cases in India was around half million. Looking at the response to the ATFE 2021 in terms of submission of research articles and request of all the participant, we have decided to conduct the ATFE 2021 through virtual format.

- **Why the meeting was virtual instead of being postponed:**
  Since, announcement of conference was take place in September 2020. Postponed of the conference demoralised the researchers as they already submit their work to the conference. Further, the postponed of the conference adversely affect the research community as the research work reach to them with delay. Such delay also affects the on-going work of research community. Considering these aspects, the conference was conducted as per schedule through virtual mode instead of being postponed.

- **Virtual Conference Dates:**
  25th – 26th March, 2021

- **Location of organisers:**
  Pandit Deendayal Petroleum University, Gandhinagar, Gujarat, India

- **Conference model: Information on plenary sessions. Amount of time that authors/presenters were given to talk:**
  Eight sessions/track were arranged during two days (Please see the ATFE-2021 Schedule document and also available at http://sot.pdpu.ac.in/atfe.html). Each session/track focused on the particular theme. Authors are given to 10 minutes time for presentation followed by 5 minutes Q and A. A session chair and co-chair along with moderator manage the Q and A session. The best paper of each session was also declared.

- **Information on discussions and Q&A sessions:**
  Authors are given to 10 minutes time for presentation followed by 5 minutes Q and A. A session chair and co-chair along with moderator manage the Q and A session. The best paper of each session was also declared.

- **Location of participants and overall participant numbers:**
  Prominent institutes like IITs, NITs, and other state and deemed universities across the India. More than 150 Participant attend the conference.

- **Technology that was used to deliver the conference:**
  Microsoft-Team (MS-Team)

- **Any reports on overall technical quality and success of delivery of the conference.**
  This two-day conference was a big success with lot of information shared through this platform. The participants thanked the organizers for knowledge transfer that took place during the two days of the event.

IOP Conf. Series: Materials Science and Engineering 1146 (2021) 011001 doi:10.1088/1757-899X/1146/1/011001

List of Committees and Editors

Patrons

Prof. Sundar Manoharan
Director General,
Pandit Deendayal Petroleum University.

Prof. Sunil Khanna
Director, School of Technology,
Pandit Deendayal Petroleum University.

Prof. Tarun Shah
Registrar,
Pandit Deendayal Petroleum University.

Editors/Conveners:

Dr. Rajesh Patel
Associate Professor,
Mechanical Engineering Department,
Pandit Deendayal Petroleum University.

Dr. Vivek Patel
Associate Professor,
Mechanical Engineering Department,
Pandit Deendayal Petroleum University.

Dr. Harshal Oza
Associate Professor,
Mechanical Engineering Department,
Pandit Deendayal Petroleum University.

Advisory Committee

Prof. Dipak Jain
Jio University, Mumbai, India.

Prof. A. R. Yildiz
Uludağ University, Turkey.

Prof. Philip Davies
University of Birmingham, UK.

Dr. Guillermo Zaragoza
Centre for Energy, Environment &
Technology Research, Spain.
Dr. Vimal Savsani

Prof. Ziad Sagir
Ryerson University, Canada.

Dr. Seshasai Srinivasan
McMaster University, Ontario, Canada.

Prof. Chao Zhu
New Jersey Institute of Technology, USA.

Prof. Santanu Bandyopadhyay
Indian Institute of Technology,
Bombay, India.
Dr. Nirajan Dhakal

IOP Conf. Series: Materials Science and Engineering 1146 (2021) 011001 doi:10.1088/1757-899X/1146/1/011001

Candore College, North Bay, Canada.

IHE Delft, Netherlands

Prof. PMV Subbarao
Indian Institute of Technology,
Delhi, India.

Prof. M.V. Rane
Indian Institute of Technology,
Bombay, India.

Prof. R.K. Shrivastva
G.B.Pant University, Uttarakhand, India.

Dr. Suphiya Khan
Bansthali University, Rajasthan, India.

Prof. T. P. Singh
Pandit Deendayal Petroleum University,
India.

Prof. S. S. Kachhwaha
Pandit Deendayal Petroleum University,
India.

Prof. Vishvesh Badheka
Pandit Deendayal Petroleum University,
India.

Organization Committee Members:

Dr. Anurag Mudgal
Associate Professor,
Mechanical Engineering Department,
Pandit Deendayal Petroleum University.

Dr. Rakesh Chaudhari
Assistant Professor,
Mechanical Engineering Department,
Pandit Deendayal Petroleum University.

Dr. Ramesh Kumar Guduru
Associate Professor,
Mechanical Engineering Department,
Pandit Deendayal Petroleum University.

Dr. Ravi Kant
Assistant Professor,
Mechanical Engineering Department,
Pandit Deendayal Petroleum University.

Dr. Jatin Patel
Associate Professor,
Mechanical Engineering Department,
Pandit Deendayal Petroleum University.

Dr. Simran Jeet Singh
Assistant Professor,
Mechanical Engineering Department,
Pandit Deendayal Petroleum University.

Dr. Garlapati Nagababu
Assistant Professor,
Mechanical Engineering Department,
Pandit Deendayal Petroleum University.

Dr. Vinay Vakharia
Assistant Professor,
Mechanical Engineering Department,
Pandit Deendayal Petroleum University.

Dr. Jaydeep Patel
Assistant Professor,
Mechanical Engineering Department,
Pandit Deendayal Petroleum University.

Dr. Vivek Kumar
Assistant Professor,
Mechanical Engineering Department,
Pandit Deendayal Petroleum University.

Dr. Jay Vora
Assistant Professor,
Mechanical Engineering Department,
Pandit Deendayal Petroleum University.

Dr. Krunal Mehta
Assistant Professor,
Mechanical Engineering Department,
Pandit Deendayal Petroleum University.

Dr. Nirav Patel

Mr. Parth Prajapati

Assistant Professor,
Mechanical Engineering Department,
Pandit Deendayal Petroleum University.

Dr. Pankaj Sahlot
Assistant Professor,
Mechanical Engineering Department,
Pandit Deendayal Petroleum University.

Dr. Pavan Kumar Gurrala
Assistant Professor,
Mechanical Engineering Department,
Pandit Deendayal Petroleum University.

Dr. Rajat Saxena
Assistant Professor,
Mechanical Engineering Department,
Pandit Deendayal Petroleum University.

Assistant Professor,
Mechanical Engineering Department,
Pandit Deendayal Petroleum University.

Mr. Abhinaya Srinivas
Assistant Professor,
Mechanical Engineering Department,
Pandit Deendayal Petroleum University.

Mr. Rahul Deharkar
Assistant Professor,
Mechanical Engineering Department,
Pandit Deendayal Petroleum University.

Dr. Anirudh Kulkarni
Assistant Professor,
Mechanical Engineering Department,
Pandit Deendayal Petroleum University.

Key note Speakers

Prof. Philip Davies
University of Birmingham, Edgbaston, Birmingham, UK

Prof. PMV Subbarao
Indian Institute of Technology, Delhi, India

Prof. M.V. Rane
Indian Institute of Technology, Bombay, India

Prof. Santanu Bandyopadhyay
Indian Institute of Technology, Bombay, India

IOP Conf. Series: Materials Science and Engineering 1146 (2021) 011002 doi:10.1088/1757-899X/1146/1/011002

Peer review declaration

All papers published in this volume of *IOP Conference Series: Materials Science and Engineering* have been peer reviewed through processes administered by the Editors. Reviews were conducted by expert referees to the professional and scientific standards expected of a proceedings journal published by IOP Publishing.

- **Type of peer review:**
  The peer review process was Single-blind. Each paper sent to two Reviewers for the review process.

- **Conference submission management system: EASYCHAIR**

- **Number of submissions received: 77**

- **Number of submissions sent for review: 63**

- **Number of submissions accepted: 56**

- **Acceptance Rate (Number of Submissions Accepted / Number of Submissions Received X 100): 72%**

- **Average number of reviews per paper: 02**

- **Total number of reviewers involved: 30**

- **Any additional info on review process:** The peer review process was Single-blind. Each paper sent to two Reviewers for the review process. Before review, paper was under go for plagiarism check. Turntin is used for plagiarism check. Paper below 15% of similarity sent for the peer review. The manuscript having plagiarism more than 15% was sent back to authors to reduce the plagiarism. The peer review process was performed with the following criteria.

  1. Technical Criteria such as scientific merit, clarity of expression, communication of ideas, readability and discussion of concepts etc.
  2. Quality Criteria Such as originality of the work, demonstration of scientific interest of the results etc.
  3. Presentation Criteria such as Title of the manuscript is appropriate for the content of the article, Information provided in Abstract, Quality of Diagram Figure, Table etc. Conclusion of the present work etc.

- **Contact person for queries: Dr Vivek K. Patel (vivekp@sot.pdpu.ac.in)**

A comparative study of effect of heater orientation on flow and heat transfer characteristics in buoyancy driven flow in cryogenic liquid

S S Ade and S K Rathore

Mechanical Engineering Department, National Institute of Technology Rourkela, Rourkela, Odisha, India 769008

E-mail: somesh20197@gmail.com; isushilrathore@gmail.com;

Abstract. The present study aims to analyse the buoyancy driven fluid flow and heat transfer characteristics using different heater orientation in liquid nitrogen for cooling superconductor. For this purpose, a localized circular heater (analogous to superconductor) generating heat at constant rate submerged in liquid nitrogen enclosure is analysed. The study is performed with two orientations of heater; one is vertical and another is horizontal. The entropy generation associated with heat transfer and fluid flow is also discussed in the present study. The Rayleigh number, average Nusselt number, maximum vertical velocity, entropy generation and Bejan number are used as comparing parameters for two orientations of heater. Under same heat generation rate, Rayleigh number is comparatively higher for vertically oriented heater whereas average Nusselt number is comparatively higher for horizontally oriented heater. The maximum vertical velocity of fluid is higher when heater is vertically oriented. The horizontally oriented heater has comparatively less value of thermal entropy generation than vertically oriented heater but horizontally oriented heater has higher value of fluid friction irreversibility. In summary, horizontally oriented heater has comparatively higher value of total entropy generation than vertically oriented heater. Bejan number is higher for horizontally oriented heater. In overall study, the value of Bejan number is less than 0.5; therefore, it can be said that fluid friction irreversibility is more dominated than heat transfer irreversibility.

1. Introduction

Superconductors are used in many applications like superconducting transformers, motors, generators, magnetic bearings, MRI and NMR machine, etc. Cryogenic fluid performs important role in convective cooling of a superconductor so that superconductor can work efficiently; but at the same time orientation of superconductor can also affect the heat transfer from its surface and flow characteristics of cryogenic liquid during convective cooling. Therefore, it is important to have an idea of how the orientation of superconductor affects the heat transfer and flow characteristics in cryogenic liquid during natural convective cooling.

Because of immersion of a localized heater, the fluid flow caused in a liquid is studied by Satpathy et al. [1]. They have noticed a characteristic 3D flow and two convective cells which are symmetric in nature. Further, they continued their study by comparing different silicon oils with liquid nitrogen for different heating conditions. Dubois et al. [2] experimentally presented the convective heat transfer process in laminar regime caused by local heater surrounded by liquid. In their study, they depicted that the motion of fluid caused by the Rayleigh-Bénard instability. They observed two revolving cells which were rotating in a counter-clockwise manner in the experimental chamber. In their investigation, they compared the data obtained from experiments conducted at room temperature for different gradients of temperature in nitrogen liquid accompanying silicon oils of various viscosities. Satpathy et al. [3] have presented a numerical analysis on cooling through natural convection in a fluid operating at cryogenic temperature. The influence of the ratio of width to the height of heater on the rate of heat transfer has been investigated in terms of Nusselt number. They have resolved that the rate of heat transfer will be more efficient when heater length is smaller. A complete study about transient natural convection and flow of fluid within an inclosure for a heat producing and heat carrying body is presented numerically by Ha et al. [4]. They have concluded that flow of fluid in the enclosure increases with increase in Rayleigh number, thereby increasing the rate of heat transfer. They have also concluded that, the shape of transient streamlines and isotherms in the enclosure is influenced by different Rayleigh number. The matter of entropy generation in rectangular cavities because of natural convective heat transfer is presented by Oliveski et al. [5]. They have concluded that the problem of entropy generation because of viscous effects increases when Rayleigh number increases under same ratio of the width to the height. They have also concluded that as the Rayleigh number increases, the total generation of entropy also increases in a linear manner. Problem of entropy generation due to natural convective heat transfer in the enclosures of right-angled triangular shapes has been studied numerically by Basak et al. [6]. They have observed that when Rayleigh number is low, the transfer of heat is mainly because of conduction and when the vertex angle at top is 15^0 the higher value of entropy generation is detected. Also, for increase of total irreversibility, the contribution of fluid friction entropy generation is higher. Ngo and Byon [7] presented a study on natural convective heat transfer inside the square cavity using finite element method. They analyzed the influence of size of heater and location of heater so that internal heat transfer due to free convection can be improved. A review on natural convective heat transfer in non-square enclosures is presented by Das et al. [8]. Sarris et al. [9] have carried out a 2D and 3D numerical study on influence of heated strips oriented horizontally at the lower wall of melting gas. In their study, the parameters that are influencing the rate of heat transfer are Rayleigh number, aspect ratio of tank and width of heated strip. They have concluded that when width of strip and aspect ratio of tank increases, fluid flow and heat transfer increases. Study on natural convective heat transfer inside an enclosure of rectangular shape filled with air was presented numerically by Cianfrini et al. [10]. In their study the bottom side of enclosure was separately heated and one side was cooled. Their study was presented with various values of the Rayleigh number, the aspect ratio of the enclosure, and size of heater. They had noticed a single roll cell in the flow structure. They had concluded that when size of heater and non-dimensional Rayleigh number increases, the effectiveness of heat removal from heater also increases. Koizumi [11] has presented an experimental and numerical study on effects of Grashof number and height of container in a rectangular container locally heated from below. Author has mentioned that thermal boundary conditions of the bottom wall can significantly influence the pattern of flow and process of transition to chaos. The author has experimentally found that when Grashof number is equal to 1800000, 3-dimensional form of roll cell changes from time dependent periodic motion to chaotic fluid flow directly. In most of the studies related to natural convection, appreciable care has been given to Rayleigh-Bénard convection from vertical enclosure or horizontal enclosure clearly and explicitly stated either with constant temperature or heat flux condition [12]; whereas, Satpathy et al. [1,3] and Dubois et al. [3] performed study which is distinctly separate from Rayleigh-Bénard convection.

As the orientation of heater (analogous to superconductor) is changed; the heat transfer characteristics of fluid flow and entropy generation in the domain will also change. It is not reported in the literature. Therefore, effect of horizontal and vertical orientation of heater on characteristics of fluid flow and heat transfer in buoyancy driven flow is presented in the present paper.

2. Problem Definition

IOP Conf. Series: Materials Science and Engineering 1146 (2021) 012001 doi:10.1088/1757-899X/1146/1/012001

For vertical orientation of heater, the 3D domain is symmetrical about *x*-axis; whereas for horizontal orientation of heater, the 3D domain is symmetrical about *y*-axis. Therefore, one-half portion of the domain is simulated to save the computational time as demonstrated in Fig. 1. The dimensions of domain which is used for computation are 100 mm (length) along *x*-axis, 100 mm (height) along y-axis and 30 mm (width) along *z*-axis. The length of heater is 50 mm and material used for heater is copper. For vertical orientation of heater, the distance between top surface of enclosure and top surface of heater is 50 mm whereas for horizontal orientation of heater, the distance between top surface of enclosure and central axis of heater is 52.5 mm. Height above the heater surface (h = 50mm) is taken as characteristics length. The diameter of heater is taken as 5mm.

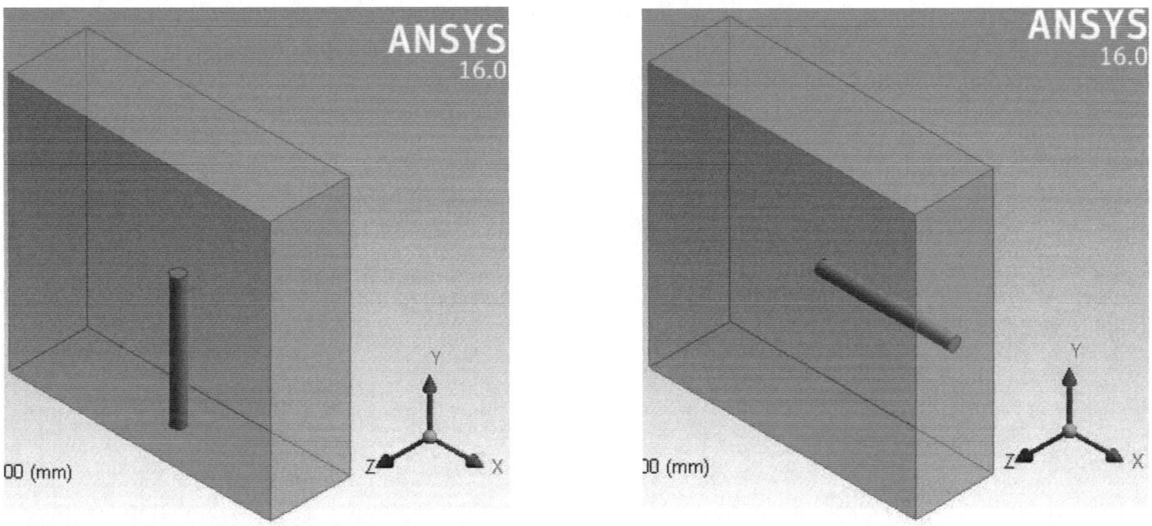

Figure 1. Computational domain for vertical orientation of heater (left) and computational domain for horizontal orientation of heater (right)

3. Governing equations

1) Governing differential equations for flow of fluid and transfer of heat in 3-Dimensional, steady state and laminar form are as follows;

a) Mass conservation equation

$$\frac{\partial u}{\partial x}+\frac{\partial v}{\partial y}+\frac{\partial w}{\partial z}=0 \tag{1a}$$

b) Momentum conservation equations

➢ Momentum conservation equation in *x*-direction

$$u\frac{\partial u}{\partial x}+v\frac{\partial u}{\partial y}+w\frac{\partial u}{\partial z}=-\frac{1}{\rho}\frac{\partial p}{\partial x}+\nu\left(\frac{\partial^2 u}{\partial x^2}+\frac{\partial^2 u}{\partial y^2}+\frac{\partial^2 u}{\partial z^2}\right) \tag{1b}$$

➢ Momentum conservation equation in *y*-direction

$$u\frac{\partial v}{\partial x}+v\frac{\partial v}{\partial y}+w\frac{\partial v}{\partial z}=-\frac{1}{\rho}\frac{\partial p}{\partial y}+\nu\left(\frac{\partial^2 v}{\partial x^2}+\frac{\partial^2 v}{\partial y^2}+\frac{\partial^2 v}{\partial z^2}\right)+g\beta(T-T_0) \tag{1c}$$

➢ Momentum conservation equation in *z*-direction

$$u\frac{\partial w}{\partial x}+v\frac{\partial w}{\partial y}+w\frac{\partial w}{\partial z}=-\frac{1}{\rho}\frac{\partial p}{\partial z}+\nu\left(\frac{\partial^2 w}{\partial x^2}+\frac{\partial^2 w}{\partial y^2}+\frac{\partial^2 w}{\partial z^2}\right) \quad (1d)$$

c) Energy conservation equation

$$u\frac{\partial T}{\partial x}+v\frac{\partial T}{\partial y}+w\frac{\partial T}{\partial z}=\alpha\left(\frac{\partial^2 T}{\partial x^2}+\frac{\partial^2 T}{\partial y^2}+\frac{\partial^2 T}{\partial z^2}\right)+\frac{Q_g}{\rho C_p} \quad (1e)$$

Here u, v, and w are the velocity components in x, y and z directions, respectively. The symbols T and T_0 stand for localized temperature and operating temperature (i.e., temperature of surrounding fluid) respectively. Furthermore, ρ is the density of liquid (kg/m^3) at temperature T_0 and p is the pressure (N/m^2), Q_g (W/m^3) is heat generation rate and g is the acceleration due to gravity (m/s^2) in negative y-direction. Moreover, α, ν and β are thermal diffusivity (m^2/s) of fluid, kinematic viscosity (m^2/s) of fluid and thermal expansion coefficient (1/K) of fluid respectively.

2) Governing equations for entropy generation are as follows

In buoyancy driven or free convection flows, entropy is generated because of presence of temperature gradients (i.e., due to heat transfer) and velocity gradients (i.e., due to fluid flow friction) [13].

$$S_T = S_{HT} + S_{FF} \quad (2a)$$

$$S_{HT} = \int_v S_h \, dv \quad (2b)$$

$$S_{FF} = \int_v S_f \, dv \quad (2c)$$

Where, S_{HT} is the generation of entropy (W/K) in the domain due to transfer of heat, S_{FF} is the generation of entropy (W/K) in the domain due to fluid flow friction and S_T is the total entropy generation (W/K) in the domain. Furthermore, S_h and S_f are localized generation of entropy due to transfer of heat and localized generation of entropy due to fluid flow friction respectively. Moreover, K_f (W/m-K) and μ (kg/m-s) are thermal conductivity and dynamic viscosity of the fluid respectively.

$$S_h = \frac{K_f}{T_0^2}\left[\left(\frac{\partial T}{\partial x}\right)^2+\left(\frac{\partial T}{\partial y}\right)^2+\left(\frac{\partial T}{\partial z}\right)^2\right] \quad (2d)$$

$$S_f = \frac{2\mu}{T_0}\left[\left(\frac{\partial u}{\partial x}\right)^2+\left(\frac{\partial v}{\partial y}\right)^2+\left(\frac{\partial w}{\partial z}\right)^2\right]+\left(\frac{\partial u}{\partial x}+\frac{\partial v}{\partial y}\right)^2+\left(\frac{\partial v}{\partial y}+\frac{\partial w}{\partial z}\right)^2+\left(\frac{\partial w}{\partial z}+\frac{\partial u}{\partial x}\right)^2 \quad (2e)$$

Bejan number (Be)

Bejan number is the ratio of generation of entropy due to transfer of heat to the total generation of entropy.

$$B_e = \frac{S_{HT}}{S_T} = \frac{S_{HT}}{S_{HT}+S_{FF}} \quad (3)$$

$If\ B_e > 0.5$, then irreversibility due to transfer of heat will be higher.
$If\ B_e < 0.5$, then fluid friction irreversibility will be higher.

4. Boundary conditions and solution methodology

For horizontal orientation of heater: Enclosure top surface is kept at a constant temperature T_0 = 75 K and free slip velocity (τ_w = 0) condition is applied. Zero heat flux and no slip velocity condition are applied to bottom wall, left wall and side walls of the enclosure. A constant heat generation rate (Q_g in W/m^3) is applied for heater. Symmetry boundary condition is applied on right wall attached to heater.

For vertical orientation of heater: Enclosure top surface is kept at a constant temperature $T_0 = 75$ K and free slip velocity ($\tau_w = 0$) condition is applied. Zero heat flux and no slip velocity condition are applied to right wall, left wall and side walls of the enclosure. A constant heat generation rate (Q_g in W/m^3) is applied for heater. Symmetry boundary condition is applied on bottom wall attached to heater.

The equations of mass, momentum and energy conservation are solved using finite-volume method with the help of commercial Ansys Fluent software. The coupling between pressure and velocity is obtained by the SIMPLE algorithm [14]. To discretize the momentum and energy equations, 1st order upwind scheme has been used. The local entropy generation equations are solved using custom field function in Ansys Fluent and the generation of total entropy is calculated through the integration of the localized generation of entropy in a computational domain. The convergence criteria for mass and momentum conservation equation are taken as 10^{-4}, whereas for energy conservation, it is taken as 10^{-9}. The operating temperature is taken as 75K.

5. Validation study and grid independence test

In the present study, the effect of heater orientation on flow characteristics of fluid and heat transfer in a buoyancy driven flow is discussed. The computational results of Satpathy et al. [1] and experimental results of Dubois et al. [2] are used to validate the present numerical results. So, we compared distribution of normalized velocity and non-dimensional temperature on the height over the surface of heater with [1] and [2]. In general, satisfactory agreement has been established between the existing simulations and the benchmark results as shown in Figure 2.

Study of grid independence is executed with three mesh sizes viz. Mesh C1, Mesh C2, Mesh C3 as shown in Table 1. It is noticed that comparative changes in the computed temperature difference ΔT ($T_H - T_0$) for vertical orientation and horizontal orientation of heater are 0.20% and 1% respectively, whereas comparative changes in computed maximum velocity of fluid ($Vmax$) for vertical orientation and horizontal orientation of heater are 2% and 0.5% respectively when moved from Mesh C2 to Mesh C3. Therefore, Mesh C2 is selected for the existing simulation in order to preserve computational cost.

Table 1. Grid independence study for vertical and horizontal orientation of heater generating heat (Qg) at a rate of 19978W/m^3

Orientation	Mesh	Number of control volumes	ΔT (K)	$Vmax$ (m/s)
	C1	6,76,101	0.1401 (75.1249-75)	0.0150
Vertical	C2	11,56,529	0.1507 (75.1507-75)	0.0165
	C3	15,60,324	0.1510 (75.1510-75)	0.0168
	C1	6,75,204	0.1092 (75.0958-75)	0.0058
Horizontal	C2	11,56,204	0.1266 (75.1266-75)	0.0064
	C3	15,52,803	0.1278 (75.1278-75)	0.006432

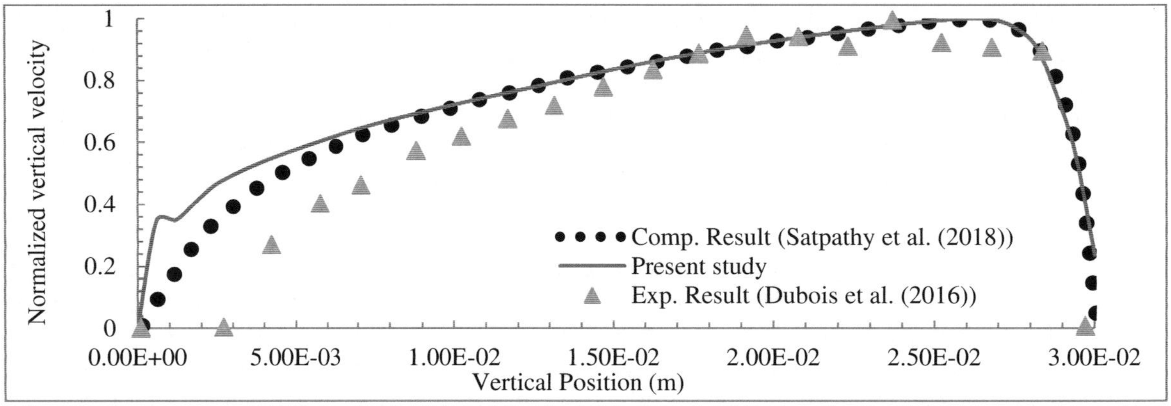

Figure 2. Normalized velocity on the height over the surface of heater

6. Dimensionless numbers

Average Nusselt number (Nu_{Avg}): It is the ratio between convective transfer of heat and the conductive heat transfer.

$$Nu_{Avg} = \frac{q_{Avg}\, h}{(T_h - T_b)\, K_f} \tag{4}$$

Rayleigh Number (Ra): It is the ratio between buoyancy force and inertia force. Higher the Rayleigh number, more will be the dominance of buoyancy force.

$$Ra = \frac{g\beta(T_h - T_b)h^3}{\nu\alpha} \tag{5}$$

7. Results and discussion

Fig. 3 (left) shows the velocity vector above heater surface for vertical orientation of heater. In Fig. 3 (left) we can see that, the fluid in contact with heater surface, gets hot, gets lighter and move towards the top surface of enclosure because of buoyancy action. When fluid reaches the top surface of enclosure, buoyancy effect ends. But because of inertia effect in fluid, the fluid gets diverted along both sides of heater and arrive at their respective top corners of enclosure. Now because of gravity action, fluid starts moving downward and hence in this way two symmetric rotating structures are formed in the fluid domain. The maximum vertical velocity of fluid when heater is vertically oriented is 0.0165 m/s. Figure 3 (right) shows the velocity vector above heater surface for horizontal orientation of heater. The fluid in contact with heater gets hot, gets lighter, moves upward, reaches top surface of enclosure. In this case also, the fluid gets diverted along both sides of heater, arrive at their respective top corners and due to gravity action again starts moving downward. But the rotating structures that are formed are not symmetric; also, their size and location is not same. One rotating structure is at left side and two rotating structures at right side of heater. The maximum vertical velocity of fluid when heater is horizontally oriented is 0.0064m/s.

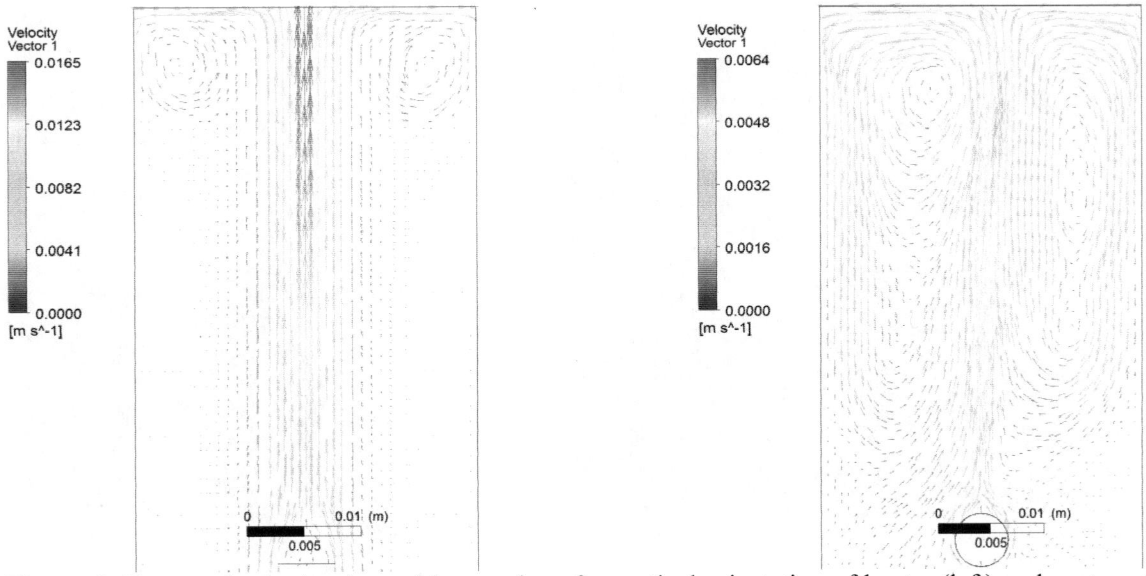

Figure 3. Vector of velocity showed in *y-z* plane for vertical orientation of heater (left) and vector of velocity in *y-z* plane for horizontal orientation of heater (right)

Figure 4 shows temperature contour in the domain when heater is vertically oriented. From Fig. 4 it is observed that the fluid portion which is in-line with vertical heater is more influenced by heat generated in heater (shown by yellow and green plume). The portion of fluid which is just below the top surface of enclosure is also affected by heat generation in heater. From Fig. 4 it is also clear that, fluid temperature which is not in-line with heater and away from top surface of enclosure is not increased i.e., it is not affected by heat generated in heater (shown by blue plume). The maximum temperature in the domain when heater is vertically oriented is 75.1507K. Figure 5 shows temperature contour in the domain when the heater is horizontally oriented. Just above the heater surface, the fluid has uniform temperature along heater length (shown by continuous red and yellow plume). The temperature is not uniform above and away from heater surface. The fluid below heater is comparatively less affected by heat and hence it has lower temperature. The maximum temperature in the domain when heater is horizontally oriented is 75.1266 K.

Figure 6 shows contour for local Bejan number distribution in the domain when heater is vertically and horizontally oriented. In the solid heater, the value of Bejan number is equal to unity (shown by red colour) and hence zero fluid friction irreversibility in the heater. In the fluid domain (i.e., outside the heater), irreversibility due to transfer of heat and fluid flow friction both comes into picture.

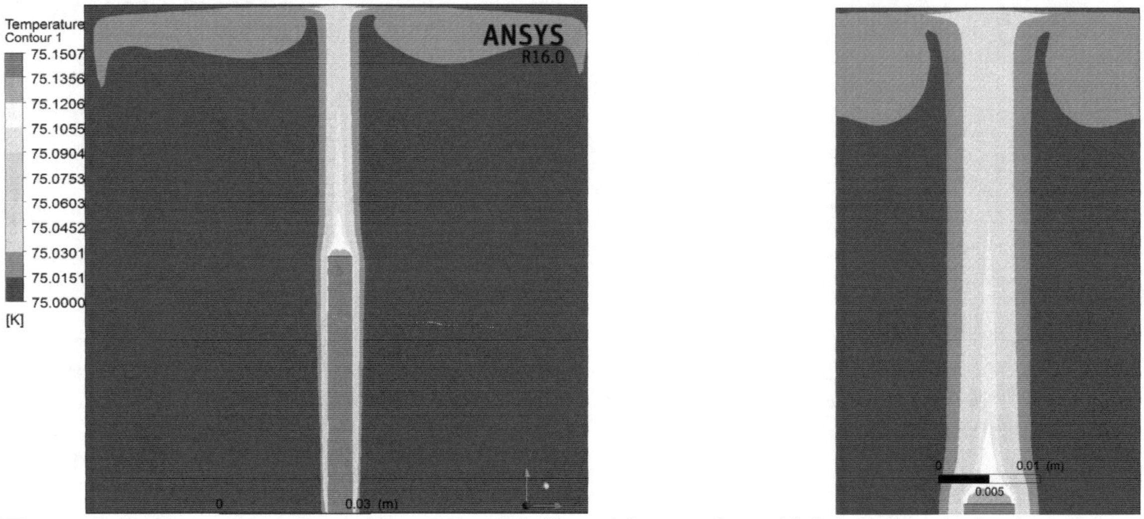

Figure 4. Temperature contour in *x*-*y* plane (left) and in *y*-z plane (right) for vertical orientation of heater

The blue plume designates dominance of irreversibility due to fluid friction; whereas red plume designates dominance of irreversibility due to transfer of heat. On the other hand, yellow plume indicates equal dominance of irreversibility due to transfer of heat and fluid flow friction. Local Bejan number effect is not significant for horizontal orientation of heater because it has lower heat transfer irreversibility. The computed values of average velocity of fluid (V_{Avg}), non-dimensional Rayleigh number (Ra), non-dimensional average Nusselt number (Nu_{Avg}), total generation of entropy or total irreversibility (S) and Bejan number (Be) are shown in Table 2. Difference between average temperature of heater surface and bulk mean temperature of fluid ($T_h - T_b$) is higher when the heater is vertically oriented. Therefore, Ra number is higher (i.e., dominance of buoyancy force in the fluid domain will be higher) when heater is vertically oriented. The average velocity V_{Avg} in fluid domain is higher when heater is horizontally oriented. Difference between average temperature of heater surface and bulk mean temperature of fluid ($T_h - T_b$) is lower when heater is horizontally oriented and hence horizontally oriented heater has higher value of average Nusselt number on its surface. Temperature gradients in the domain are higher when heater is vertically oriented. Therefore, heat transfer irreversibility and hence Bejan number are higher when heater is vertically oriented. Velocity gradients in the domain are higher when heater is horizontally oriented, which results in comparatively higher irreversibility due to fluid friction and lower Bejan number.

Table 2. Analysis on heater orientation

Orientation	$(T_h - T_b)$ K	V_{Avg} (m/s)	Ra	Nu_{Avg}	S_T (W/K)	Be
Vertical	0.13553	0.000508	51935519.53	60.1851	0.009743	0.14820
Horizontal	0.10377	0.000859	39764988.28	78.6055	0.041486	0.02979

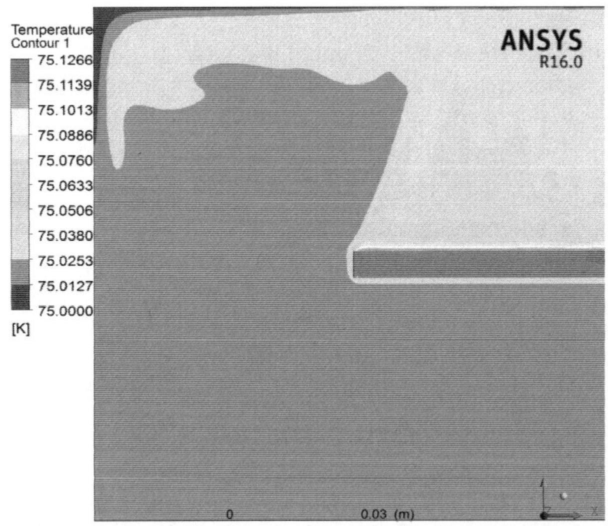

Figure 5. Temperature contour in x-y plane for horizontal orientation of heater

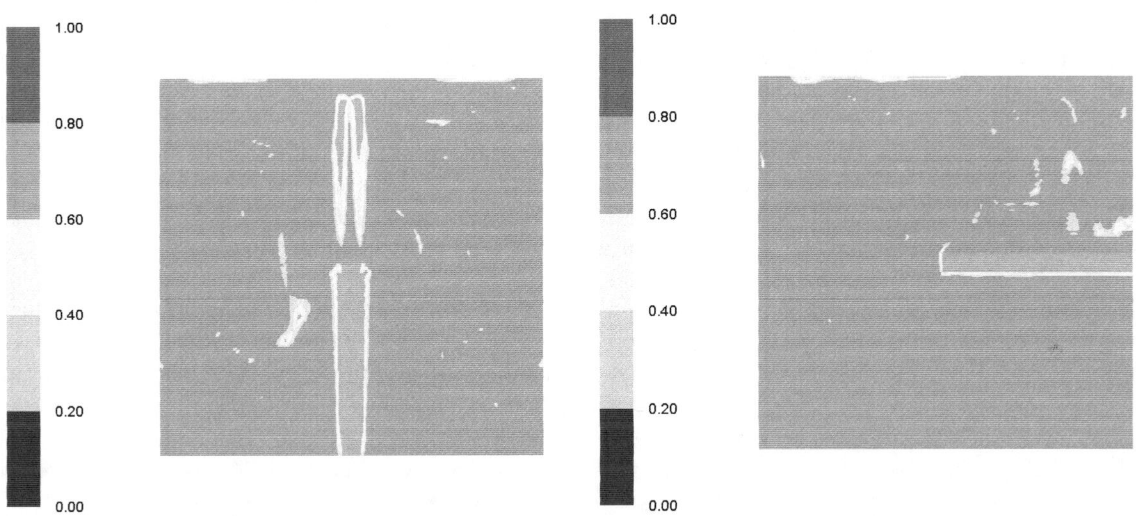

Figure 6. Local Bejan number distribution in x-y plane for vertical orientation of heater (left) and local Bejan number distribution in x-y plane for horizontal orientation of heater (right)

8. Conclusion

3D computational study on effect of heater orientation on flow characteristics of fluid and heat transfer in a buoyancy driven flow using cryogenic liquid is presented. Two symmetric rotating structures of approximately same size are formed in the fluid domain when heater is vertically oriented; whereas, non-symmetric rotating structures of non-identical size are formed in the fluid domain when heater is horizontally oriented. The maximum velocity of the fluid is found to be decreased when heater orientation is changed from vertical to horizontal. When heater is vertically oriented, only some portion of fluid is affected by heat generation in heater; whereas, when heater is horizontally oriented, complete portion of fluid is influenced by heat generation in heater. Rayleigh number (dominance of buoyancy force) is higher in the fluid domain when heater is vertically oriented, because difference between average temperature of heater surface and bulk mean temperature of fluid is higher for vertical orientation of heater. Higher average velocity of the fluid is obtained when heater orientation is changed from vertical to horizontal. Velocity gradients in the domain are higher when heater is horizontally oriented and hence fluid friction irreversibility in the domain is higher (i.e., lower *Be*). Temperature gradients in the domain are higher when heater is vertically oriented and hence heat transfer irreversibility is higher (i.e., higher *Be*). Difference between average temperature of heater surface and bulk mean temperature of fluid is lower when heater is horizontally oriented. Therefore, Nu_{Avg} on heater surface is higher when heater is horizontally oriented. The vertical orientation of heater involves lower total entropy generation in the domain.

Nomenclature

T_b Bulk mean temperature of fluid
T_H Maximum temperature in the domain
h Characteristic length
Ra Rayleigh number
q_{Avg} Average heat flux from heater surface (W/m^2)
Be Bejan Number
S_T Total entropy generation
T_h Average temperature of heater surface
V_{Avg} Average velocity of fluid
Nu_{Avg} Average Nusselt number
K_f Thermal conductivity of fluid (W/m K)

References

[1] Satpathy K, Duchesne A, Dubois C, Fagnard J-F, Caps H, Vanderbemden P and Vanderheyden B 2018 Study of Buoyancy driven heat transport in silicone oils and in liquid nitrogen in view of cooling applications *Int. J. Heat Mass Transf.* **118** 538-550
[2] Dubois C, Duchesne A, Vanderheyden B, Vanderbemden P and Caps H 2016 Locally induced laminar convection in liquid nitrogen and silicone oils *Eur. Phys. J. E* **39** 1-8
[3] Satpathy K, Dubois C, Duchesne A, Fagnard J-F, Caps H, Vanderbemden P and Vanderheyden B 2017 Studies convective cooling of cryogenic fluids towards superconducting applications *WIT TRANS ENG SCI* **118** 95-106
[4] Ha M Y, Jung M J and Kim Y S 1999 Numerical study on transient heat transfer and fluid flow of natural convection in an enclosure with a heat generating conducting body *Numerical Heat Transfer A* **35** 415-433
[5] Oliveski R D C, Macagnan M H and Copetti J B 2009 Entropy generation and natural convection in rectangular cavities *Appl. Thermal Eng.* **29** 1417-1425
[6] Basak T, Anandalakshmi R and Gunda P 2015 Role of entropy generation during convective thermal processing in right-angled triangular enclosures with various wall heatings ***Chem*** *Eng Res Des* **90** 1779-1799
[7] Ngo I-L and Byon C 2015 Effects of heater location and heater size on the natural convection heat transfer in a square cavity using finite element method *J. Mech. Sci. Tech.* **29(7)** 2995-3003
[8] Das D, Roy M and Basak T 2017 Studies on natural convection within enclosures of various (non-square) shapes- A review *Int. J. Heat Mass Transf.* **106** 356-406
[9] Sarris I E, Lekakis I and Vlachos N S 2004 Natural convection in a rectangular tanks heated locally from below *Int. J. Heat Mass Transf.* **47** 3549-3563
[10] Cianfrini C, Corcione M, Habib E and Quintino 2015 Convective Transport in Rectangular Cavities Partially Heated at the Bottom and Cooled at One Side *J. Thermal Sci.* **22**, 55-63
[11] Koizumi H 2007 Flow pattern formation and the transition to chaos in a confined container heated locally from below *Int. J. Thermal Sci.* **46** 953-962
[12] Öztop H F, Estellé P, Yan W M, Al- Salem K, Orfi J and Mahian O 2015 A brief review of natural convection in enclosures under localized heating with and without nanofluids *Int. Commu. Heat Mass Transfer* **60**
[13] Bejan A 1979 A study of entropy generation in fundamental convective heat transfer *J. Heat Transfer ASME* **101** 718-725
[14] Patankar S V 1980 *Numerical Heat Transfer and Fluid Flow* New York: Hemisphere

Analysis of interfacial behavior in two-phase flow using image processing

Digpriya[1], S Saini[1], J Banerjee[1]
[1]Department of Mechanical Engineering, S V National Institute of Technology, Surat, Gujrat-395007, India.

E-mail: digpriya22@gmail.com; jaish0220@gmail.com; jbaner@med.svnit.ac.in;

Abstract. Two-phase flow has a wide range of application from petroleum to space industry. Gas-liquid two-phase flow in horizontal pipe commonly occurs in petro-chemical, nuclear and geothermal industry, where simultaneously gas and liquid are transported through pipelines. During the transportation of gas-liquid flow simultaneously, due to the interaction of two phases, slug forms from the stratified-wavy flow. Slug flow consists of strong aeration at the slug front which is responsible for pipe failure and hazards in industries. Thus, it is required to understand the phenomena of slug formation in industrial pipeline to avoid the catastrophic pipe failure. In the present research, image processing technique is used to accurately analyze the interfacial behavior of two-phase flow. The interfacial behavior plays crucial role, as it is the plane where the two phases interact and interaction of the phases leads to the transition of different two-phase pattern. Experimentally captured images are processed using image processing algorithm, to study the variation of liquid film thickness. With the help of liquid film thickness transition of the interface is analyzed for different superficial Reynolds number of liquid and gas. Further, a PDF analysis is reported to understand the mechanism of the growth of the wavy interface.
Keywords: Stratified-wavy; Slug; Image processing; Liquid film thickness.

1. Introduction

Two-phase flow is widely seen in oil (petroleum and chemical industry) and gas (nuclear power plant and nuclear industry) industry during transportation of oil and natural gas through pipelines. In transportation of two-phase flow from offshore to onshore and further process facility the flow regime changes due to change in fluid flow parameters (inlet flow rate of gas and liquid, density of both the phases and local phase velocity of gas and liquid ([1]-[7])) in the pipeline. Two-phase flow is the simultaneous flow of two distinct phases in a channel or pipe having common interfaces [25], which may show large number of flow patterns (separated flow, intermittent flow and dispersed flow ([1]-[4])) depending on the variation in fluid flow properties of the phases. Some of the flow regime of two-phase flow like slug (type of intermittent flow in which liquid slug flows alternatively between the elongated gas bubbles associated with strong aeration at the slug front ([6]-[7])) leads to safety issues and severe damage in terms of corrosion ([8]-[10]) and leakage of transportation pipeline ([11]-[12]).

Considering this, a lot of research has been reported in this field ([1]-[4], [7], [13]-[19]) and studies in gas-liquid horizontal two-phase flow remains an active area of research due to buoyancy, asymmetric distribution of gas phase and interfacial instability such as K-H instability [15]. Earlier studies reported by ([6], [14]-[19]) analyzed in detail the transition of stratified-wavy flow to slug flow regime in horizontal pipe flow. Kordyban and Ranov [15], are first who identified the transition of stratified-wavy flow to slug flow pattern in horizontal channel. They reported that high velocity air flows over the wave crest which leads to suction pressure higher than the downward gravity force. This phenomenon leads to Bernoulli Effect on the wavy flow and flow transits into slug flow pattern. Saini and Banerjee [18], analyzed the roll waves using subjective methodology of flow visualization technique. They suggested that beyond the critical wave height, waves undergo the plunging kind of wave breaking which leads to slug formation in pipe flow. In addition, Saini et al. [19], reported the critical liquid height for transition of wavy flow to slug flow using Laser Doppler Velocimetry (LDV).

In-spite of the extensive research, the influence of fluid flow properties on gas-liquid interface is not yet explicitly understood. In this direction several researchers proposed analysis of the gas-liquid interface using image processing technique. Dinh et al. [20] presented a new application of image processing technique to two-phase bubbly/slug flow in vertical pipe for identification and measurement of bubble parameters. Do Amaral et al. [21] have also developed a method to extract quantitative parameters through image processing algorithms and measured bubble dimensions, velocity and frequency etc. Image processing technique is used for interfacial analysis of wavy flow sub-patterns ([22]-[23]). Liu et al. [24] used image processing technique to swirling flow regime in vertical channel. They have obtained liquid film thickness for different swirling flow regime, as the liquid film in swirling flow plays an important role in mass, momentum and heat transfer.

In addition, researchers have used multiple techniques to analyze the interfacial characteristics of two-phase flow regime such as capacitance probe, conductance probe, particle image velocimetry and image processing etc. Among all these techniques, image processing technique shows more potential because it is a non-intrusive technique and interface of two-phase flow can be easily analyzed by developing a suitable image processing algorithm. There is a wide scope for development of suitable algorithm for image processing so as to accurately analyze the interfacial behavior of two-phase flow. In the present work, visualization and characterization of two-phase flow is reported using image processing technique. Further analysis of wavy flow is done by introducing probability distribution of wave profile. Transition of stratified to wavy flow and wavy flow to slug flow is also analyzed by increasing liquid superficial Reynolds number. In this work, image processing technique is used for accurate analysis interfacial properties of two-phase flow.

2. Experimental setup

The experiments for gas (air)-liquid (water) are carried out on existing two-phase flow test rig [10]. The test facility mainly includes, two-phase flow test rig system and high-speed photography system (HSPS). The two-phase flow test rig facility includes some major components like air-flow loop, water flow loop, SCADA (Supervisory control and data acquisition), test section for two-phase flow measurement etc. The test facility contains three different pipe diameters of 12 mm, 25 mm and 50 mm made of Perspex glass and each of 14 m length as shown in figure 1. The two-phase flow experiments reported here are performed on pipe of 50 ±0.20 mm inner diameter. The liquid (water) is pumped from the 200 L tank using centrifugal pump which passes through the water filter, heat bath exchanger (for maintaining the liquid temperature constant), Coriolis mass flow meter and Pneumatic valve before entering in the test section. Simultaneously, air is introduced in the test section which supplied from a screw compressor at 4 bar pressure and ambient temperature. Before entering the test section, air passes through a filter (which remove the condensed air and produces the dry air), Coriolis mass flow meter and pneumatic value. At the outlet of the test section a separation tank is placed in which air vent to outside the chamber and water collects in the supply tank. Hence, water is recirculated in the test section as shown in figure 1.

Figure 1. Two-phase flow test rig system.

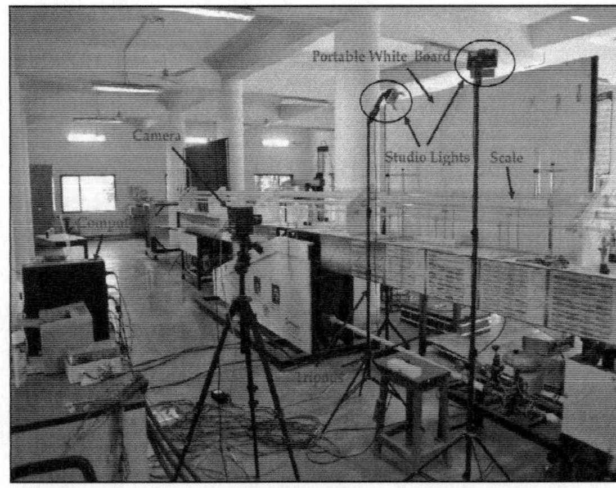

Figure 2. High-speed Photography System.

Table 1. Summary of the experimental condition

Pipe diameter	50±0.20 mm
Two-phases	Air and water
Liquid density	1000 kg/m^3
Gas density	1.2 kg/m^3
Dynamic viscosity of liquid	$7.7733*10^{-4}$ kg/ms
Dynamic viscosity of gas	$1.868*10^{-5}$ kg/ms
Range of liquid superficial Reynolds number (Re_{SL})	5000 to 12700
Range of gas superficial Reynolds number (Re_{SG})	2270

A high-speed photography is conducted for the flow visualization of the two-phase flow patterns. For high-speed camera made of Photron FASTCAM is used for capturing the images of obtained flow patterns in pipe flow. The images are captured at the 2000 fps (frame rate per sec) and 1280×616 resolution. The high-speed photography system (shown in figure 2) consists of a Photron made FASTCAM camera with specification, speed: 30 to 100000 FPS, resolution: 1280×1280. The other components with high-speed camera are tripod, workstation, portable white board, alight and a scale etc.

3. Methodology

The technique adopted here for the study of interfacial behavior of two-phase flow is image processing technique. In this technique, digital images (captured using high-speed photography system) are processed using an algorithm to eliminate noise. Digital image is made of an element called pixel, which gives information of color intensity at a point of an image. We use this property of digital image for image processing. In this work using high-speed camera in order to capture the maximum number of frames per second, so that the interfacial phenomenon of two-phase flow can be captured with higher order of accuracy with each frame.

Different flow patterns of two-phase flow are recorded using of FASTCAM Photron camera. The raw images are taken from PFV (PHOTRON FASTCAM VIEWER APPLICATION). These raw images are further processed in the MATLAB software. In MATLAB software there are sequential processes for processing an image, the software contains an image processing tool box, which have inbuilt function to read and process the image further. In image processing technique, we process the raw image to get the most contrast and less noise image.

IOP Conf. Series: Materials Science and Engineering 1146 (2021) 012002 doi:10.1088/1757-899X/1146/1/012002

The raw (RGB) image is first read in MATLAB software by 'imread' inbuilt function. The function reads the RGB image in form of three-layer matrices with respect to Red, Green and Blue planes. For analysis of particular area of interest of raw image, 'imcrop' function is used to crop the image. The cropped RGB image is then converted into binary image to visualize the interface clearly.

In order to get the detectable interface, 'imfill' function is used, which is used to fill the holes of dispersed air-water area. The final processed binary image has clearly visible and detectable air-water interface. Now the data for the interface can be extracted. Further, some mathematical MATLAB functions like 'polyfit' and 'polyval' are used for smoothening of the irregular interface. With the help of attached scale on the pipe and measurement of pipe diameter, pixel values can be converted into true coordinate values. The true coordinate values can be used to calculate interfacial properties like liquid film thickness of two-phase flow. The flow chart of image processing algorithm to get the interface of two-phase flow is shown figure 3.

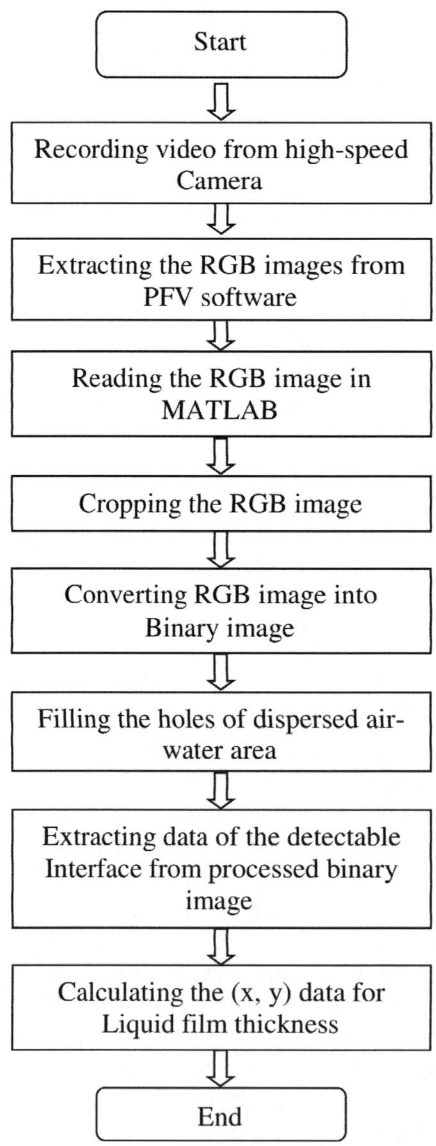

Figure 3. Flowchart for image processing algorithm.

4. Results

4.1. Characterization of two-phase flow

The two-phase flow has different flow patterns, as it is an interactive flow of two distinct phases. So whenever fluid properties of any phase or both the phases varies, then it is reflected as different flow patterns [25]. Interface of these flow patterns can be precisely visualized and analyzed by using the image processing technique. In the present work, characterization of two-phase flow is done by analyzing the liquid film thickness of different flow patterns with axial distance and time, with the help of image processing algorithm.

4.1.1. Stratified smooth flow. The Stratified smooth flow regime is captured at Re_{SG}, 2270 and Re_{SL}, 5000 shown in figure 4. The interface between the two phases of fluid, is smooth and no fluctuation observed at the gas-liquid interface as shown in figure 5. The obtained results of liquid film thickness varying with time is obtained from analysing 400 sequenced images as shown in figure 6.

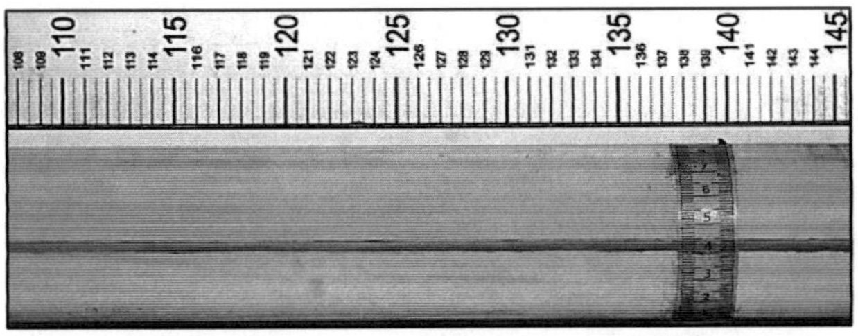

Figure 4. Raw image of stratified smooth flow pattern.

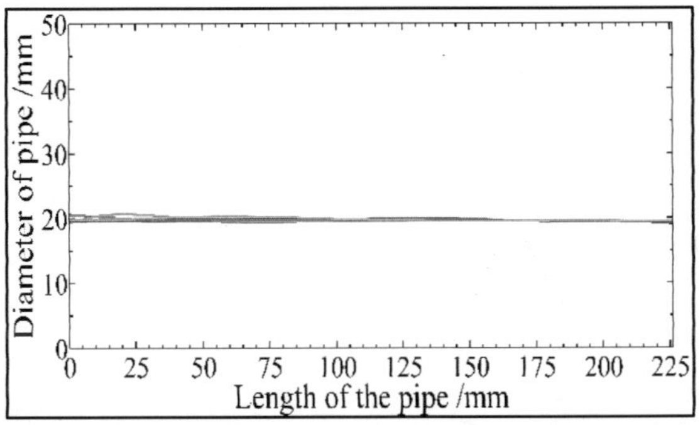

Figure 5. Stratified smooth interface along axial length for sequenced image frames.

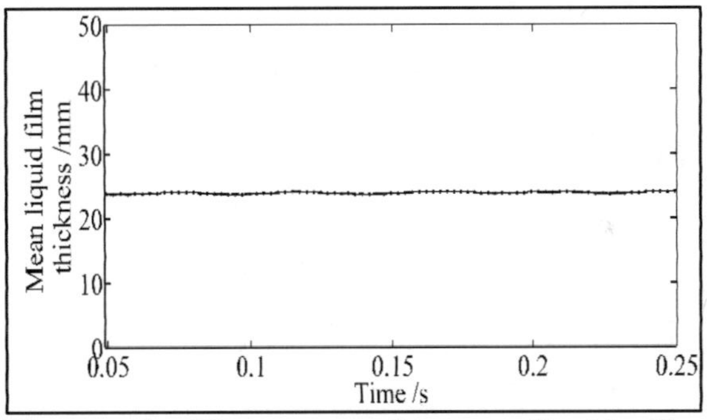

Figure 6. Time-trace of mean liquid film thickness of stratified smooth flow pattern.

It is observed that the mean height of liquid film thickness is nearly constant along with time. Thus, this flow pattern falls under stratified smooth flow pattern. Also, very little fluctuation in liquid film thickness is observed from the processed images. The plot for the mean liquid film thickness shows that the liquid film thickness for stratified smooth flow remains undisturbed with time.

4.1.2. Wavy flow. As the Re_{SL} is increased from 5000 to10900, keeping the Re_{SG} constant (2270), the flow pattern alters from stratified smooth to stratified wavy flow as captured in image shown in figure 7. Increase in Re_{SL}, disturbs the smooth interface and it leads to wave formation due to pressure fluctuation over the interface (as shown in figures 8). By the visualization only, we can infer that the interface is wavy.

Alteration in liquid film thickness of wavy interface with axial distance and time-trace of mean liquid film thickness is shown in figure 8 and figure 9 respectively. The fluctuations of the liquid film thickness with axial distance is of the order of 5 mm, and it is calculated by using image processing algorithm.

4.1.3. Slug flow. With further increase in Re_{SL} to 12700, by keeping Re_{SG} constant as 2270, the wavy flow pattern altered into slug flow pattern. When the Re_{SL} is increased, the waves with large amplitude grows and it touches to the upper wall of the pipe and blocks the path of air flow, this phenomenon leads to slug formation. The pipe fills with liquid momentary in a small time interval and we can visualize the periodic flow of liquid slug and elongated gas bubble. In raw image (figure 10), the nose of the formed slug can be seen and further in figure 11 and 12, we can depict that the interface height reaches up to 50 mm, which is equal to the diameter of the pipe.

The interface of the slug flow pattern is highly unstable which is shown in raw image (figure 10). Thus, processing the interface with the image processing algorithm doesn't give a smooth interface, shown in figure 11. The disturbed red line shows the processed interface of slug nose without smoothening of the curve. Further a smooth blue curve is obtained by developed algorithm, which helps in examine the flow pattern.

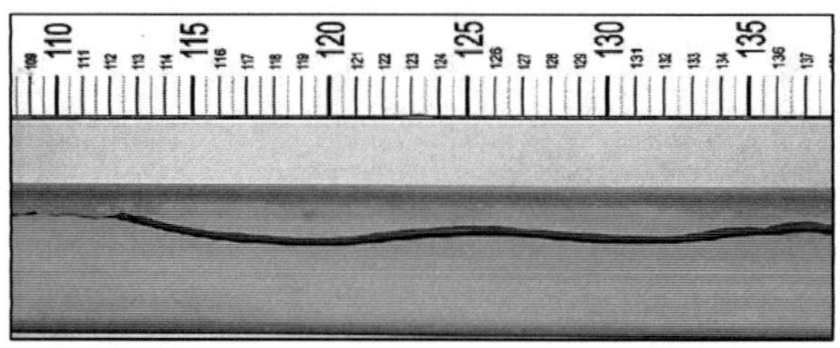

Figure 7. Raw image of wavy flow pattern.

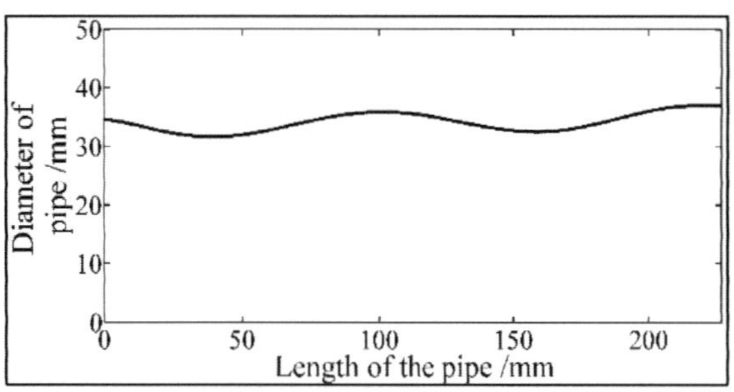

Figure 8. Wavy interface along axial length.

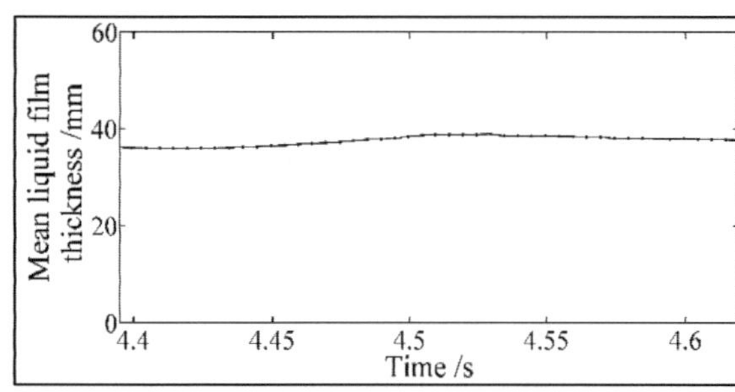

Figure 9. Time-trace of mean liquid film thickness of wavy flow pattern.

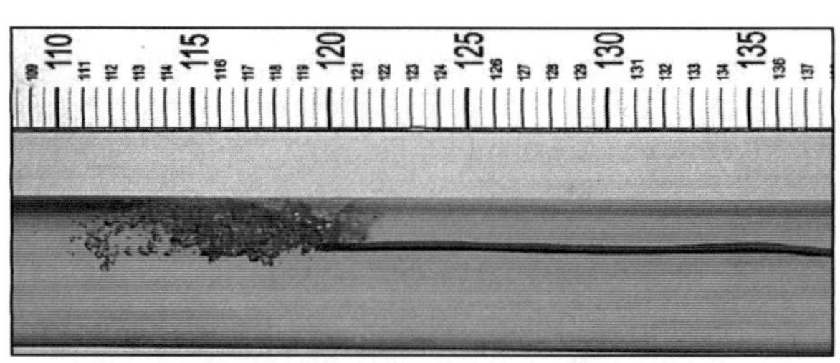

Figure 10. Raw image of slug flow pattern.

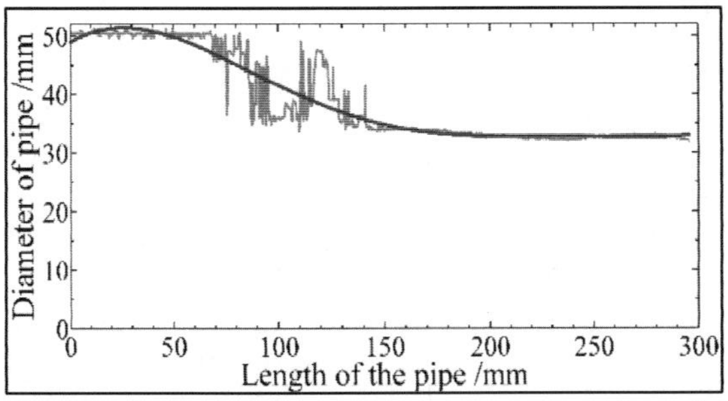

Figure 11. Interface of slug nose with axial length.

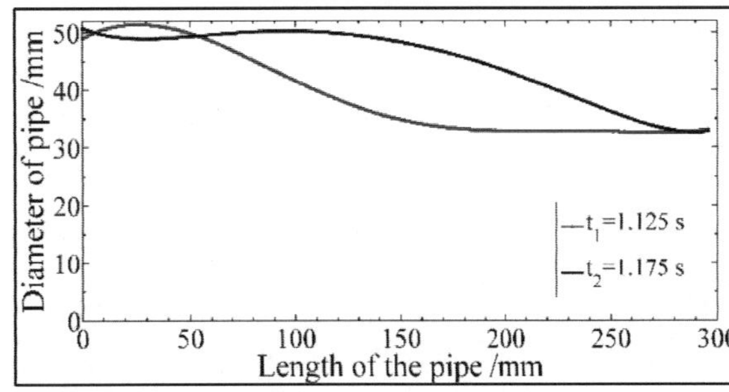

Figure 12. Smoothen slug interface along axial length for sequenced image frames at different time.

In figure 12, the interface of two sequenced image frames at a gap of 20 milli second is drawn to visualize the movement of interface of the slug along axial length with time. Here, (in figure 12) the blue curve is for image frame 1 at time t_1 and black color curve is next image for frame 2 at time t_2.

4.2. Interfacial analysis of wavy flow

In literature, different types of wavy flow patterns have been found in two-phase flow in a horizontal pipe ([22]-[23]-[27]) i.e. 2D small amplitude waves, 2D large amplitude waves, 3D small amplitude waves, 3D large amplitude waves and roll waves etc. In present work (at superficial Reynold number of water 10900 and that of air 2270), we observed that 2D large amplitude waves are not uniform non-sinusoidal. The wave properties such as amplitude and wavelength of the wavy interface are changing at the same Re_{SL}. In figure 13, at same Re_{SL}, we can see the two different interfacial wave at time t_1 and time t_2.

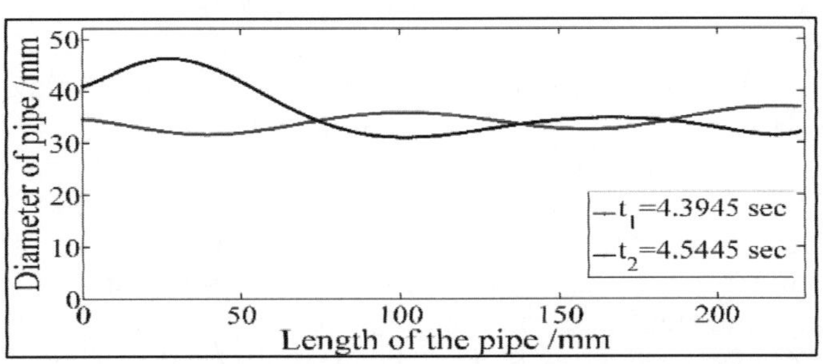

Figure 13. Wavy interface of two sequenced image frames.

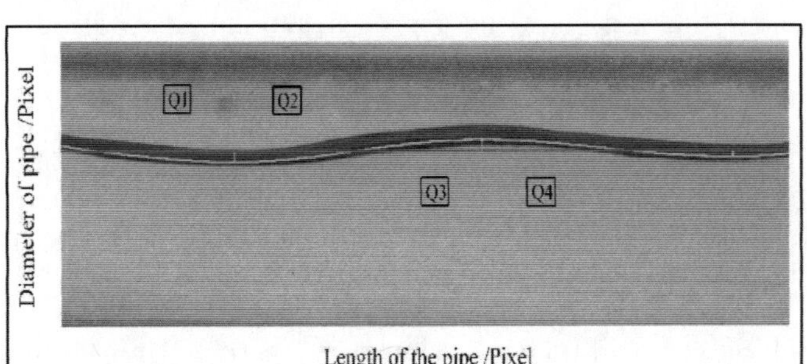

Figure 14. Quarter division of a complete wave cycle.

To analyze the asymmetric behavior of wavy interface in more detail, the probability distribution analysis is carried out for 110 sequenced image frames for 55 milli seconds. A complete wave cycle of wavy interface is taken and divided into four quadrants [26] as shown in figure 14. The crest and through of the wave are divided by zero crossing line (undisturbed position of interface). The length and height of the four quadrants for each different image frame is calculated with the help of image processing algorithm. Further, the non-uniformity of the interfacial wave is shown with the probability distribution of wave quarters, which is calculated from the image processing algorithm.

In order to calculate probability distribution of quarter wave length and height, the x-axis is divided into number of uniform bins and probability of quarter length and height is calculated for each bin. The length and height of the wave quarters Q1, Q2, Q3 and Q4 lies in different range of bins, shown in figures 15 and 16 respectively. Similarly, the upstream and downstream side of quarters also fall under different range of length. From the probability distribution, it is depicted that the length and height of crest and trough is not uniform. Thus, by PDF analysis, it is reported that the 2D large amplitude wave is non-uniform (and asymmetric) with space and time.

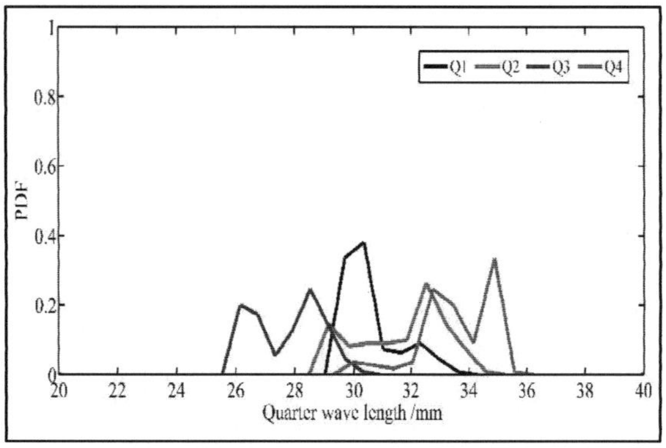

Figure 15. Probability distribution of quarter wave length.

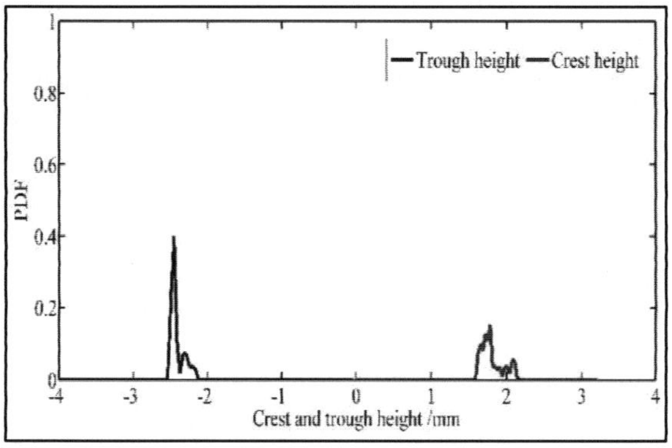

Figure 16. Probability distribution of quarter wave height.

By probability distribution analysis, it is observed that length of the wave quarters Q1, Q2, Q3 and Q4 are not equal. The crest length which is the sum of length of Q3 and Q4 are different. The difference in crest wave quarters Q3 and Q4 is due to the presence of different magnitude of pressure on windward side and leeward side as was demonstrated in terms of sheltering mechanism by Harold Jeffrey [28]. It states that the pressure on the wave slope facing the air will be greater than those away from it. Similarly, the trough quarters Q1 and Q2 also differ in length due to the sheltering effect.

The variation in wave height in the wave quarters depicts the wave growth due to given perturbation. In literature, the wave growth phenomenon is referred as Kelvin Helmholtz instability [15], which states that due to perturbations, stratified smooth interface changes into wavy interface and the waves grows and one of the peak rises and approaches to the pipe wall. The growth (response) of the interface observed based on image processing here shows that it is not sinusoidal, as the wave length of quarters of the crest (Q3 and Q4) and that of trough (Q1 and Q4) have different values. This can be attributed to the sheltering effect.

4.3. Transition of stratified smooth to slug flow pattern

With an increase in Re_{SL}, at constant Re_{SG}, the stratified smooth flow regime transits into wavy flow. As Re_{SL} is further increased, the waves start growing and it touches the upper wall of the pipe and undergoes the plunging kind of wave breaking [18]. This phenomenon leads to slug formation in pipe flow with vigorous aeration in liquid body. As shown in figure 17, transition of stratified smooth to slug flow has been reported with image processing technique by showing the increment in the liquid film thickness from stratified smooth to slug flow pattern. The interface of stratified smooth, wavy and slug flow regime shown in figure 17, is obtained by using the algorithm. As the Re_{SL} is increased from 5000 to 12700 with constant Re_{SG} 2270, the liquid film thickness increased from 24 mm (for stratified flow) to 50 mm (for slug flow in which liquid slug occupies the complete cross-section area of the pipe) while flow regime changed from stratified smooth to slug.

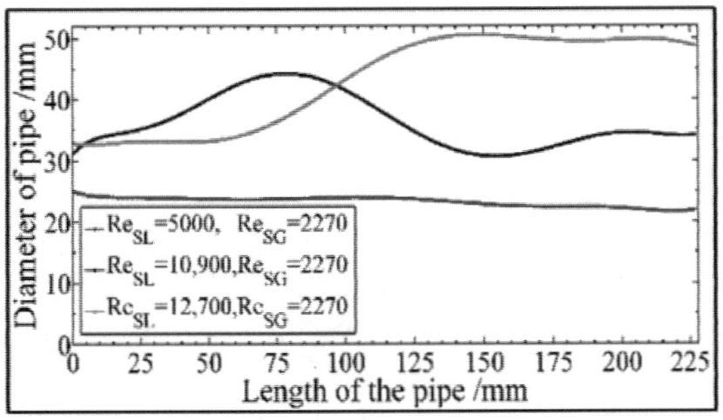

Figure 17. Transition of liquid film thickness with flow pattern

With the above results, it can be seen that the liquid film thickness is linked with the flow pattern. As the flow pattern changes from stratified smooth to wavy flow pattern, the liquid film thickness increases and passage for air-flow reduces, resulting in decrease in pressure over the interface which leads to the wave growth. Further, when the liquid film increases (due to increase in Re_{SL}), one of the peaks of wavy interface grows significantly and touches the upper wall of the pipe, leading to the blockage of air-flow. It increases the pressure at the windward side of wave-slope, thus the air-flow pushes the water in the flow direction and forms liquid slug.

5. Closure

In the present work, a method is developed to extract the interface of different two-phase flow pattern, through image processing algorithm. The interfacial analysis of two-phase flow is reported by using experimental flow visualization as well as image processing technique. In the present work, the two-phase flow pattern is analyzed in terms of interface fluctuation and the alteration in liquid film thickness is done by varying the Re_{SL} at fixed Re_{SG}. The probability distribution of wavy interface depicted that the wavy interface is not uniform and non-sinusoidal. By the PDF analysis, it could be concluded that for the growth of the wavy interface, sheltering and K-H instability mechanism both are responsible. The transition of the interface from stratified smooth to slug flow pattern is analyzed with increase in Re_{SL} and it is observed that the liquid film thickness increases with increase of mass flow rate of any of the phase, air or water.

6. References

[1] Mandhane, J., Gregory, G., and Aziz, K. (1974). A flow pattern map for gas-liquid flow in horizontal pipes. *International journal of multiphase flow*, **1(4)**:537–553.

[2] Barnea, D., Shoham, O., Taitel, Y., and Dukler, A. (1980). Flow pattern transition for gas-liquid flow in horizontal and inclined pipes. Comparison of experimental data with theory. International *Journal of Multiphase Flow*, **6(3)**:217–225.

[3] Ruder, Z. and Hanratty, T. (1990). A definition of gas-liquid plug flow in horizontal pipes. *International journal of multiphase flow*, **16(2)**:233–242.

[4] Netto, J. F., Fabre, J., and Peresson, L. (1999). Shape of long bubbles in horizontal slug flow. *International Journal of Multiphase Flow*, **25(6-7)**:1129–1160.

[5] Ghajar, A. J. and Tang, C. C. (2007). Heat transfer measurements, flow pattern maps, and flow visualization for non-boiling two-phase flow in horizontal and slightly inclined pipe. *Heat Transfer Engineering*, **28(6)**:525–540.

[6] Vaze, M. and Banerjee, J. (2011). Experimental visualization of two-phase flow patterns and transition from stratified to slug flow. Proceedings of the Institution of Mechanical Engineers, Part C: *Journal of Mechanical Engineering Science*, **225(2)**:382–389.

[7] Thaker, J. and Banerjee, J. (2015). Characterization of two-phase slug flow sub-regimes using flow visualization. *Journal of Petroleum Science and Engineering*, **135**:561–576.

[8] Sun, J.-Y., Jepson, W., et al. (1992). Slug flow characteristics and their effect on corrosion rates in horizontal oil and gas pipelines. In SPE Annual Technical Conference and Exhibition. *Society of Petroleum Engineers.*
[9] Ahmed, W. H., Bello, M. M., El Nakla, M., Al Sarkhi, A., and Badr, H. M. (2014). Experimental investigation of flow accelerated corrosion under two-phase flow conditions. *Nuclear Engineering and Design*, **267**:34–43.
[10] Thaker, J. and Banerjee, J. (2016a). Influence of intermittent flow sub-patterns on erosion-corrosion in horizontal pipe. *Journal of Petroleum Science and Engineering,* **145**:298–320.
[11] Ossai, C. I., Boswell, B., and Davies, I. J. (2015). Pipeline failures in corrosive environments–a conceptual analysis of trends and effects. *Engineering Failure Analysis*, **53**:36–58.
[12] Bordalo, S. N. and Morooka, C. K. (2018). Slug flow induced oscillations on subsea petroleum pipelines. Journal of Petroleum Science and Engineering, **165**:535–549.
[13] Kong, R., Kim, S., Bajorek, S., Tien, K., and Hoxie, C. (2018a). Effects of pipe size on horizontal two-phase flow: Flow regimes, pressure drop, two-phase flow parameters, and drift-flux analysis. *Experimental Thermal and Fluid Science*, **96**:75–89.
[14] Kong, R., Rau, A., Kim, S., Bajorek, S., Tien, K., and Hoxie, C. (2018b). Experimental study of horizontal air-water plug-to-slug transition flow in different pipe sizes. *International Journal of Heat and Mass Transfer*, **123**:1005–1020.
[15] Kordyban, E. S. and Ranov, T. (1970). Mechanism of slug formation in horizontal two-phase flow. *Journal of Basic Engineering*, **92(4):** 857–864.
[16] Wallis, G. B. and Dodson, J. E. (1973). The onset of slugging in horizontal stratified air-water flow. *International Journal of Multiphase Flow*, **1(1):**173–193.
[17] Andritsos, N., Williams, L., and Hanratty, T. (1989). Effect of liquid viscosity on the stratified-slug transition in horizontal pipe flow. *International Journal of Multiphase Flow*, **15(6)**:877–892.
[18] Saini, S. and Banerjee, J. (2021). Physics of aeration in slug: Flow visualization analysis in horizontal pipes. *Journal of Visualization*, pages **1–14**.
[19] Saini, S., Thaker, J., and Banerjee, J. (2021). Analysis of interfacial dynamics in stratified and wavy-stratified flow using laser Doppler velocimetry. *Experimental and Computational Multiphase Flow*, pages **1–14**.
[20] Dinh, T. B., Kim, B. S., & Choi, T. S. (1999). Application of image processing techniques to air/water two-phase flow. In Applications of Digital Image Processing XXII (Vol. **3808**, pp. 725-730). *International Society for Optics and Photonics.*
[21] Do Amaral, C. E., Alves, R. F., da Silva, M. J., Arruda, L. V., Dorini, L., Morales, R. E., & Pipa, D. R. (2013). Image processing techniques for high-speed videometry in horizontal two-phase slug flows. *Flow Measurement and Instrumentation*, **33**, 257-264.
[22] Kuntoro, H. Y., Hudaya, A. Z., & Dinaryanto, O. (2015). Experimental study of the interfacial waves in horizontal stratified gas-liquid two-phase flows by using the developed image processing technique. *arXiv preprint arXiv*:**1511**.06087.
[23] Hudaya, A. Z., Widyatama, A., Dinaryanto, O., & Juwana, W. E. (2019). The liquid wave characteristics during the transportation of air-water stratified co-current two-phase flow in a horizontal pipe. *Experimental Thermal and Fluid Science*, **103**, 304-317.
[24] Liu, L., Wang, K., & Bai, B. (2020). Comparative investigation of liquid film thickness and interfacial wave properties of swirling gas-liquid flows. *Chemical Engineering Science*, **213**, 115407.
[25] Faghri, A., & Zhang, Y. (2006). Transport phenomena in multiphase systems. Elsevier.
[26] Birvalski, M., Tummers, M. J., Delfos, R., & Henkes, R. A. W. M. (2014). PIV measurements of waves and turbulence in stratified horizontal two-phase pipe flow. *International journal of multiphase flow*, **62**, 161-173.
[27] Chen, X. T., Cal, X. D., & Brill, J. P. (1997). Gas-liquid stratified-wavy flow in horizontal pipelines. Journal of Energy Resources Technology, **119 (4)**: 209-216.
[28] Jeffreys, H. (1925). On the formation of water waves by wind. *Proceedings of the Royal Society of London*. Series A, Containing Papers of a Mathematical and Physical Character, **107(742)**, 189-206.

IOP Conf. Series: Materials Science and Engineering 1146 (2021) 012003 doi:10.1088/1757-899X/1146/1/012003

Automated Product Inspection in Industry 4.0 Environment

M B Kiran[1]

[1]Associate Professor, Department of Mechanical Engineering,
Pandit Deendayal Energy University, Gandhinagar, Gujarat, INDIA

E-Mail: MB.Kiran@sot.pdpu.ac.in

Abstract. The emergence of Industry 4.0 technologies demands new techniques of measurement which would facilitate seamless integration with other devices in the wireless IoT network. Many of the existing product inspection methods cannot be deployed directly in the IoT environment. Thus, there is a need for innovative inspection techniques in the Industry 4.0 environment. In this context, the proposed inspection technique assumes special significance. Surface irregularities observed in product manufacturing can be due to chatter, vibration, worn-out cutting tools, condition of the machine tools, etc. Evaluation of surface texture helps in predicting a product's functionality. In this work, an attempt has been made to identify the surface texture images acquired from Shaping, Milling, Electric discharge machining (E.D.M.) and Sand Blasting processes during online inspection. In addition to surface texture identification, the proposed method will also measure surface roughness and component dimensions. Thus, entire product inspection can be done online, and in a single setup. This is also an 100% online inspection method. The main contribution of this proposed research work is that all types of inspection are completed in a single set up, resulting in significant savings.

1.Introduction

Surface texture may be defined as the superposition of roughness, waviness, and error in the form. Surface irregularities having shorter wavelengths are known as roughness. Surface irregularities having medium wavelength are known as waviness and surface irregularities having large wavelength are known as form error. Surface texture evaluation helps in predicting the functionality of a machined component. Many industries such as automobile and machine tool manufacturing normally do the surface inspection of machined components. This is required for ensuring customer satisfaction. During the component inspection, they check not only the surface finish but also the processor texture of the machined surface. This is because the same roughness can be achieved by the different manufacturing processes. Thus, even the component drawing in addition to specifying dimensions will specify both surface finish and texture information. Many researchers have been working on texture evaluation. But as far as the knowledge of the author goes not a single paper exists today which addresses (i) dimensional inspection (ii) surface roughness inspection and (iii) process identification or texture classification. In this work, an attempt has been made to identify the surface textures acquired from Shaping, Milling, Electric discharge machining (EDM) and Sand Blasting processes. Subsequently, both dimensional and surface roughness assessments are done. The findings presented in this article are useful to academicians, practitioners, and researchers.

2. Literature survey

Texture evaluation methods can be classified into -statistical, structural, and morphological methods. Statistical methods extract a set of features from a given image [1]. These features are used for classifying a given image by using statistical pattern recognition techniques. The statistical methods can be classified as follows.

- Gray level difference method [2]. In this technique, the probability density function is calculated for the variation in image intensities.
- The spatial grey level dependence method [3]. In this method, joint gray level distribution for two gray levels at distance 'd' and angle 'θ 'is estimated. These are called first and second-order statistics.
- The grey-level run-length method [4, 5]. A set of connected pixels having the same gray value is called an identical run. This method computes the number of identical runs.

2.1 *Gray level Cooccurrence Matrix (GLCM)*

GLCM was proposed by Haralick et al. [3]. Many researchers have used this method for classifying images [3]. The main problem with this method is to define both 'd' and angle 'θ '. In this work, an attempt has been made to identify the surface textures produced from Shaping, Milling, Electric discharge machining (E.D.M.) and Sand Blasting processes. In this research, the experiment was repeated at different values of 'd' and angle 'θ'. Surface texture classification is useful for determining whether a product becomes successful when it is put into service. Several researchers have been working in the area of classification.

2.2 Structural methods

These methods Haralick et al. [3] try to identify a set of primitives from a given image. The method then tries to define texture *as* an arrangement of such primitives by different placement rules. A major problem with this method is to define the shape and size of the area where texture features should be collected.

2.3 Multiresolution filtering method

Multiresolution filtering method (e.g., Gabor filtering) is very much similar to the way the human visual system works. Gabor filter work with texture in spatial -frequency domain s [6, 7, 8, 9, 10, 11]. Thus, it *is* clear from above that not a single paper exists today which addresses (i) dimensional inspection (ii) surface roughness inspection and (iii) process identification or texture classification, *in* a single setup. With the advent of Industry 4.0, there is a need to design and develop novel techniques, as many of the existing inspection techniques cannot be deployed directly in Industry 4.0 environments [12,13,14]. Also, nowadays, there is a great demand for 100% inspection of machined components. Especially, in automobile and machine tool companies. Thus, it is clear from above that not a single paper exists today which addresses (i) dimensional inspection (ii) surface roughness inspection and (iii) process identification or texture classification, in a single setup. This is very much required in industrial inspection of machined components.
In this work, an attempt has been made to identify the surface textures acquired from Shaping, Milling, Electric discharge machining (E.D.M.) and Sand Blasting processes. Subsequently, both dimensional and surface roughness assessments are done. The findings presented in this article are useful to academicians, practitioners, and researchers.

3. Methodology
Research methodology consists of -Experimental set-up, Specimen preparation, Surface texture inspection, Surface roughness inspection and Dimensional inspection.

3.1 Experimental set-up.

The experimental set-up (Figure 1) consists of a charge coupled device (CCD) camera, cables, a frame grabber, an advanced Image processing board, and a high-end computer. The workstation uses the Windows operating system. In-house software is developed using C++, for computing the roughness parameters, dimensional inspection, and texture assessment.

3.2 Specimen preparation

Test specimens are made of EDM, Shaping, Milling, and Sand Blasting processes. Table 1 shows the sample specimen details of milled specimens.

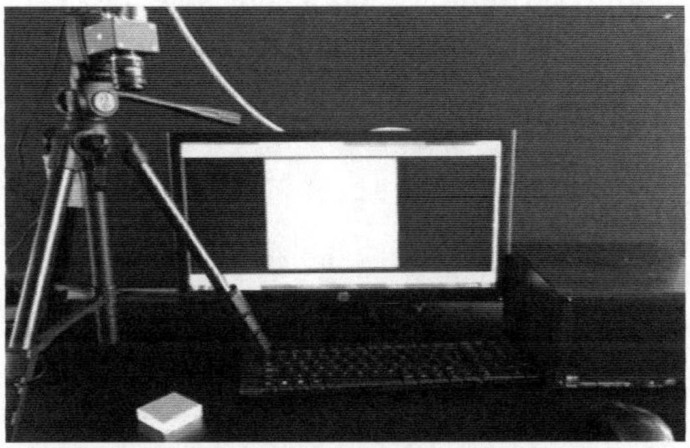

Figure 1. Experimental set-up

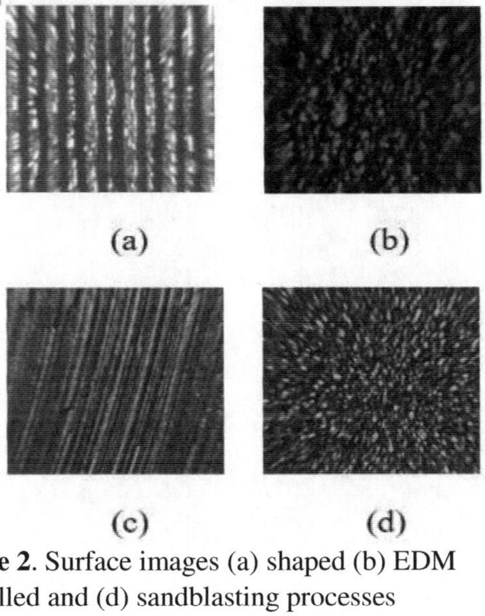

Figure 2. Surface images (a) shaped (b) EDM (c) milled and (d) sandblasting processes

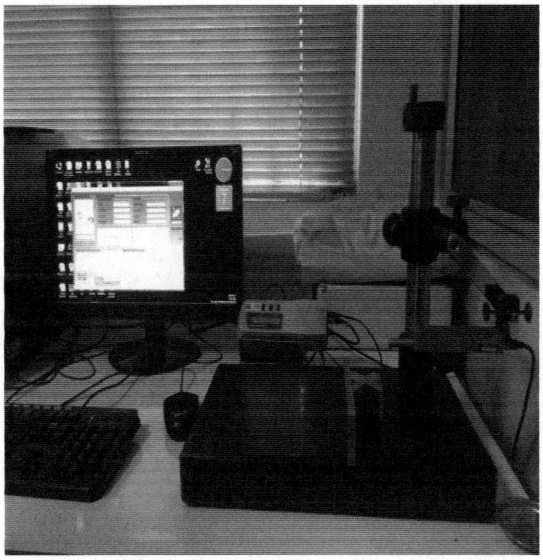

Figure 3 Stylus instrument measurement setup

Table 1 Milled specimen details

Speed (rpm)	Depth of Cut(mm)	Feed (mm/min)	StylusR_a(μm)
280	0.4	12.5	1.31
280	0.4	31.5	1.53
280	0.4	80.0	3.18
280	0.4	100.0	3.30
280	0.4	200.0	2.80
280	0.4	250.0	3.70

3.3 Surface texture inspection

Figure 2 (a) to (d) shows the surface *images* of shaping, EDM, milling, and sandblasting processes respectively, captured by using a CCD camera connected to the vision system. Table 2 shows the digital image of a milled surface.

3.4 Texture Identification

Texture identification process consists of the following steps.

3.4.1 *Grey level co-occurrence matrix (GLCM):* Digital images obtained in the previous step are used for computing the co-occurrence matrix. each entry in the co-occurrence table shows the joint distribution of gray levels *i* and *j*, separated at distance '*d*' and angle 'θ '. Table 3, shows a sample co-occurrence matrix. co-occurrence matrix tries to capture texture details from a given image. this it does, by using image intensity values.

Table 2: Digital Image of a Milled surface

200	138	236	225	180	167	74
182	127	235	181	171	137	65
160	147	232	161	157	122	55
143	173	201	133	165	100	52
131	217	164	141	178	105	46
122	229	190	146	229	128	75
122	202	213	145	236	132	81

Table 3: Co-occurrence matrix (GLCM)

12	15	0	0	0	0	0
20	4	1	0	0	0	0
0	4	0	0	0	0	0
12	15	0	0	0	0	0
12	15	0	0	0	0	0
12	15	0	0	0	0	0
12	15	0	0	0	0	0

3.4.2. Texture features extraction from GLCM
Texture features are calculated by using following formulae.

Contrast	$\sum_{n=0}^{G-1} n^2 \left\{ \sum_{i=1}^{G} \sum_{j=1}^{G} P(i,j) \right\}, \lvert i-j \rvert = n$	(1)
Correlation	$\sum_{i=0}^{G-1} \sum_{j=0}^{G-1} \frac{\{i \, X \, j\} \, X \, P(i,j) - \{\mu_x \, X \, \mu_y\}}{\sigma_x \, X \, \sigma_y}$	(2)
Cluster Prominence	$\sum_{i=0}^{G-1} \sum_{j=0}^{G-1} \{i + j - \mu_x - \mu_y\}^4 \, X \, P(i,j)$	(3)
Dissimilarity	$\sum_{i=0}^{G-1} \sum_{j=0}^{G-1} P(i,j) \lvert i-j \rvert$	(4)
Energy	$\sum_{i=0}^{G-1} \sum_{j=0}^{G-1} \{P^2(i,j)\}$	(5)
Entropy	$-\sum_{i=0}^{G-1} \sum_{j=0}^{G-1} P(i,j) X log(P(i,j))$	(6)
Homogeneity	$\sum_{i=0}^{G-1} \sum_{j=0}^{G-1} \frac{P(i,j)}{1 + \lvert i-j \rvert}$	(7)
Maximum Probability	$max(p_{i,j})$	(8)
Sum entropy	$-\sum_{i=0}^{2G-2} P_{x+y}(i) \, log\left(P_{x+y}(i)\right)$	(9)
Difference Variance	$\sum_{i=0}^{G-1} \sum_{j=0}^{G-1} (i-\mu)^2 \, P(i,j)$	(10)
Difference entropy	$-\sum_{i=0}^{G-1} P_{x+y}(i) \, log\left(P_{x+y}(i)\right)$	(11)
Inverse difference moment normalized	$\sum_{i=0}^{G-1} \sum_{j=0}^{G-1} \frac{1}{1 + (i-j)^2} P(i,j)$	(12)
Inverse difference normalized	$\sum_{i=0}^{G-1} \sum_{j=0}^{G-1} \frac{P_{i,j}}{(1 + \frac{\lvert i-j \rvert}{N})^2}$	(13)

Sample texture features of EDM, shaped, milled and sandblasted surface images are shown in Table 4, 5, 6, and 7 respectively.

3.4.3. Feature set reduction: To improve computational efficiency, only those features that are required for classification are selected by using the singular value decomposition method is used. So that all redundant features are eliminated.

3.4.4. Method of texture classification: After selecting a set of required features from the set of available features, the K- nearest neighbour classification scheme is used for classifying images. Out of the total number of samples, one sample (t1) is taken out for classification. The method computes the Euclidean distance between the test sample (t1) and the training sample (t2). The method then assigns the test sample to that class, where the Euclidean distance is minimum. Table 8 shows the results of the classification experiment (Confusion matrix). 15 samples, belonging to different classes, were used in the classification experiment. From Table 8 it is clear the classification rate is 100% in case of EDM (E), milling (M), shaping (S) and sand blasting specimen images. However, in the case of Ground (G) specimens, out of 15 specimen images, 12 were classified correctly as belonged to the class of Ground specimens. But, 2 specimens were classified as Sand Blasting (SB) and one was classified as EDM (E) specimen.

Table 4 Texture features values of EDM surface

Feature No.	E1	E2	E3	E4	E5	E6	E7	E8	E9	E10	E11	E12	E13	E14	E15
F1	0.75	0.74	0.68	0.63	0.64	0.68	0.62	0.78	0.67	0.70	0.64	0.79	0.79	0.69	0.68
F2	0.70	0.76	1.09	1.40	1.20	1.15	1.57	0.57	1.19	1.04	1.38	0.54	0.54	1.07	1.12
F3	0.75	0.74	0.68	0.63	0.64	0.68	0.62	0.78	0.67	0.70	0.64	0.79	0.79	0.69	0.68
F4	2.10	2.13	2.28	2.48	2.46	2.39	2.54	2.02	2.47	2.33	2.47	1.88	1.88	2.26	2.34
F5	0.50	0.49	0.40	0.29	0.36	0.48	0.29	0.63	0.48	0.51	0.36	0.57	0.57	0.40	0.44
F6	17.0	17.0	18.0	20.0	20.0	19.0	20.0	17.0	20.0	19.0	19.0	15.0	15.0	17.0	19.0
F7	5.52	5.52	5.61	5.88	5.95	5.74	5.91	5.47	5.93	5.67	5.78	5.14	5.14	5.50	5.68
F8	1.62	1.64	1.67	1.75	1.76	1.76	1.79	1.63	1.82	1.74	1.77	1.50	1.50	1.66	1.72
F9	0.91	0.93	1.05	1.15	1.08	1.08	1.20	0.84	1.10	1.05	1.15	0.82	0.82	1.06	1.07
F10	3.35	3.81	5.93	4.49	4.43	7.86	4.70	6.24	8.07	8.05	5.97	4.34	4.34	5.22	6.82
F11	18.0	21.0	35.0	28.0	28.0	49.0	30.0	38.0	52.0	51.0	36.0	23.0	23.0	30.0	43.0
F12	0.99	0.99	0.98	0.98	0.98	0.98	0.98	0.99	0.98	0.98	0.98	0.99	0.99	0.98	0.98

Table 5 Sample texture features obtained from shaped surface

Feature No.	S1	S2	S3	S4	S5	S6	S7	S8	S9	S10	S11	S12	S13	S14	S15
F1	0.72	0.72	0.72	0.70	0.73	0.72	0.71	0.75	0.75	0.73	0.67	0.78	0.72	0.72	0.75
F2	0.86	1.02	1.08	0.84	0.84	1.01	1.21	0.75	0.77	1.03	1.30	0.64	0.86	1.15	0.85
F3	0.72	0.72	0.72	0.70	0.73	0.72	0.71	0.75	0.75	0.73	0.67	0.78	0.72	0.72	0.75
F4	2.76	2.53	2.46	2.68	2.70	2.57	2.62	2.45	2.58	2.50	2.74	2.29	2.77	2.46	2.26
F5	0.80	0.68	0.67	0.75	0.78	0.73	0.69	0.77	0.80	0.74	0.70	0.78	0.78	0.69	0.67
F6	57.0	47.0	44.0	52.0	54.0	48.0	48.0	46.0	51.0	47.0	53.0	43.0	56.0	45.0	40.0
F7	9.21	8.40	8.11	8.90	9.03	8.42	8.44	8.33	8.81	8.28	8.82	8.02	9.21	8.12	7.76
F8	2.22	1.94	1.89	2.14	2.17	2.00	2.01	1.99	2.12	1.96	2.09	1.88	2.21	1.91	1.77
F9	0.98	1.05	1.06	0.96	0.97	1.04	1.10	0.93	0.94	1.04	1.14	0.88	0.98	1.08	0.97
F10	15.0	7.0	12.0	10.0	11.0	15.0	15.0	14.0	15.0	20.0	12.0	15.0	12.0	18.0	11.0
F11	157	65	85	98	120	118	119	108	138	157	124	107	132	131	73
F12	0.99	0.99	0.98	0.99	0.99	0.99	0.98	0.99	0.99	0.98	0.98	0.99	0.99	0.98	0.99

Table 6 Texture features values of milled surface

Feature No.	M1	M2	M3	M4	M5	M6	M7	M8	M9	M10	M11	M12	M13	M14	M15
F1	0.83	0.68	0.62	0.70	0.86	0.72	0.62	0.76	0.90	0.68	0.68	0.83	0.83	0.65	0.81
F2	0.36	0.82	1.27	0.87	0.31	0.68	1.12	0.58	0.21	0.81	0.84	0.37	0.34	0.95	0.40
F3	0.83	0.68	0.62	0.70	0.86	0.72	0.62	0.76	0.90	0.68	0.68	0.83	0.83	0.65	0.81
F4	1.63	2.30	2.56	2.28	1.43	2.02	2.40	1.82	1.03	2.21	2.21	1.48	1.31	2.29	1.58
F5	0.49	0.39	0.30	0.39	0.47	0.30	0.22	0.25	0.36	0.31	0.28	0.36	0.21	0.30	0.37
F6	31.5	29.9	27.2	28.2	31.1	27.6	24.6	25.7	27.8	24.6	22.1	22.7	23.7	20.6	18.8
F7	6.89	7.00	6.79	6.81	6.70	6.65	6.45	6.34	6.12	6.42	6.14	5.91	5.88	5.98	5.53
F8	1.37	1.70	1.79	1.68	1.20	1.51	1.66	1.36	0.89	1.61	1.60	1.22	1.06	1.64	1.29
F9	0.67	0.89	1.09	0.97	0.64	0.86	0.99	0.83	0.51	0.90	0.93	0.69	0.65	0.93	0.69
F10	0.51	0.86	1.89	2.22	0.70	0.47	1.03	0.67	0.28	0.79	0.62	0.41	0.09	1.55	0.22
F11	3.17	9.27	15.1	13.6	2.48	4.64	8.34	3.89	1.21	6.82	6.61	3.32	1.27	9.82	2.68
F12	0.99	0.99	0.98	0.99	1.00	0.99	0.98	0.99	1.00	0.99	0.99	0.99	0.99	0.99	0.99

Table 7 Texture features values of a sandblasted surface

Feature No.	SB1	SB2	SB3	SB4	SB5	SB6	SB7	SB8	SB9	SB10	SB11	SB12	SB13	SB14	SB15
F1	0.85	0.81	0.82	0.87	0.88	0.77	0.81	0.87	0.80	0.77	0.82	0.88	0.82	0.78	0.88
F2	0.29	0.41	0.38	0.26	0.26	0.48	0.41	0.28	0.42	0.50	0.38	0.23	0.37	0.51	0.24
F3	0.85	0.81	0.82	0.87	0.88	0.77	0.81	0.87	0.80	0.77	0.82	0.88	0.82	0.78	0.88
F4	1.66	1.86	1.85	1.51	1.58	2.04	1.77	1.48	1.98	2.17	1.70	1.33	1.81	2.13	1.37
F5	0.64	0.59	0.61	0.63	0.67	0.59	0.50	0.55	0.64	0.67	0.51	0.56	0.60	0.64	0.55
F6	7.44	8.36	7.31	8.52	7.99	8.21	8.33	9.56	8.18	9.00	9.05	9.61	7.09	7.68	8.11
F7	3.93	4.16	3.96	4.05	3.97	4.20	4.13	4.21	4.18	4.34	4.22	4.13	3.91	4.07	3.89
F8	1.46	1.56	1.57	1.33	1.39	1.68	1.46	1.28	1.67	1.79	1.42	1.17	1.54	1.75	1.19
F9	0.60	0.71	0.69	0.57	0.58	0.74	0.71	0.60	0.71	0.76	0.69	0.55	0.68	0.78	0.56
F10	0.12	1.33	1.13	0.85	0.72	1.24	0.55	0.22	1.00	2.12	0.83	0.51	0.77	2.06	0.06
F11	4.63	9.41	9.02	6.28	7.03	10.76	5.54	3.13	9.90	16.90	5.58	3.02	6.75	15.34	2.79
F12	1.00	0.99	0.99	1.00	1.00	0.99	0.99	1.00	0.99	0.99	0.99	1.00	0.99	0.99	1.00

Table 8 Confusion Matrix

Training						Testing					
G	S	M	SB	E	Classified as	G	S	M	SB	E	Classified as
15	0	0	0	0	G	12	0	0	2	1	G
0	15	0	0	0	S	0	15	0	0	0	S
0	0	15	0	0	M	0	0	15	0	0	M
0	0	0	15	0	SB	0	0	0	15	0	SB
0	0	0	0	15	E	0	0	0	0	15	E

Table 9. Comparison between Vision and CMM readings

CMM reading(mms)	Vision reading (mms)
35.42	35.4345
40.44	40.4786
45.61	45.6654
50.42	50.4986
55.63	55.6543
60.43	60.4564
65.62	65.6543
70.42	70.4342

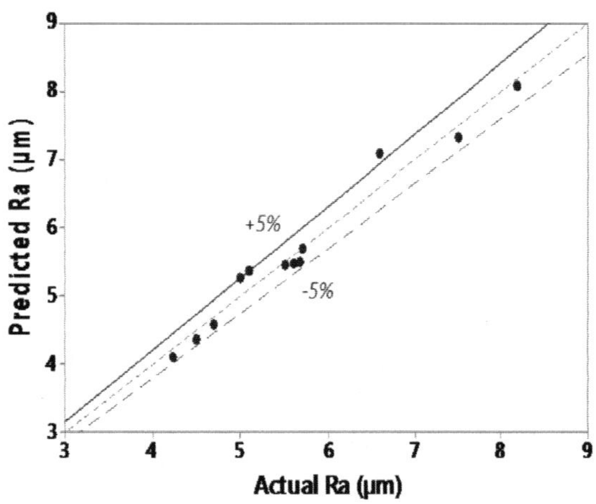

Figure 4 Correlation between Vision and Stylus roughness

Surface roughness inspection

Table 4, 5, 6 and 7 shows sample texture feature values obtained for EDM, shaped, milled and sandblast surface image, respectively. Texture features calculated for the different surface images are then used for computing the correlation with the roughness values (Actual Ra) obtained by the stylus method. The readings obtained from the stylus method (Figure 3), is considered as standard and are used for validating vision roughness parameters. Figure 4 shows the correlation between the contrast value (predicted) with the roughness (Actual Ra) obtained from the stylus instrument. Since there is a good correlation between contrast (f1) and Ra value obtained from the stylus instrument. Thus, contrast value (f1) can be used as a vision roughness parameter.

3.5. Dimensional inspection

Table 9 shows the results obtained after the dimensional inspection. Coordinate measuring machine (CMM) readings are used as the standard for calibrating the readings obtained from the vision system.

4. Conclusion

Surface texture classification is carried out on machined surfaces produced by machining processes viz. shaping, sand-blasting, milling, and EDM. The surface texture was identified using a combination of gray level co-occurrence matrix and classifier. The classifiers gave an accuracy of 96%. The method can also be used for performing roughness measurements. It was observed that there was a good correction between vision roughness and that of Ra value obtained from the stylus method. The method can also be used in performing the dimensional inspection. It was found that the minimum accuracy of dimensional inspection obtained is 96%. Thus, the current research demonstrates that the entire product inspection can be done in a single set-up and thus saves the cost of the inspection. Also, the method can be used in the online inspection of machined components and products.

Summary:

- Method presented in this work is capable of identifying images having textures of Shaping, Milling, Grinding, EDM and Sandblasting processes.
- Method can be used for measuring component dimensions e.g., length, diameter, etc.
- Method is also useful in measuring surface roughness of manufactured components.
- By this method 100% inspection is possible.
- Method is non-contact in nature and hence suitable for online inspection.
- Method cannot be used in the texture identification of smooth surfaces (e.g., Polished surface).

5. References

[1] Devdas Shetty and Clement Imbert 1984 Computer Identification of machined surfaces, *Journal of Testing and Evaluation*, **12** 6 375-379

[2] Wesczka J.S and Rosenfeld A 1976 An application of Texture analysis to materials inspection, Pattern Recognition **8** 195-199

[3] Haralick R.M. Shanmugam K. and Dinstein I. 1973 Texture features for Image classification *IEEE Transactions on Systems, Man and Cybernetics* **3** 610-621

[4] Galloway M.M 1975 Texture analysis using grey level run lengths, *Computer Vision Graphics Image Processing* **4** 172-179

[5] Chu A., Sehgal C.M., and GreenLeaf J.F. 1990 Use of grey level value distribution of run lengths for texture analysis Letters **11** 415-420

[6] Francesco Bianconi, Antonio Fernández 2007 Evaluation of the effects of Gabor filter parameters on texture classification *Pattern Recognition* **40** 12

[7] Ramchandra Manthalkar, P.K Biswas, B.N Chatterji 2003 Rotation invariant texture classification using even symmetric Gabor filters *Pattern Recognition Letters* **24** 12 2061-2068

[8] Mahamadou Idrissa, Marc Acheroy 2002 Texture classification using Gabor filters *Pattern Recognition Letters* **23** 9, Pages 1095-1102

[9] Nam Chul Kim, Hyun Joo So, Directional statistical Gabor features for texture classification, *Pattern Recognition Letters* **112** 18-26

[10] Somayeh Molaei, Mohammad Ebrahim Shiri Ahmad Abadi 2020 Maintaining filter structure: A Gabor-based convolutional neural network for image analysis *Applied Soft Computing* **88**,105960

[11] Arivazhagan, S., Ganesan,L., Padam Priyal,S 2006 Texture classification using Gabor wavelets-based rotation-invariant features *Pattern Recognition Letters*, **27** 16 1976-1982

[12] Yuhang Pan, Ping Zhou, Ying Yan, Anupam Agrawal, Yonghao Wang, Dongming Guo, Saurav Goel 2021 New insights into the methods for predicting ground surface roughness in the age of digitalisation *Precision Engineering* **67** 393-418

[13] Lee BY, Tarng YS.2001 Surface roughness inspection by computer vision in turning operations *International Journal of Machine Tools and Manufacture* **41** 9 1251-1263

[14] Franco A, Otaduy D, Barreda A I, Fernández-Luna J L, Merino S, González F, Moreno F 2019 Optical inspection of manufactured nanohole arrays to bridge the lab-industry gap *Optics & Laser Technology* **116** 48-57

Acknowledgments
The author would like to express his sincere thanks to the management of Pandit Deendayal Energy University, for providing the necessary infrastructure and timely support.

Basic characterization and carbon capture study of an indigenous activated carbon

Gautam[1*], Satyabrata Sahoo[1]

[1]Department of Mechanical Engineering, Indian Institute of Technology (Indian School of Mines), Dhanbad-826004, India

[1*]E-mail: gautam.s02668@gmail.com; satya@iitism.ac.in;

Abstract. The current research article deals with the fundamental characterization and carbon capture study of an indigenous activated carbon (AC) named MODISORB PL 4, derived from bituminous coal. Physical characterization to study the porous properties is carried out using N_2 adsorption/desorption at 77 K on a 3 Flex Micromeritics Setup. The activated carbon sample has a BET surface area of 1213.016 m²/g, a pore volume of 0.285 cm³/g, and a pore size of 33.44 Å. The structural analysis carried out by powder XRD confirms the presence of graphite lattice in the adsorbent AC. The surface morphology is studied by FE-SEM analysis, which revealed that adsorbent activated carbon is microporous. The CO_2 adsorption study is carried out for a relative pressure range of 0 to 1 and temperature varying from 298-333 K. The experimental data confirm the exothermic and physisorption behavior of the process. Isotherm models; Toth and Sips are fitted with the experimental CO_2 adsorption data, in which both resulted in a good fit for the temperature and pressure range taken. The corresponding isotherm constants are illustrated in the article. Further, a comparison study is carried out with the existing literature to compare the CO_2 uptake at ambient conditions, revealing that the assorted AC sample has immense potential for gas sorption, energy storage and carbon capture.

Keywords: *Carbon capture, CO_2 adsorption, adsorption isotherm, activated carbon*

1. Introduction

Adsorption systems are among the most promising clean technologies in the modern world, dealing with many environmental issues. The rate of CO_2 emission into our atmosphere is increasing and is expected to reach 37-40 Gt by 2030 [1]. Carbonaceous materials like activated carbons (ACs) are the most favorable candidates for carbon capture and carbon dioxide-based sorption systems, as they are cheap, thermally stable, hydrophobic in nature, and have an affinity towards CO_2 [2]. Moreover, ACs are up-and-coming candidates in other fields like petrochemical, environmental, food & beverages, automobile and chemical sectors [3]. Serafin et al. [4] prepared and characterized different activated carbons derived from various biomass materials, having higher CO_2 adsorption capacity. The ACs were prepared by KOH activation and carbonization. It was noted that micropores in the 0.30-0.86 nm range were responsible for efficient CO_2 adsorption at 1 bar and 273 K. Maximum CO_2 uptake and selectivity (6.03 mmol/g at 1 bar and 273 K) was obtained for pomegranate peel based activated carbon. Gomez-Delgado et al. [5] prepared various ACs via KOH activation from Pinus canariensis cones. Effects of various parameters such as impregnating KOH/carbonized cone ratio (IR) and temperatures (873-1073 K) were examined. The study showed that the activated carbon prepared with

IR=3 at 973 K resulted in a BET surface area of 1900 m^2/g and a maximum CO_2 uptake at 273 K of 6.4 mmol/g. Rashidi and Yusup [6] prepared palm shell kernel-based ACs via single-stage physical activated technique. The goal of the study was to examine the CO_2 uptake and product yield at room temperature and pressure under the impact of different input parameters, such as gas flow rate, temperature, and dwell time. At 850 °C, a gas flow rate of 450 cm^3/min, 25.15 percent of product yield, holding time of 60 min, the maximum CO_2 uptake of 2.13 mmol/g was obtained. The produced ACs are economical as per the industrial applications, as they showed similar CO_2 uptakes compared to the ACs prepared by double-stage activation techniques. Pal et al. [7] reported a benchmark study on CO_2 uptakes onto various Biomass-derived ACs (BACs) (mangrove and waste palm trunk). The BACs were prepared from waste palm trunk and mangrove, and were activated at 900 °C and carbonized at 500 and 600 °C. The maximum CO_2 absorption of 2.175 g/g (absolute adsorption) by waste palm trunk AC carbonized at 500 °C, and at an adsorption temperature and pressure of 25 °C and 5.04 MPa, respectively, was shown in the study.

Furthermore, many studies are available in the literature that deals with the adsorption isotherm study of few ACs samples indigenous and other foreign grades [8-10]. Nevertheless, studies related to fundamental morphological characteristics of indigenous ACs and their CO_2 capture ability under ambient conditions are still very limited in the literature. In the present analysis, an indigenous AC named 'MODISORB PL 4' is taken for the study. Basic morphological characterization viz... BET analysis, XRD, EDAX, FE-SEM and CO_2 adsorption capacity are tested for the temperature varying from 298 K to 333 K and for pressure ranging from 0 to 1 bar.

2. Experimental section

The adsorbent AC is provided by M M Corporation, Ahmedabad, India. Different physical properties of the adsorbent activated carbon such as BET surface area, pore size and volume, pore size distribution
(PSD), structural information and surface morphology are studied by BET analysis, XRD, FE-SEM and EDAX, respectively. The CO_2 adsorption isotherm is carried out using a 3 Flex Micromeritics setup at 298, 313, and 333 K for P/P_0=0 to 1. The detailed analysis of all the aforementioned tests is discussed in the upcoming sections.

2.1. Physical Characterization

The BET surface area, PSD, pore-volume and diameter are analyzed using 3 Flex Micromeritics Version 3.02 using the N_2 adsorption/desorption technique. The analysis is done at -196 °C and a relative pressure range of 0 to 1. The AC sample is degasified overnight at 300 °C before the test. The tested activated carbon has a BET surface area of 1213.016 m^2/g, a pore volume of 0.285 cm^3/g, and an average pore diameter of 33.44 Å. Nitrogen adsorption/desorption isotherms onto the Modisorb PL 4 pellet along with the BET report is shown in Figure 1. The isotherm curves of N_2 adsorption and desorption almost follow the same path indicating that the processes are reversible in nature. The packing density of the AC sample is 0.43-0.52 g/cm^3, which is provided by the manufacturer. Figure 2 shows pore size distribution (PSD) analysis of the activated carbon pellet. The PSD analysis is carried out using NLDFT given by the software package of 3 Flex Micromeritics. It clear from the figure that the AC pellet is microporous as the pore volume is well below 20 Å. Moreover, the pore size distribution is more concentrated in the pore width range of 4-16 Å.

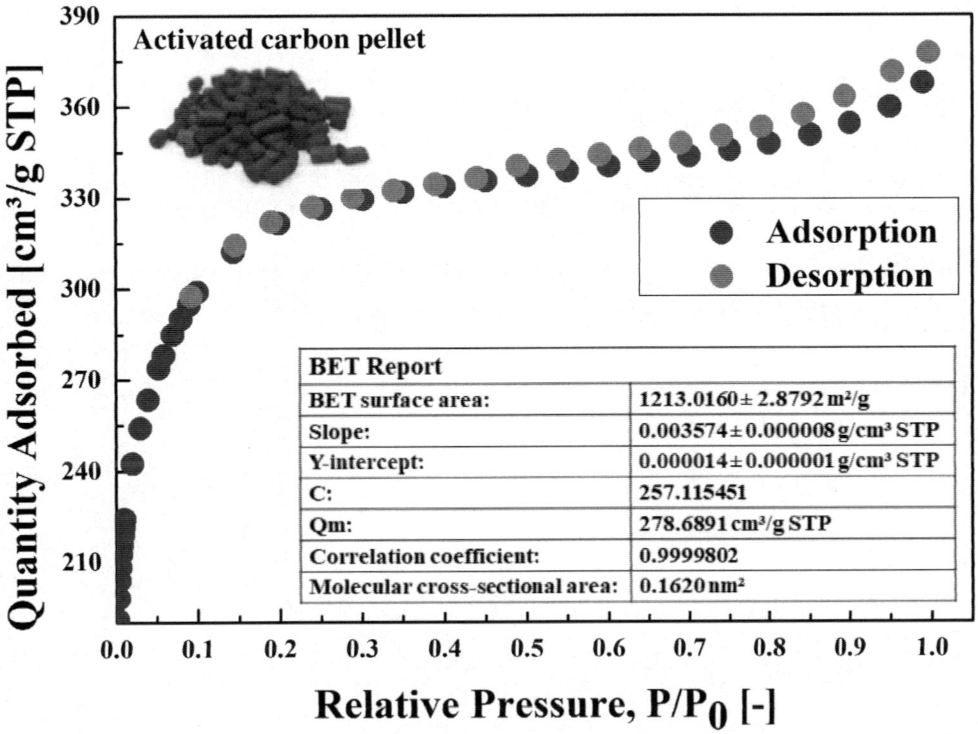

BET Report	
BET surface area:	1213.0160 ± 2.8792 m²/g
Slope:	0.003574 ± 0.000008 g/cm³ STP
Y-intercept:	0.000014 ± 0.000001 g/cm³ STP
C:	257.115451
Qm:	278.6891 cm³/g STP
Correlation coefficient:	0.9999802
Molecular cross-sectional area:	0.1620 nm²

Figure 1. Adsorption/desorption isotherms of nitrogen on AC Modisorb PL 4 at -196 °C

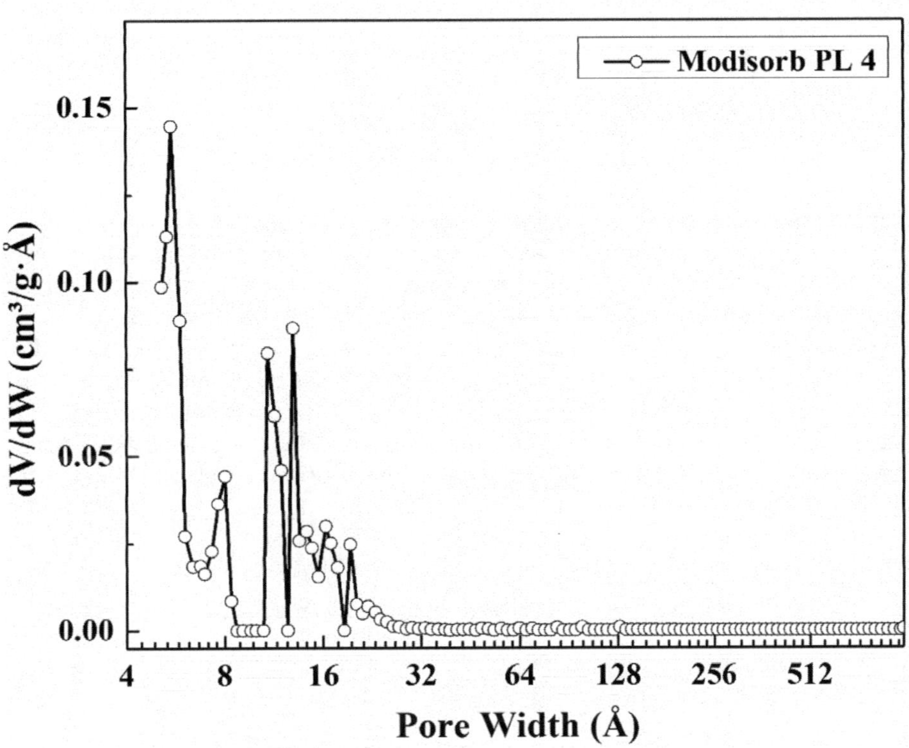

Figure 2. Pore size distribution of Modisorb PL 4

X-ray powder diffraction (XRD) on the Rigaku Smart lab instrument is used to analyze the structural properties of the activated carbon sample using CuKa radiation with λ= 1.54 Å. For the analysis, 40 kV and 40 mA of tube voltage and current are maintained. The diffraction patterns with a step size of 0.02° and a scanning speed of 2 °/ min are taken for 2θ ranging between 10 to 80°. Figure 3 shows the XRD patterns of the activated carbon sample. Five sharp diffraction peaks are observed at 2θ=26.603°, 31.558°, 36.113°, 42.465°, and 44.670°. Further interpretations of XRD analysis are carried out using X'Pert High Score Plus software. The peaks at 2θ=26.603°, 42.465°, and 44.670° denoted the presence of graphite with (002), (100), and (101) planes, respectively. The peaks at 2θ=31.558° and 36.113° may be due to some impurities (which may be present because of the pelletization of the AC).

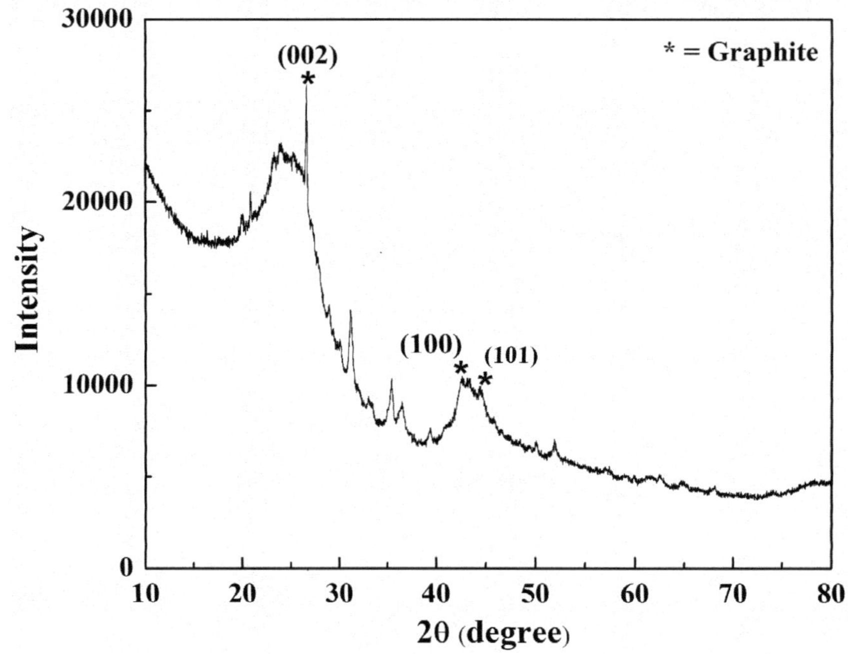

Figure 3. XRD analysis of MODISORB PL 4

The surface morphology and microstructure of the adsorbent activated carbon are studied by FE-SEM using a Supra 55 Zeiss at a magnification of 5 KX at a 5 kV accelerating voltage. The FE-SEM photograph of the activated carbon is shown in Figure 4. It is depicted from the figure that the AC sample consists of smaller flakes and a rock-type hybrid structure and is a microporous adsorbent, which confirms the suitability for CO_2 adsorption. Area-wise Energy Dispersive X-Ray Analysis (EDAX) is carried out on powdered activated carbon to explore the elemental composition. The results of the EDAX analysis are shown in Figure 5 and Table 1. The analysis confirmed the highest carbon percentage in the activated carbon sample. The study also revealed the presence of residual mineral ions in the AC sample, which can be related to the peaks observed at 2θ=26.603°, 42.465°, and 44.670° during the XRD analysis.

Table 1. Elemental composition of the AC obtained from the EDAX analysis

Element	Weight %	Element	Weight %
C K	80.83	Fe K	0.74
O K	9.90	Mo L	0.90
Mg K	0.44	Sb L	1.94
Al K	1.19		
Si K	1.86		
Ca K	2.19		

Figure 4. FE-SEM image of Modisorb PL 4 at 5 KX magnification

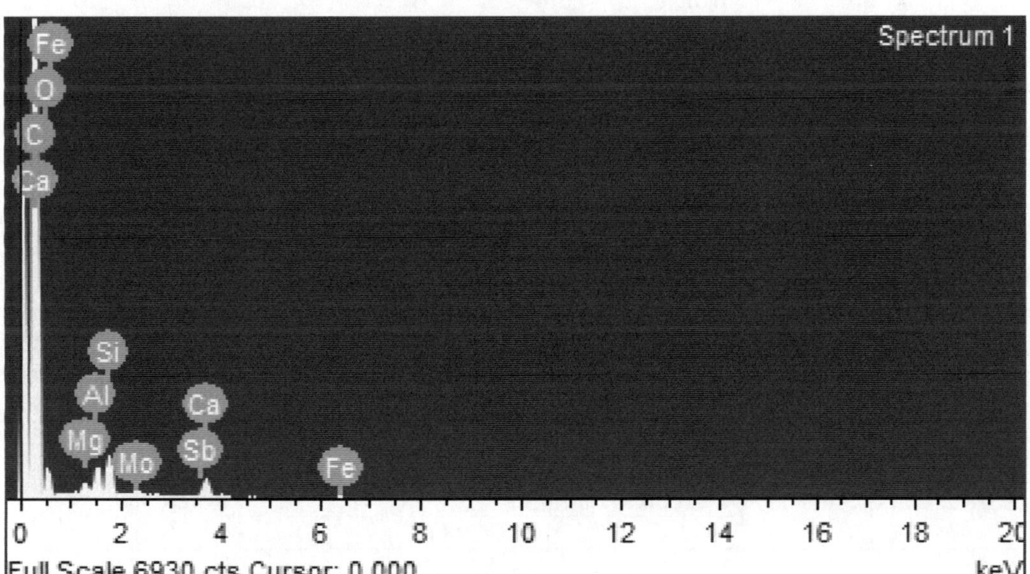

Figure 5. EDAX analysis of AC Modisorb PL 4

2.2. CO_2 adsorption
The carbon capture study is conducted using a 3 Flex Micromeritics Setup at varying temperatures ranging from 298 to 333 K for a relative pressure range (P/P_0) of 0 to 1. The results are illustrated in Figure 6. The figure clearly indicates that with a rise in temperature, the adsorption of CO_2 decreases due to the decrease in binding strength between the adsorbate and the adsorbent. In addition, the decrease in the ability of CO_2 adsorption at higher temperatures suggests that the sorption process that occurs is exothermic and a physisorption process.

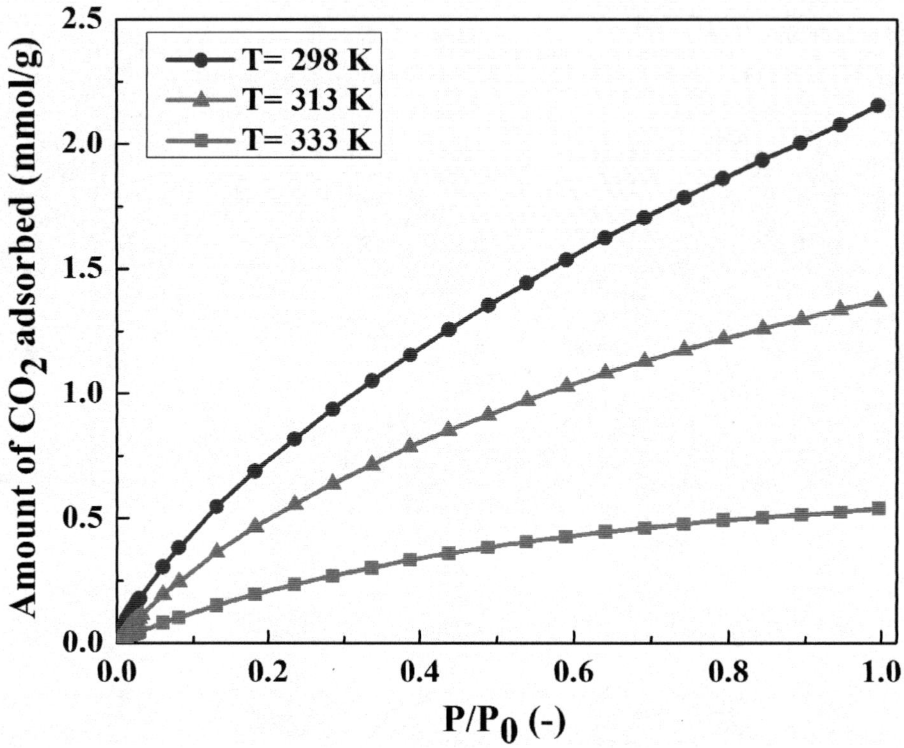

Figure 6. Experimental equilibrium data of CO_2 adsorption onto AC at 298-333 K

2.2.1. Comparison of CO_2 uptakes
Recent studies show that BACs have a significant potential for carbon capture under ambient conditions. Table 2 shows comparative values of CO_2 uptakes of various BACs taken from the open literature and current work results. The selected activated stands well with the other BACs under similar conditions. In fact, the CO_2 uptake of the selected AC is better than BACs derived from palm fruit bunch, which is quite appreciable. Therefore it is noteworthy to mention that the selected AC is a suitable candidate for gas adsorption/separation, energy storage, adsorption based heat pumps and carbon capture.

Table 2. Comparison of CO_2 uptakes onto ACs under ambient conditions

Activated Carbon Precursor	CO_2 uptake (mmol/g)		Reference
	T=273 K	T=298 K	
Fern leaves	4.52	4.12	[4]
Pomegranate peels	6.03	4.11	[4]
Carrot peels	5.64	4.18	[4]
Coffee ground	6.89	4.00	[11]

Lignin	-	2.8	[12]
Wood Camellia japonica	3.0	-	[12]
Fungi	5.5	3.5	[13]
Coconut shell	5.6	3.9	[14]
Coffee grounds	4.7	3.0	[15]
African palm stones	3.7	-	[16]
Coconut shell	3.7	-	[16]
Starch	6.1	4.8	[17]
Palm fruit bunch	1.3	1.12	[18]
Bituminous coal	-	2.14	Present work

2.3. Isotherm modeling

All the experimental data at various temperatures are fitted with the Sips and Toth model, given by Eq. (1) & (2), respectively. The corresponding isotherm constants are given in Table 3. It is depicted from Figure 7 shows that both the isotherm model provides an excellent fit for all the temperature and pressure ranges taken in the study. Both the isotherm models are suitable for heterogeneous adsorbents like activated carbon and depict monolayer adsorption. Moreover, the increasing trends of the heterogeneity factor 't' with an increase in temperature refers to higher adsorption energies. Also, decreasing the values of the limiting uptake 'q_m' shows the exothermic behavior of CO_2 adsorption with increasing temperature.

$$q = \frac{q_m (KP)^{1/t}}{1+(KP)^{1/t}} \tag{1}$$

$$q = \frac{q_m (KP)}{\left(1+(KP)^t\right)^{1/t}} \tag{2}$$

Table 3. Isotherm constants for CO_2-MODISORB PL 4 pair

Isotherm model	Sips			Toth		
Temperature (K)	298	313	333	298	313	333
q_m (mmol/g)	11.27064	3.32619	0.9497	96.64889	6.01578	1.10384
K (1/mbar)	0.00015	0.00066	0.00133	0.00014	0.00077	0.00146
t (-)	0.76358	0.86636	0.94236	0.25417	0.51708	0.77890

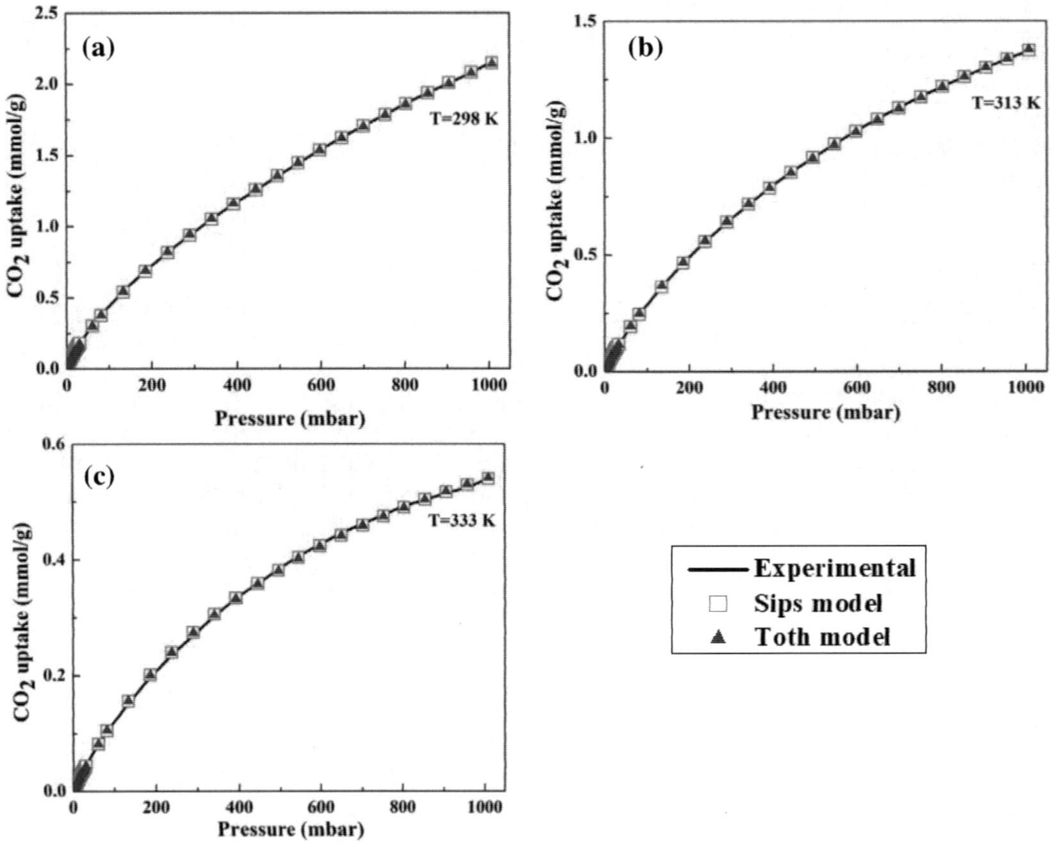

Figure 7. Isotherm data by linear regression method for CO_2 uptake at 298, 313, and 333 K

3. Conclusion

In the present work, basic characterization and CO_2 adsorption study are carried out on bituminous coal-based activated carbon called MODISORB PL 4. The BET analysis is carried out on the sample AC sample using the N_2 adsorption/desorption technique, which revealed a surface area of 1213.016 m^2/g with a pore volume of 0.285 cm^3/g. The CO_2 adsorption study is carried out for a relative range of pressure (P/P_0=0 to 1) and different temperatures (298 K, 313 K, 333 K) using a 3 Flex Micromeritics Setup. Experimental results proved that the adsorption behavior was physisorption as increment and decrement in CO_2 adsorption capacity are noticed with an increase in pressure and temperature, respectively. A comparison is made for CO_2 uptake at ambient conditions with various grades of biomass-derived ACs, which shows that the selected AC has potential for gas sorption and carbon capture. Toth and Sips isotherm models are fitted with the obtained experimental data of CO_2 uptake, and both of them resulted in a good fit for the whole pressure and temperature range. The isotherm data of the selected AC is significant and can be used to compare other indigenous commercially available activated carbons. Moreover, studies like TGA and FTIR will further enable us to describe the surface morphology and adsorbent properties of the assorted activated carbon, which is included in our future work.

Acknowledgments
The authors would like to acknowledge TEQIP-III IIT (ISM) Dhanbad for the financial assistance provided to carry out the present research work. The authors would also like to acknowledge the Central Instrument Facility (CRF) IIT (ISM) Dhanbad for providing the testing facility.

References

[1] Gautam, Sahoo S. Effects of geometric and heat transfer parameters on adsorption–desorption characteristics of CO_2-activated carbon pair. *Clean Technologies and Environmental Policy*. 2020 Jun 6:1-21. https://doi.org/10.1007/s10098-020-01866-3.

[2] Gautam, Kumar G, Sahoo S. Performance improvement and comparisons of CO_2 based adsorption cooling system using modified cycles employing various adsorbents: a comprehensive study of subcritical and transcritical cycles. *International Journal of Refrigeration*. 2020 Apr 1;**112**:136-54.

[3] Bagheri S, Muhd Julkapli N, Bee Abd Hamid S. Functionalized Activated Carbon Derived from Biomass for Photocatalysis Applications Perspective. *International Journal of Photoenergy*. 2015 Feb 23; 1-30. http://dx.doi.org/10.1155/2014/218743

[4] Serafin J, Narkiewicz U, Morawski AW, Wróbel RJ, Michalkiewicz B. Highly microporous activated carbons from biomass for CO_2 capture and effective micropores at different conditions. *Journal of CO_2 Utilization*. 2017 Mar 1;**18**:73-9.

[5] Gomez-Delgado E, Nunell G, Cukierman AL, Bonelli P. Tailoring activated carbons from Pinus canariensis cones for post-combustion CO_2 capture. *Environmental Science and Pollution Research*. 2020 Feb 8; **27**:13915–29.

[6] Rashidi NA, Yusup S. Production of palm kernel shell-based activated carbon by direct physical activation for carbon dioxide adsorption. *Environmental Science and Pollution Research*. 2019 Nov; **26**:33732-46.

[7] Pal A, Uddin K, Saha BB, Thu K, Kil HS, Yoon SH, Miyawaki J. A benchmark for CO_2 uptake onto newly synthesized biomass-derived activated carbons. *Applied Energy*. 2020 Apr 15;**264**:114720.

[8] Singh VK, Kumar EA, Saha BB. Adsorption isotherms, kinetics and thermodynamic simulation of CO_2-CSAC pair for cooling application. *Energy*. 2018 Oct 1;**160**:1158-73.

[9] Singh VK, Kumar EA. Thermodynamic analysis of single-stage and single-effect two-stage adsorption cooling cycles using indigenous coconut shell based activated carbon-CO_2 pair. *International Journal of Refrigeration*. 2017 Dec 1;**84**:238-52.

[10] Singh VK, Kumar EA. Experimental investigation and thermodynamic analysis of CO_2 adsorption on activated carbons for cooling system. *Journal of CO_2 Utilization*. 2017 Jan 1;**17**:290-304.

[11] Travis W, Gadipelli S, Guo Z. Superior CO_2 adsorption from waste coffee ground derived carbons. *RSC advances*. 2015;**5(37)**:29558-62.

[12] Coromina HM, Walsh DA, Mokaya R. Biomass-derived activated carbon with simultaneously enhanced CO 2 uptake for both pre and post combustion capture applications. *Journal of Materials Chemistry A*. 2016;**4(1)**:280-9.

[13] Wang J, Heerwig A, Lohe MR, Oschatz M, Borchardt L, Kaskel S. Fungi-based porous carbons for CO_2 adsorption and separation. *Journal of Materials Chemistry*. 2012;**22(28)**:13911-3.

[14] Ello AS, de Souza LK, Trokourey A, Jaroniec M. Coconut shell-based microporous carbons for CO_2 capture. *Microporous and Mesoporous Materials*. 2013 Nov 1;**180**:280-3.

[15] Plaza MG, González AS, Pevida C, Pis JJ, Rubiera F. Valorisation of spent coffee grounds as CO_2 adsorbents for postcombustion capture applications. *Applied Energy*. 2012 Nov 1;**99**:272-9.

[16] Vargas DP, Giraldo L, Silvestre-Albero J, Moreno-Piraján JC. CO_2 adsorption on binderless activated carbon monoliths. *Adsorption*. 2011 Jun;**17(3)**:497-504.

[17] Sevilla M, Fuertes AB. Sustainable porous carbons with a superior performance for CO_2 capture. *Energy & Environmental Science*. 2011;**4(5)**:1765-71.

[18] Parshetti GK, Chowdhury S, Balasubramanian R. Biomass derived low-cost microporous adsorbents for efficient CO_2 capture. *Fuel*. 2015 May 15;**148**:246-54.

IOP Conf. Series: Materials Science and Engineering 1146 (2021) 012005 doi:10.1088/1757-899X/1146/1/012005

Biodiesel Production from Waste Cooking Oil Using Sequential Process Intensification Technique (Ultrasound and Microwave)

Himanshi Gupta[1], Kartikkumar Thakkar[1], S.S. Kachhwaha[1], Pravin Kodgire[2]

[1]Department of Mechanical Engineering, Pandit Deendayal Petroleum University, Gujarat, India
[2]Department of Chemical Engineering, Pandit Deendayal Petroleum University, Gujarat, India
E-mail: himanshi.gmtmt18@sot.pdpu.ac.in; kartik.tphd16@sot.pdpu.ac.in; Surendra.Singh@sot.pdpu.ac.in; Pravin.Kodgire@sot.pdpu.ac.in;

Abstract. The use of biodiesel as a renewable fuel has increasingly become more attractive due to its environmental advantages such as non-toxicity and biodegradability. While biodiesel has the capability to mitigate the energy crisis, its production is hindered by the availability of feedstock. The general drawbacks of the transesterification-based synthesis include high energy consumption, high operating cost, high separation times, and low production efficiency. The present study demonstrates an innovative and industrially viable biodiesel production process from waste cooking oil (WCO) as feedstock and KOH as a catalyst using sequential ultrasound (22kHz + 35kHz) and microwave (0 – 1000W) reactor. The sequential approach of ultrasound and microwave removes the barrier of mass transfer (high-velocity liquid jets and turbulence induced by ultrasound) and enhances the heat transfer (rapid dipolar moment induced by microwave) by their synergism while significantly improving the reaction efficiency and overall economics of the process. The reaction conditions that can attain a maximum yield of 96.9% were methanol/oil molar ratio, 7:1, catalyst concentration, 0.9 wt%, and reaction temperature, 62°C in 5 minutes of reaction time for a sample size of 4500 mL.

1. Introduction

Currently, global heating and environmental deterioration are threatening the world. The gases emitted by burning fossil fuels are the major source of pollution. Therefore, alternative energy sources need to be identified that are feasible, ecologically friendly, and commercially sustainable. In addition, the world-wide exhaustion of petroleum has also encouraged the search for alternative sources.

Biodiesel as an alternative fuel holds great potential. It is a biodegradable, renewable, non-toxic fuel, and a reducer of carbon monoxide emissions that can be recycled by photosynthesis. Besides, biodiesel has the benefit of good fuel properties such as good lubricity, reduced exhaust gas emissions, low sulfur, and aromatic contents, carbon-neutral and lower carbon dioxide emissions in the atmosphere, a cetane number and cloud point that is largely dependent on the feedstock and a high flash point that tends to make it volatile and easy to handle [1].

For biodiesel productivity, edible oils are categorized as first-generation feedstock. Nonetheless, due to the clash with the accessibility of food and waste disposal in the environment, there is currently a switch towards the use of low-grade feedstock (second-generation) such as inedible oils and waste cooking oil (WCO) to produce biodiesel [2].

Various characteristics of edible oil are modified by frequent frying. Therefore, controlling the quality of the oil is necessary to avoid using deteriorated oil for cooking purposes. The Food Safety and Standards Authority of India (FSSAI) stipulates a threshold of 25% for Total Polar Compounds (TPC), above which vegetable oil is not appropriate for use [3]. Vegetable oil with a TPC of more than 25% is referred to as waste cooking oil.

The properties of WCO vary globally. WCO is collected from large-scale food processing industries, households, and business establishments. The usage of WCO as a substitute feedstock has the potential to mitigate the environmental concern of waste disposal [4]. The other benefits of using waste cooking oil over edible vegetable oil include easy availability, better oxidation stability, and a greater percentage of saturated fatty acids [5]. Currently, 850 crore litres of diesel is used in India every month. The government intends to blend 5% of biodiesel into diesel by 2030 in line with the National Policy on Biofuels -2018 [3]. There is therefore a requirement for 500 crore litres of biodiesel per annum. The current consumption of vegetable oil in India is 2700 crore litres, of which 140 crore litres of WCO may be obtained from the majority of consumers like restaurants, canteens, and hotels, which could provide approximately 110 crore litres of biodiesel annually [3].

Approximately 75% of the biodiesel's production price corresponds to the price of virgin vegetable oil [6]. On an average, WCO is considered to be 2 to 3 times cheaper than that of virgin cooking oil [7]. The higher concentration of free fatty acid (FFA) makes WCO less expensive.

Biodiesel fuels are generally produced through a transesterification process, a method in which triglyceride (oil/fat) reacts with an alcohol involving a catalyst to produce esters (biodiesel) and glycerol (waste by-product). Transesterification comprises three successive reversible reaction steps, with each step resulting in one ester molecule [8] as shown in equations (1), (2), and (3).

$$Triglyceride + R'OH \leftrightarrow Diglyceride + R'COOR \quad (1)$$

$$Diglyceride + R'OH \leftrightarrow Monoglyceride + R'COOR \quad (2)$$

$$Monoglyceride + R'OH \leftrightarrow Glycerol + R'COOR \quad (3)$$

The traditional approach of mechanical stirring for biodiesel production from WCO as a feasible feedstock has been reported by several researchers in their studies [9][10]. This method has many drawbacks like low product conversion, an excess concentration of reactants, slow and inefficient heating, and higher by-product yield [11]. Thus, resulting in excess power consumption for the production of biodiesel.

The concept of process intensification refers to the implementation of innovative techniques and/or equipment that can achieve substantially higher benefits or yields compared to the existing techniques and/or equipment in operation [12]. In the perspective of the production of biodiesel, process intensification refers to the rise in heat and mass transfer rates between reaction products at various stages of biodiesel production such as purification and/or separation and/or transesterification and/or extraction. Various techniques have been proposed to improve biodiesel productivity like reactive distillation [13], membrane reactor [14], microwave, and ultrasonic irradiation (independently or in a combination) [15][7]. Among them, the most noticeable results were observed when microwave or ultrasonic-assisted methods were employed for process intensification.

Several kinds of research involving microwave and ultrasound as techniques for intensifying the processes have demonstrated increased biodiesel yields, reduced energy consumption, better biodiesel quality, and reduced reaction times owing to the thermal and non-thermal effects of these radiations. In particular, microwaves produce fast and appropriate heating but have a mass transfer limit [16] while ultrasound induces intense mechanical mixing through cavitation but has heat transfer limitations [17]. These restrictions of heat and mass transfer can be resolved simultaneously by integrating the two separate methods into a single apparatus leading to a balanced operation that could lead to energy-efficient biodiesel productivity. Ultrasound will produce the mechanical mixing that microwave needs to boost mass transfer and minimize localized overheating; and microwaves will produce thermal energy that is lacking in ultrasound, hence enhancing the process and increasing the benefits [18].

The studies conducted by Yan et al. [19], Thakkar et al. [20], Yu et al. [21], Safieddin Ardebili et al. [22], Gole et al. [23], and Hsiao et al. [18], [24] showed use combined microwave-ultrasound and achieved the yield up to 98%. Studies on waste cooking oil using combined microwave-ultrasound were reported by Martinez-Guerra et al. [12], [22] reported the yield around 95%. The production of biodiesel by this method meets the standards for biodiesel.

While the impacts of ultrasound and microwave irradiations on the process of transesterification have been studied separately in several studies but their combined effect on biodiesel productivity has been stated in relatively few studies. The purpose of this research was to investigate the sequential impact of ultrasound followed by microwave on the biodiesel productivity from waste WCO and potassium hydroxide (KOH) as feedstock and catalyst respectively. The optimum reaction conditions of various process parameters such as methanol: oil molar ratio, catalyst content, reaction temperature, and time were also examined.

2. Materials and Methods

2.1. Overview of Experimental Configuration

Figure 1 demonstrates the system developed for the batch synthesis of biodiesel employing sequential ultrasound and microwave. The developed sequential reactor set-up comprises an oil tank, water tank, heater, stirrer, rotameter, flow control valves, pump, ultrasound generator, ultrasonic reactor, microwave, by-pass valves, and control panel. The system is composed of ultrasound and microwave connected in series. It is possible to use the ultrasound and the microwave separately, or both in series. The output of the ultrasound is supplied to the microwave in the series connection.

At first, a known quantity of oil is fed into the tank. It is heated to a required temperature utilizing a hot water jacket. A mixture of catalyst and solvent is supplied into the tank after the oil attains the required temperature, and then this mixture is stirred by means of a stirrer. This mixture then passes to the ultrasound through a pump. It passes to the microwave from ultrasound and then back to the tank. The mixture also keeps on rotating inside the reactor for a stipulated amount of time after which it is collected employing a drain valve at the bottom of the tank.

2.2. Reagents and Materials Used

WCO was collected from the canteen of PDPU, methanol (Emplura Grade, 99% purity) was used as a solvent and potassium hydroxide (KOH) flakes (Emplura Grade, 85% purity) were utilized as a homogeneous base catalyst.

2.3. Procedure for Experimental Run

The process of transesterification was performed in an oil tank (capacity: 20 liters) equipped with a condenser and thermocouple using a batch size of 4500 mL. A precise quantity of pre-mixed catalyst and methanol mixture was introduced into the tank. US power, MW power, and stirrer speed are 600W, 1000W, and 500 rpm respectively were fixed during all the reaction runs while varying the catalyst content from 0.9 to 1.2 wt%, methanol: oil molar ratio from 4.5:1 to 12:1, temperature from 52 to 62 °C, and reaction time varying from 5 to 15 minutes. Two phases of reaction mass, i.e., the top phase of biodiesel and the bottom phase of glycerol involving a small amount of unreacted methanol and oil, were found upon completion of the reaction. The extra methanol in the liquid product was recovered. After segregation of glycerol layer and recovery of methanol, the synthesized biodiesel was washed four times with distilled water to eliminate traces of the catalyst. The phase rich in biodiesel was washed again with distilled water to eliminate any remaining oil traces and then heated to □82 °C to extract any moisture.

2.4. Analysis and Characterization of Biodiesel

The samples of biodiesel were analyzed on GC (Agilent Technologies, 5975C) with HP-5MS capillary column possessing a flame ionization detector (FID) by employing the GC method defined for in-house laboratory use. The FAME identification [25] in the samples was done by comparison with the standard retention times. Biodiesel yield was determined using Equation (4):

$$Yield\ (\%) = \frac{Biodiesel\ (grams)\ x\ FAME\ (\%)}{Raw\ Oil\ (grams)}\ x\ 100$$

(4)

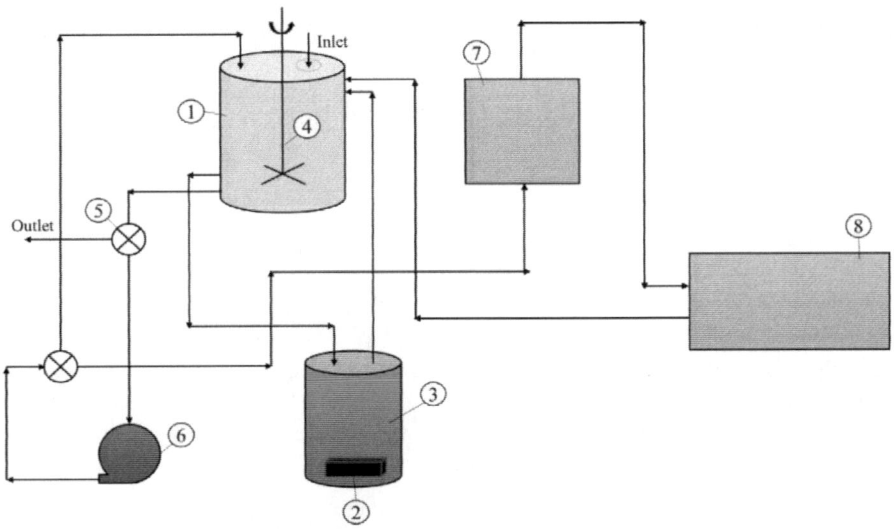

Figure 1. Schematic illustration of biodiesel production employing sequential ultrasound and microwave. 1: Oil tank, 2: Heater, 3: Water tank, 4: Stirrer, 5: By-pass valve, 6: Pump, 7: Ultrasound, 8: Microwave

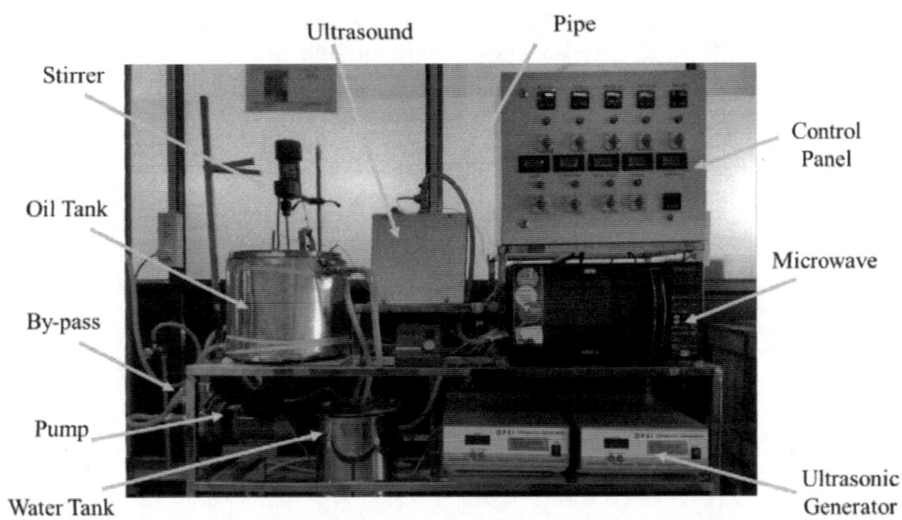

Figure 2. General view of the pilot-scale unit.

2.5. Energy Analysis

Energy consumed during biodiesel production is analyzed during the following steps: heating of waste cooking oil to extract moisture, conversion of raw oil to biodiesel in the reactor, extraction of extra methanol utilizing rotary evaporator, and heating of washed biodiesel to eliminate any trapped moisture.

Specific energy consumption (SEC), efficiency factor (η), and energy use index (EUI) [26]–[28] are the various performance parameters used in the energy analysis of the experiments. Energy can be categorized into supplied energy (Q_{MW+US}) and dissipated energy (Q_{th}).

$$Q_{MW+US} = P_{MW+US}\, t \tag{5}$$

$$Q_{th} = mC_p\Delta T \tag{6}$$

where Q_{MW+US} is the energy input to the reactor, P_{MW+US} is available reactor power (W), t is the time duration for which power is supplied (sec), Q_{th} is the energy needed to attain a particular temperature,

m is the reactant mass (kg), C_p is the heat capacity of the reactants (J/kg°C) and ΔT is the difference between mixture's initial and final temperature (°C).

Efficiency factor (η) as shown in Equation (7) can be determined as the proportion of dissipated energy to the supplied energy:

$$\eta = \frac{Q_{th}}{Q_{MW+US}} \tag{7}$$

Specific Energy Consumption (SEC) for the synthesis of biodiesel is determined by Equation (8):

$$SEC = \frac{\text{Total Energy Consumption (Wh)}}{\text{Biodiesel produced (kg)}} \tag{8}$$

The energy consumption of laboratory equipment is the focus of energy analysis in this study. However, a comprehensive study of energy flow needs to be conducted in order to optimize and control the energy employed in the process to improve the efficiency of biodiesel production.

3. Results and Discussions

3.1. Waste Cooking Oil Characterization

The fatty acid structure of waste cooking oil (WCO) was examined utilizing gas chromatography (Table 1). Palmitic acid, oleic acid, and linoleic acid are the main components of WCO according to this analysis. The acid value of WCO has been identified to be 2.25 mg KOH/g that corresponds to a free fatty acid concentration of 1.13%, considered suitable for single-step conversion (direct transesterification without esterification) of WCO to biodiesel. The measured moisture content was less than 0.05%.

Table 1. Fatty acids and their composition in WCO.

Fatty Acids	Composition (%)
Palmitoleic acid (C16:1)	0.96
Oleic acid (C18:1)	24.68
Linoleic acid (C18:2)	40.89
Palmitic acid (C16:0)	28.66
Stearic acid (C18:0)	2.56
Linolenic acid (C18:3)	1.27
Myristic acid (C14:0)	0.98

3.2. Assessment of Process Parameters

3.2.1 Impact of Methanol: WCO Molar Ratio

The method of transesterification involves reacting the triglycerides (fat/oil) with alcohol utilizing a catalyst for the formation of alkyl esters and crude glycerol. Stoichiometrically, 1 mole of triglyceride and 3 moles of alcohol are required for a complete transesterification reaction. In practice, this ratio must be greater than the stoichiometric ratio to attain a maximum ester yield. As this process is reversible, surplus alcohol is employed to transfer the balance to the synthesis of esters (product) [8]. The impact of various M: O molar ratios i.e., 5:1, 6:1, and 7:1, on the transesterification reaction at a pre-set temperature of 62 °C and catalyst content of 0.9 wt% was evaluated. As shown in Figure 3(a), the yield of biodiesel increases with increasing M: O molar ratio and time. A molar ratio of 7:1 was determined to achieve a maximum yield of 96.9% in 5 minutes. The biodiesel yield improved marginally as the reaction time increased further.

3.2.2 Impact of Catalyst

The catalysts utilized in the production of biodiesel are usually acids or bases (homogeneous or heterogeneous) due to which their properties are associated with different advantages and disadvantages [29], [30]. The reaction rates induced by base catalysts are higher than those by acid-catalyzed reactions which makes them the most preferred catalyst in biodiesel production. Sodium hydroxide (NaOH) and

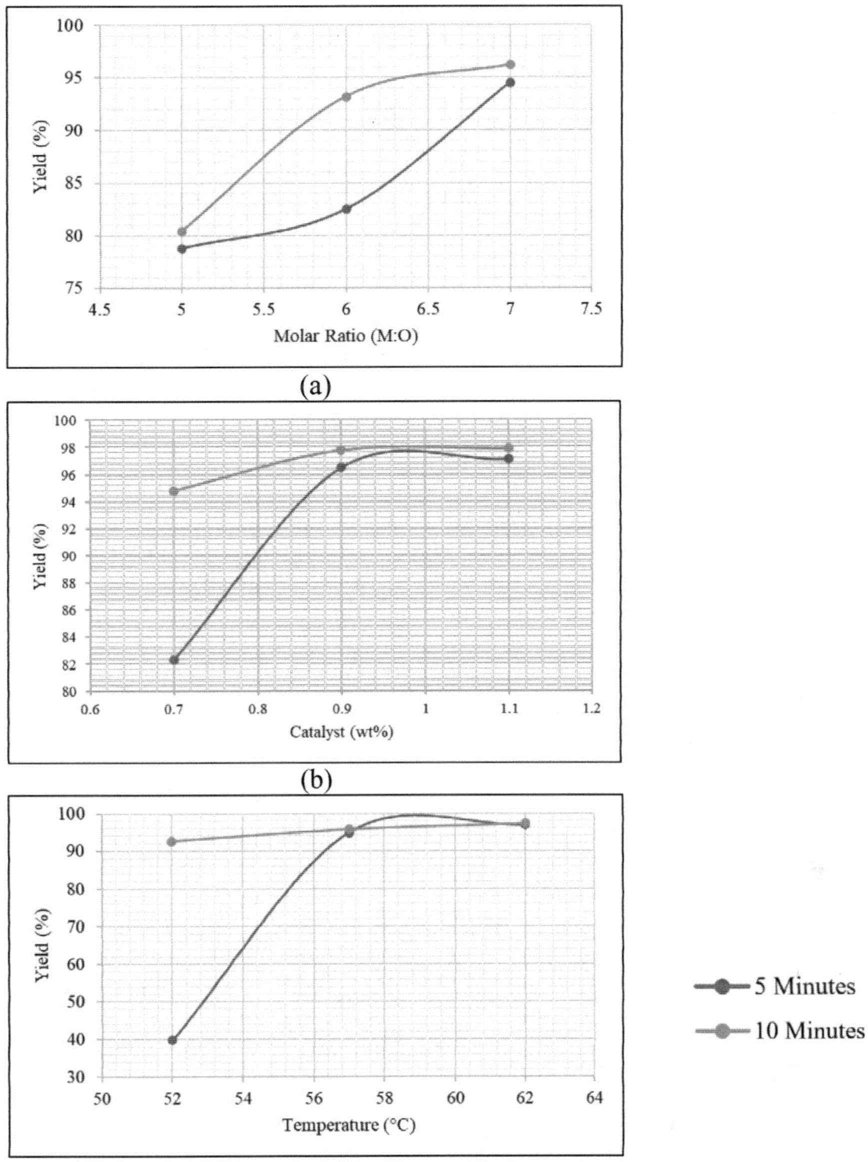

Figure 3. Impact of process parameters on yield (%): (a) M: O molar ratio at 0.9 wt% of KOH catalyst and 62 °C temperature; (b) catalyst concentration (wt%) at 7:1 M:O molar ratio and 62 °C temperature; and (c) reaction temperature at 7:1 M:O molar ratio and 0.9 wt% of KOH catalyst, on ultrasound and microwave irradiated sequential transesterification.

potassium hydroxide (KOH) flakes are among the extensively utilized base catalysts in the biodiesel industry. Here, we have selected KOH which is the most frequently used homogeneous base catalyst. The impact of the catalyst content (wt%) on the yield of biodiesel can be seen in Figure 3(b). For these experiments, the temperature was set at 62 °C and the methanol: oil molar ratio at 7:1. A catalyst content of 0.9 wt% was determined to achieve a biodiesel yield of 97.9% in 10 minutes. A lower amount of catalyst at less reaction time was not sufficient to accelerate the reaction forward, possibly due to a reduction in the mass transfer induced by methanol resulting in lower yields. Increasing the catalyst content from 0.7 to 0.9 wt%, increased the biodiesel yield, while increasing catalyst concentration to 1.1 wt% reduced the biodiesel yield. Two explanations can be given to explain why a lower conversion rate was reached with the catalyst amount at 0.7 and 1.1 wt%. Some amount of the catalyst was neutralized with traces of FFA present in WCO during the process of transesterification. Thus, increasing the catalyst content will increase the biodiesel yield. This may justify why the biodiesel yield was only 82.3% in 5 minutes of reaction time when 0.7 wt% of the catalyst was used. Increases in the catalyst amount from 0.7 to 0.9 wt% improved the biodiesel yield. Second, a saponification reaction could be triggered by too much of the catalyst concentration. This could explain why 0.9 wt% of the catalyst was more effective than 1.1 wt%. In short, the catalyst must be used in an optimal amount.

3.2.3 Impact of Reaction Temperature

Temperature is a vital factor in biodiesel production since it affects the rate and yield of transesterification. [31]. At higher reaction temperature, a reduction in reaction time and an upsurge in reaction rate are observed due to a decrease in oil viscosity. It also minimizes the limitation of mass transfer leading to increased yield of methyl ester and FFA conversion. However, higher reaction temperatures (above the optimum level) promote triglycerides saponification and methanol vaporization which results in decreased biodiesel yield. To inhibit the vaporization of alcohol, transesterification must occur at a temperature lower than the boiling point of alcohol. The optimum temperature limit for the reaction can fluctuate between 50°C and 60°C, depending on the type of oil used. Figure 3(c) shows the relation between reaction temperature and biodiesel yield. Constant process parameters were methanol: oil molar ratio of 7:1 and catalyst content of 0.9 wt%. The reaction temperatures used were: 52, 57, and 62 °C. It was observed that the biodiesel yield increased with the temperature rise. An optimal temperature of 62°C was determined to achieve a biodiesel yield of 96.9% in 5 minutes. The biodiesel yield improved marginally as the reaction time increased further.

3.3. Biodiesel Fuel Properties

Biodiesel fuel properties like calorific value, density, flash point, oxidation stability, and kinematic viscosity are calculated in compliance with ASTM standards. Table 2 shows the evaluated biodiesel properties synthesized using waste cooking oil, compared to ASTM standards.

Table 2. Comparison of WCO-produced biodiesel properties with ASTM standards.

Properties	ASTM Standard Values	ASTM Standard	This Study
Density (kg/m^3)	860 - 900	ASTM D4052	876
Flash point (°C)	130 (min)	ASTM D93	179
Kinematic viscosity (mm^2/sec)	1.9 - 6.0	ASTM D445	3.9
Oxidation stability (h)	3 hrs (min)	EN 14112	7.6
Calorific value (MJ/kg)	35 - 45	ASTM D240	37.6

3.4. Energy Analysis

The energy consumption during the biodiesel production process is shown in Table 3. Equations (6) and (7) were used to calculate energy dissipation in liquid mixtures and efficiency factor respectively. A system energy meter installed in the control panel of the reactor was used to measure the energy delivered to the set-up.

The energy requirement is driven by variables such as the sample volume, ultrasound/ microwave radiation intensity, nature of the solvent, and the reaction volume depth. The energy requirements for the various heating and mixing methods are compared in Table 4. The observations demonstrate that the energy required for processing and separation by the sequential approach has been reduced multifold compared to individual and simultaneous operation. The outcome of the sequential ultrasound and microwave approach has a substantial impact on the overall energy savings and production of biodiesel with greater economic prospects.

Table 3. Energy analysis of sequential reactor for 4500 mL batch size.

Run Order	M:O, Catalyst, Temp, Time	Measured Parameters			Performance Parameters			
		Heating of Oil (Wh)	Sequential Reactor (Wh)	Rotary Evaporator (Wh)	Q_{MW+US} (Wh)	Q_{th} (Wh)	η (%)	SEC (Wh/kg)
1	4.5:1, 0.9, 57, 5	625	218	2000	843	204	24.2	876
2	4.5:1, 0.9, 57, 15	625	538	2000	1163	204	17.5	954
3	7:1, 0.9, 52, 15	521	527	2000	1048	199	19	799
4	7:1, 0.9, 62, 5	799	263	2000	1062	249	23.4	766
5	7:1, 1.2, 57, 5	619	257	2000	876	199	22.7	733
6	7:1, 1.2, 57, 15	619	554	2000	1173	199	17	801

Table 4. Comparison of energy consumption by different methods of biodiesel production.

Technique	Experimental settings	Energy consumed (kJ/g)	Reference
Heating	Camelina sativa oil (9 g of oil, 15 min, 500 W)	50	[32]
Ultrasonic-assisted reactor	Vegetable oil (52 g of oil, 30.7 min, 130 W)	4.6	[33]
Microwave-assisted reactor	Nagchampa oil (160 W, 30 min, 20 g oil)	6.7	[23]
Simultaneous microwave/ultrasound	Waste vegetable oil (18 g of oil, 2 min,100 MW/ 100 US)	1.3	[34]
Sequential ultrasound and microwave	Waste cooking oil (4122 g of oil, 15 min, 300 US/1000 MW)	2.66	This study

4. Conclusion

The present study analyzed the transesterification of waste cooking oil (WCO) using KOH as a catalyst employing sequential ultrasound and microwave approach for biodiesel production. A substantial degree of process intensification is provided by the synergistic relationship between the mechanical effects of micro-emulsification and the acoustic propagation of ultrasound combined with the quick heating and dipolar rotation of molecules caused by microwaves. Results show that under the optimum process parameters of 7:1 methanol: oil molar ratio, 0.9 wt% catalyst content, and 62°C reaction temperature, the biodiesel yield achieved was 96.9% in 5 minutes of reaction time for a sample size of 4500 mL with energy consumption of 2.66 kJ/g. This study offers an insight into the progress of biodiesel production technologies to not only take the advantage of synergistic effect and potential energy and cost savings, but also in terms of time, robustness, safety, and environmental effects.

Acknowledgment
The authors are thankful to PDPU's Centre for Biofuels and Bioenergy Studies (CBBS) for the lab facilities provided for the experiments.

References

[1] Z. Yaakob, M. Mohammad, M. Alherbawi, Z. Alam, and K. Sopian, "Overview of the production of biodiesel from Waste cooking oil," *Renew. Sustain. Energy Rev.*, vol. 18, pp. 184–193, Feb. 2013, doi: 10.1016/J.RSER.2012.10.016.

[2] A. E. Atabani, A. S. Silitonga, I. A. Badruddin, T. M. I. Mahlia, H. H. Masjuki, and S. Mekhilef, "A comprehensive review on biodiesel as an alternative energy resource and its characteristics," *Renewable and Sustainable Energy Reviews*, vol. 16, no. 4. pp. 2070–2093, May 2012, doi: 10.1016/j.rser.2012.01.003.

[3] FSSAI, "Repurpose Used Cooking Oil," 2018. [Online]. Available: https://fssai.gov.in/ruco/ruco_booklet.php.

[4] T. Issariyakul and A. K. Dalai, "Biodiesel from vegetable oils," *Renew. Sustain. Energy Rev.*, vol. 31, pp. 446–471, Mar. 2014, doi: 10.1016/J.RSER.2013.11.001.

[5] Y. C. Sharma, B. Singh, and S. N. Upadhyay, "Advancements in development and characterization of biodiesel: A review," *Fuel*, vol. 87, no. 12, pp. 2355–2373, Sep. 2008.

[6] T. Murayama, Y. Fujiwara, and T. Noto, "Evaluating waste vegetable oils as a diesel fuel."

[7] A. Talebian-Kiakalaieh, N. A. S. Amin, and H. Mazaheri, "A review on novel processes of biodiesel production from waste cooking oil," *Applied Energy*, vol. 104. Elsevier Ltd, pp. 683–710, 2013, doi: 10.1016/j.apenergy.2012.11.061.

[8] Y. Asakuma, K. Maeda, H. Kuramochi, and K. Fukui, "Theoretical study of the transesterification of triglycerides to biodiesel fuel," *Fuel*, vol. 88, no. 5, pp. 786–791, May 2009, doi: 10.1016/J.FUEL.2008.10.045.

[9] J. M. Marchetti, V. U. Miguel, and A. F. Errazu, "Techno-economic study of different alternatives for biodiesel production," *Fuel Process. Technol.*, vol. 89, no. 8, pp. 740–748, Aug. 2008, doi: 10.1016/J.FUPROC.2008.01.007.

[10] M. G. Kulkarni and A. K. Dalai, "Waste Cooking OilAn Economical Source for Biodiesel: A Review," *Ind. & Eng. Chem. Res.*, vol. 45, no. 9, pp. 2901–2913, Mar. 2006, doi: 10.1021/ie0510526.

[11] G. L. Maddikeri, A. B. Pandit, and P. R. Gogate, "Intensification Approaches for Biodiesel Synthesis from Waste Cooking Oil: A Review," *Ind. & Eng. Chem. Res.*, vol. 51, no. 45, pp. 14610–14628, Nov. 2012, doi: 10.1021/ie301675j.

[12] V. G. Gude, "Synergism of microwaves and ultrasound for advanced biorefineries," *Resour. Technol.*, vol. 1, no. 2, pp. 116–125, Dec. 2015, doi: 10.1016/j.reffit.2015.10.001.

[13] Y. S. Pradana, A. Hidayat, A. Prasetya, and A. Budiman, "Biodiesel production in a reactive distillation column catalyzed by heterogeneous potassium catalyst," *Energy Procedia*, vol. 143, pp. 742–747, Dec. 2017, doi: 10.1016/J.EGYPRO.2017.12.756.

[14] M. A. Dubé, A. Y. Tremblay, and J. Liu, "Biodiesel production using a membrane reactor," *Bioresour. Technol.*, vol. 98, no. 3, pp. 639–647, Feb. 2007, doi: 10.1016/J.BIORTECH.2006.02.019.

[15] X. Fan, X. Wang, and F. Chen, "Ultrasonically assisted production of biodiesel from crude cottonseed oil," *Int. J. Green Energy*, vol. 7, no. 2, pp. 117–127, Mar. 2010, doi: 10.1080/15435071003673419.

[16] J. Milano *et al.*, "Optimization of biodiesel production by microwave irradiation-assisted transesterification for waste cooking oil-Calophyllum inophyllum oil via response surface methodology," *Energy Convers. Manag.*, vol. 158, pp. 400–415, Feb. 2018, doi: 10.1016/J.ENCONMAN.2017.12.027.

[17] E. Martinez-Guerra and V. G. Gude, "Continuous and pulse sonication effects on transesterification of used vegetable oil," *Energy Convers. Manag.*, vol. 96, pp. 268–276, May 2015, doi: 10.1016/J.ENCONMAN.2015.02.073.

[18] M. C. Hsiao, C. C. Lin, Y. H. Chang, and L. C. Chen, "Ultrasonic mixing and closed

microwave irradiation-assisted transesterification of soybean oil," *Fuel*, vol. 89, no. 12, pp. 3618–3622, Dec. 2010, doi: 10.1016/j.fuel.2010.07.044.

[19] J. Yan, Y. Zhao, K. Li, H. Zhang, L. Fan, and Z. Lu, "Efficient production of biodiesel from ionic liquid catalyzed esterification using ultrasonic-microwave combined intensification," *Chem. Eng. Process. - Process Intensif.*, vol. 149, Mar. 2020, doi: 10.1016/j.cep.2020.107870.

[20] K. Thakkar, K. Shah, P. Kodgire, and S. S. Kachhwaha, "In-situ reactive extraction of castor seeds for biodiesel production using the coordinated ultrasound – microwave irradiation: Process optimization and kinetic modeling," *Ultrason. Sonochem.*, vol. 50, pp. 6–14, Jan. 2019, doi: 10.1016/J.ULTSONCH.2018.08.007.

[21] M. Ulfah, Firdaus, S. Octavia, H. Suherman, and Subagjo, "Biodiesel Production Through Waste Cooking Oil (WCO) Esterification Using Sulfated Alumina as Catalyst," in *IOP Conference Series: Materials Science and Engineering*, 2019, vol. 543, no. 1, doi: 10.1088/1757-899X/543/1/012007.

[22] S. M. Safieddin Ardebili, T. T. Hashjin, B. Ghobadian, G. Najafi, S. Mantegna, and G. Cravotto, "Optimization of biodiesel synthesis under simultaneous ultrasound-microwave irradiation using response surface methodology (RSM)," *Green Process. Synth.*, vol. 4, no. 4, pp. 259–267, Aug. 2015, doi: 10.1515/gps-2015-0029.

[23] V. L. Gole and P. R. Gogate, "Intensification of synthesis of biodiesel from non-edible oil using sequential combination of microwave and ultrasound," *Fuel Process. Technol.*, vol. 106, pp. 62–69, Feb. 2013, doi: 10.1016/j.fuproc.2012.06.021.

[24] M.-C. Hsiao, "Optimization of the Production of Biodiesel Assisted by Ultrasonic and Microwave," *Int. J. Energy Power Eng.*, vol. 2, no. 2, p. 54, 2013, doi: 10.11648/j.ijepe.20130202.14.

[25] A. Sharma, P. Kodgire, and S. S. Kachhwaha, "Biodiesel production from waste cotton-seed cooking oil using microwave-assisted transesterification: Optimization and kinetic modeling," *Renew. Sustain. Energy Rev.*, vol. 116, Dec. 2019, doi: 10.1016/j.rser.2019.109394.

[26] A. R. Gupta, S. V. Yadav, and V. K. Rathod, "Enhancement in biodiesel production using waste cooking oil and calcium diglyceroxide as a heterogeneous catalyst in presence of ultrasound," *Fuel*, vol. 158, pp. 800–806, Jun. 2015, doi: 10.1016/j.fuel.2015.05.064.

[27] M. Nüchter, B. Ondruschka, W. Bonrath, and A. Gum, "Microwave assisted synthesis – a critical technology overview," *Green Chem.*, vol. 6, no. 3, pp. 128–141, Mar. 2004, doi: 10.1039/b310502d.

[28] A. Sharma, P. Kodgire, and S. S. Kachhwaha, "Investigation of ultrasound-assisted KOH and CaO catalyzed transesterification for biodiesel production from waste cotton-seed cooking oil: Process optimization and conversion rate evaluation," *J. Clean. Prod.*, vol. 259, Jun. 2020, doi: 10.1016/j.jclepro.2020.120982.

[29] M. Agarwal, G. Chauhan, S. P. Chaurasia, and K. Singh, "Study of catalytic behavior of KOH as homogeneous and heterogeneous catalyst for biodiesel production," *J. Taiwan Inst. Chem. Eng.*, vol. 43, no. 1, pp. 89–94, Jan. 2012, doi: 10.1016/J.JTICE.2011.06.003.

[30] M. Kouzu and J. Hidaka, "Transesterification of vegetable oil into biodiesel catalyzed by CaO: A review," *Fuel*, vol. 93, pp. 1–12, Mar. 2012, doi: 10.1016/J.FUEL.2011.09.015.

[31] M. Mathiyazhagan and A. Ganapathi, "Factors Affecting Biodiesel Production," *Res. Plant Biol.*, vol. 1, no. 2, pp. 1–05, 2011, [Online]. Available: www.resplantbiol.com.

[32] P. D. Patil, V. Gnaneswar Gude, L. Mar Camacho, and S. Deng, "Microwave-Assisted Catalytic Transesterification of Camelina Sativa Oil," *Energy & Fuels*, vol. 24, no. 2, pp. 1298–1304, Nov. 2009, doi: 10.1021/ef9010065.

[33] B. Salamatinia, H. Mootabadi, I. Hashemizadeh, and A. Z. Abdullah, "Intensification of biodiesel production from vegetable oils using ultrasonic-assisted process: Optimization and kinetic," *Chem. Eng. Process. Process Intensif.*, vol. 73, pp. 135–143, 2013, doi: https://doi.org/10.1016/j.cep.2013.07.010.

[34] E. Martinez-Guerra and V. G. Gude, "Synergistic effect of simultaneous microwave and ultrasound irradiations on transesterification of waste vegetable oil," *Fuel*, vol. 137, pp. 100–108, Aug. 2014, doi: 10.1016/j.fuel.2014.07.087.

IOP Conf. Series: Materials Science and Engineering 1146 (2021) 012006 doi:10.1088/1757-899X/1146/1/012006

Brackish ground water and dairy wastewater treatment using electrodialysis system

Dipak Ankoliya[1*], Anurag Mudgal[2*], Manish Kumar Sinha[3], Vivek Patel[4], Jatin Patel[5]

[1] Senior Research Fellow, INDIA H_2O Project, Department of Mechanical Engineering, Pandit Deendayal Energy University, Gandhinagar (Guj.) – 382007, India
[2,4,5] Department of Mechanical Engineering, Pandit Deendayal Energy University, Gandhinagar (Guj.) – 382007, India
[3] Department of Chemical Engineering, Pandit Deendayal Energy University, Gandhinagar (Guj.) – 382007, India
Email:[2*]anurag.mudgal@sot.pdpu.ac.in, [1*]dipak.personal77@gmail.com

Abstract. For brackish water desalination, there is necessity to reduce the volume of saline waste water generated in existing processes due to scarcity of water and to cope up with increasing water demand. The electrodialysis process typically known for its higher water recovery rate with competitive energy consumption per unit volume of clean water produced. Theoretical and experimental investigation of electrodialysis in the feed water composition of Indian ground water would make this technology more adaptable at commercial level. With the capability of bipolar membrane to split water into hydroxide ions and protons, the electrodialysis with bipolar membrane process can produce acid and base solution from the corresponding salt solution. This onsite acid and base production capacity of electrodialysis with bipolar membrane process reduces the need of purchasing these chemicals and also reduces the saline wastewater generation in many industries including dairy industry. This research is aimed to study the electrodialysis system at pilot plant scale installation along with long term testing would make this technology one step ahead in desalination and wastewater treatment.

1. Introduction

Water remains vital for life and crucial common freedoms, for example, livelihood and satisfactory food. Regardless of having covered majority of the Earth's surface by water, large proportion of population remain inaccessible to the clean water. A portion of the variables adding to developing water shortage are populace development, urbanization, climate change including restricted clean water reserves. It is imperative that we investigate new innovations and take advantage from elective water sources, for example, seawater for our expanding water demands. Despite the fact that seawater is broadly accessible water source, it has high concentration of salts which should be first removed by means of desalination. Desalination is the process to remove dissolved salts and other minerals from the brackish and seawater.

Initially, thermal energy driven processes like multi-effect distillation and multi-stage flash remain market leader for long period of time then very famous membrane technology, reverse osmosis, covers the majority of the desalination market. The growing demand and preference of these pressure intensive membrane processes are due to its simple operation, less space, chemicals and energy requirement and further improvement in materials of membrane [1].

Electrodialysis (ED) is an emerging innovation that can be utilized in the creation of drinking water resource by brackish water desalination. ED can be easily run by the solar energy produced by photovoltaic (PV) panel because ED consumes direct current (DC) for its desalting operation [2]. Due to high recovery ratio, less chemical and energy requirement and high membrane selectivity, ED can become a savvy strategy for the brackish water desalination and wastewater treatment. As an electromembrane process, ED can only allow charged ions to migrate through the special selective membranes under the influence of electric potential and generates two streams one is became rich in ions and other became dilute. ED is widely used in the drinking water production from the brackish water source and table salt production from the seawater [3].

Initial feed water concentration of solution and type of membrane decide the electrical voltage difference required across the ED stack and consequently the energy consumption of the process. The speed at which salt removed from the feed water solution is directly proportional to the electric current. There is also energy requirement to pump the feed water into the ED cell as well as in the electrode compartment. The ED process efficiency strongly depend upon the feed water quality, types of ion exchange membranes and quantity of salt removed from the initial concentration. For any particular feed water, the voltage difference across ED membrane cell vary with the conductivity of membrane and accordingly decides the system efficiency [1]. The use of ED stays restricted by the significant capital expense of system which includes ion exchange membranes, electrodes and replacement cost due to the degradation of membrane polymer.

Electromembrane technologies, for example, electrodialysis and bipolar membrane electrodialysis (BMED) have been generally applied for increasing recovery and valorisation of seawater reverse osmosis reject stream which is called hyper saline stream because of high concentration of NaCl present. The BMED process can covert this NaCl into the acid (HCl) and base (NaOH) by use of electrical energy [4].

BMED is a combination of two processes, one is ordinary electrodialysis and another is splitting water molecule through bipolar membrane. It can make salt transformation and produce H^+ and OH^- ions without second salt contamination. Due to the economic benefit, technical development and environment friendliness of BMED process, it is considered in vast applications from everywhere in the world, for example, applications in bio food preparation, chemical processes, environmental preservation etc. [5]. As discussed earlier in this section for the ED process, the efficiency of a BMED process is influenced by some of the parameters like fluid flow velocity, current density, cell design, stack layout, spacers and membrane properties.

2. Method

Electromembrane Processes consist of those technologies that apply the electric potential gradient as the driving force of ion transport. They are increasingly important group of separation methods for removal of charged components from solutions such as for producing potable water from brackish water [6].

The main electromembrane processes are:
Electrodeionization (EDI)
Electrodialysis (ED)
Electrodialysis reversal (EDR)
Electrodialysis with bipolar membrane (BMED)
Capacitive deionization (CDI)

Lately these processes have started to be applied into vast applications for example energy generation, food processing, personal and health care products, high grade water production, separation of biomolecules and organic materials, sensors, industrial effluent treatment, etc. The basic elements of the Electromembrane technologies can be ion-exchange or bipolar membrane. Ion migration is taking place across these charged membranes which are placed in the electric field.

2.1. Electrodialysis (ED):

In the ED process, when we applied the direct current (DC) voltage across the membrane stack then charged dissolved ions are move through ion exchange membranes. The selectivity of ED process is defined by the electrical charge of feed water contamination. This process allows charged dissolved solids via its ion exchange membrane under the influence of electrical potential while non-charged particles may remain untouched in the brackish water. Due to advances of technology in ion exchange membrane properties and material of construction as well as innovation of polarity reversal, ED process is gaining popularity rapidly [7,8]. The schematic diagram for the working of electrodialysis is illustrated in Figure 1.

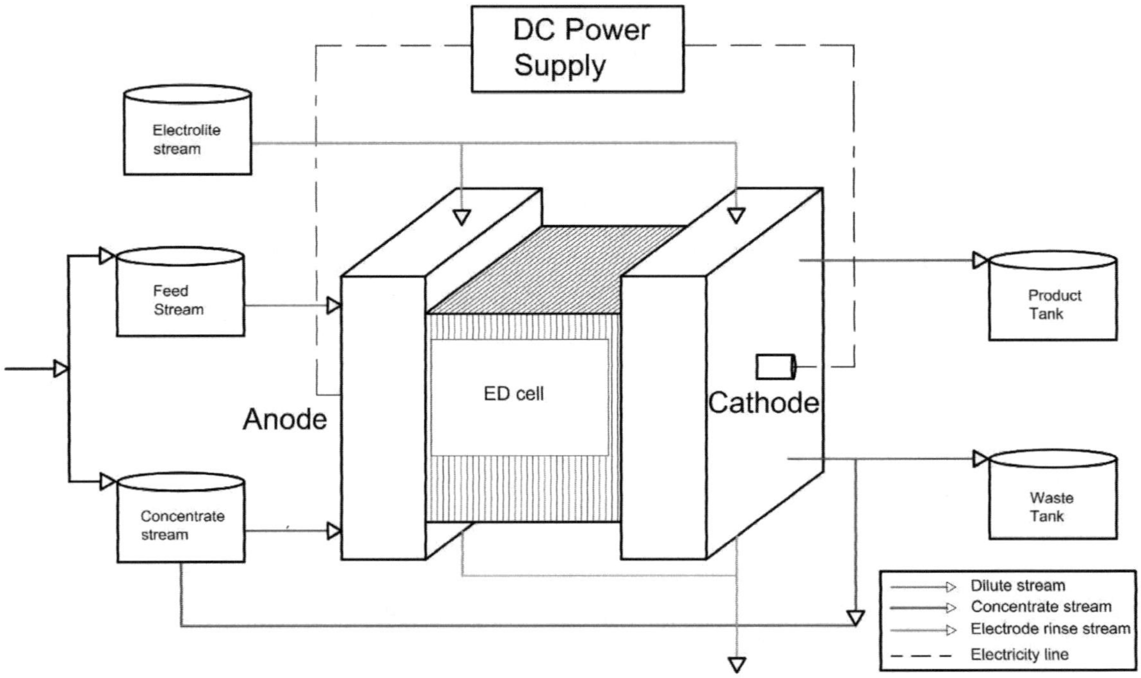

Figure 1. Schematic diagram of electrodialysis system.

Brackish water desalination for inland water has its own limitation to create it as water resource for example, the amount of concentrate stream water with more salinity than feed water coming from the desalination process would not be a cost effective and environment friendly solution if disposed inland. The higher recovery brackish water desalination system is required to reduce waste management and cost of disposal [9–11].

Electrodialysis process can achieve low volume and high concentration of brine by targeting high (up to 95%) recovery rates [12]. To achieve high water recovery, it is necessary to control the precipitation of sparingly soluble salt like $CaSO_4$. It is investigated that electrodialysis reversal can achieve higher tolerable $CaSO_4$ saturation level compared to the RO process due to system's electrochemical behavior [13,14].

2.1.1. Design. Electrodialysis process has following important governing input variables:

C^{fd} Concentration at inlet of dilute stream
C^{d} Concentration at exit of dilute stream
R Recovery ratio
Δ Thickness of cell pair
u Linear flow velocity
Q_d Product water flow rate (Production capacity)

As all these process parameters are interrelated between each other, it is necessary to choose the chronology between these design parameters and fixing some inputs to calculate all design parameters. Based on these input variables, the aim is to find out length, membrane area, current, voltage and pressure drop across ED stack. By knowing feed water salinity and setting salinity value for the pure output water, we can get salinity value of concentrate stream at outlet by deciding at what recovery ratio ED plant should be run. Here the input salinity of concentrate stream is same as the feed water salinity. By using mass balance equation for both the dilute and concentrate stream, we can find all concentration values [15].

$$C^c = \frac{C^{fd} - R\,C^d}{1-R} \tag{1}$$

Here, the C^c is the concentration at outlet of reject stream, C^{fd} and C^d are the concentration of dilute stream at inlet and outlet respectively and R is recovery ratio. Current density applied is limiting parameter in electrodialysis because it can decide the efficiency of the system. Higher current density can lead to the unwanted phenomenon like water splitting and current leakage which can increase the overall energy consumption of the system. So it is necessary to calculate the limiting current density of the given system which cannot be exceeded at any point of operation of the system to reduce energy loss at the same time increase the full potential of the system. The empirical relation is defined by the Hong-Joo Lee et al. [15] for the limiting current density which is dependent on the flow velocity and outlet concentration in dilute compartment and the geometry of stack. They have also derived formula for the calculation of the membrane area requirement. After calculating the current density, membrane area requirement can be calculated by using concentration, volume flow rate and other constants.

$$i_{prac} = s\,a\,C^d u^b \tag{2}$$

$$A_{prac} = \frac{\left[ln\left(\frac{C^c C^{fd}}{C^d C^{fc}}\right) + \frac{\Lambda(\rho_A+\rho_C)(C^{fd}-C^d)}{\Delta}\right] z\,F\,C^d\,Q^d}{\left[\frac{C^d}{C^c}+1+\frac{\Lambda\,C^d(\rho_A+\rho_C)}{\Delta}\right] i_{prac}\,\varsigma} \tag{3}$$

Where, s is safety factor, a and b are constants and u is flow velocity. The Δ is the cell pair thickness, ς is current utilization and Λ is the equivalent conductance of solution. Current flowing in stack is calculated by using the derived values of current density and membrane area. The electrical potential drop across the cell pair can be derived by using current flow and cell pair resistance in stack. Energy requirement in stack is multiplication of current and voltage applied in stack.

$$I_{st} = \frac{z\,F\,Q^d\,C^{\Delta}}{\varsigma} \tag{4}$$

$$U_{st} = \frac{\Delta\,i_{prac}}{\Lambda}\left[\frac{1}{C^c} + \frac{1}{C^d} + \frac{\Lambda(\rho_A+\rho_C)}{\Delta}\right] \tag{5}$$

$$E_{des} = \frac{I_{st}\;x\;U_{st}}{1000\;Q^d \times 3600} \tag{6}$$

The current flow is denoted by I_{st}, the U_{st} is the electrical potential drop and E_{des} is the specific energy consumption of the unit.

2.2. Bipolar membrane Electrodialysis (BMED):

Bipolar membrane electrodialysis process is used to produce acids and bases from a respective salt solution by water separation phenomenon occurred in bipolar membrane by applying electrical potential.

The basic working principle and arrangement of BMED process is shown in Figure 2 having bipolar membranes and cation-exchange and anion exchange membranes which is arranged parallel to form single unit cell between two electrodes. This unit cell consists three compartment and this unit cell is repeated multiple times between two electrodes to form a whole BMED stack. Suppose a feed solution consisting salt is entered in the middle compartment then under influence of electrical potential applied via two electrodes, the cations and anions in this salt solution will move towards the cathode and anode respectively. Now ions will permeate through the cation and anion-exchange membrane and enter into the adjacent compartment in which hydroxide ions and protons are present which is generated by the bipolar membrane. These anions and cations form an acid and a base by combining with the proton and hydroxide ions respectively. The whole process is resulted into the generation of an acid and a base in BMED cell from the corresponding salt solution.

Technical feasibility and the economics of the BMED process is very much depends upon the performance of bipolar membrane which becomes critical component of this process [16].

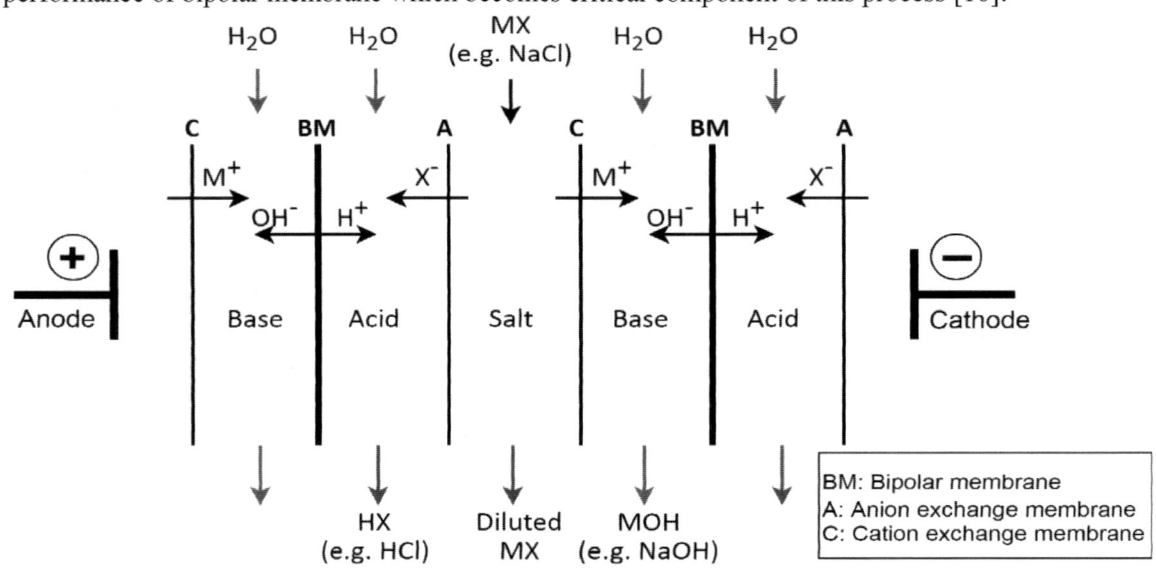

Figure 2. Working principle of bipolar membrane electrodialysis stack.

To achieve optimum working point in this electrodialysis process, membrane resistance should be minimum to reduce the energy requirement while current density should be maximum to reduce membrane area requirement.

The design of BMED process is depend on the various parameters like current, voltage drop and membrane area requirement as per the model equations given below [16]:

$$I = \frac{Q_p F \,(c_s^f - c_s^p)}{\xi} \tag{7}$$

$$A = \frac{Q_p F \,(c_s^f - c_s^p)}{i\,\xi} \tag{8}$$

$$\Delta U_{cell} = i\left[\Delta\left(\frac{ln\frac{c_s^f}{c_s^p}}{\Lambda_s\,(c_s^f - c_s^p)} + \frac{ln\frac{c_a^f}{c_a^p}}{\Lambda_a\,(c_a^f - c_a^p)} + \frac{ln\frac{c_b^f}{c_b^p}}{\Lambda_b\,(c_b^f - c_b^p)}\right) + r^{tr} + r^{bml} + r^{am} + r^{cm}\right] + 0.814 \tag{9}$$

Where, the symbols used in above equations are described in the section 3 with the nomenclature of each symbol in Table 4.

2.2.1. Application. As shown in Table 1, many applications of BMED process were identified under the category of pollution control or resource recovery.

IOP Conf. Series: Materials Science and Engineering 1146 (2021) 012006 doi:10.1088/1757-899X/1146/1/012006

The most established application of BMED process in industrial steel plant is to recover the HF and HNO_3 after pickling bath where the waste flow of KF and KNO_3 is came out. Washington Steel plant in Pennsylvania is the first commercial plant started using BMED process in 1987 [17]. Because of difficulty in the extraction of organic acids after fermentation operation, the application of BMED process in this case becomes very promising [18].

Table 1. Application of bipolar membrane electrodialysis system [17].

Sr No	Type	Application
1	Sulphate recovery	Acid recovery in battery production line
		Converting sodium sulphate waste in rayon manufacturing
2	Acid recovery	Recovery of acid in steel plant
		Acid recovery from spent aluminium pot linings
		Fluorosilicic acid conversion to HF, SiO_2
		Control of fluoride emission in chemical plant
3	Nitrate recovery	Ammonium nitrate conversion from the uranium processing
		Conversion of KNO_3
4	Pulp & Papers	Recycling of sodium alkali
5	Amine recovery	Recovery of catalyst in Al coating moulds
6	Flue gas desulfurization	SO_2 recovery in Soxal™ process

2.2.2. Benefit. To generate H^+ and OH^- ions out of the water by routine method, it requires electrolysis process in which H_2 and O_2 gas is produced which consumes about half of the total applied voltage so the electrical energy of the whole process. These days, double layered ion-exchange membranes called bipolar membranes are available for producing H^+ and OH^- ions from the water without generating unwanted gases. This bipolar membrane with conventional cation- and/or anion-exchange membranes is used to form BMED stack which is used to convert salts into their respective acids and bases [17]. It has low running cost as well as low space requirement.

2.2.3. Limitation. A significant hindrance in the vast commercial application of BMED process is thought to be as per the following [19]:

a) Drastic decrease of the perm selectivity of the membrane due to reduced Donnan exclusion at high product concentrations, which resulted into huge salt spillage via membrane into acid and base.
b) Precipitation of sparingly soluble salt like di- or multivalent salt and organic material at high pH alkaline solution.
c) For very high acid and base concentration, the stability of the anion and cation exchange layer of bipolar membrane becomes very poor.

2.3. Dairy wastewater characterisation:

The quality and quantity of effluent produced in the dairy plant are majorly depend on processing plant parameters like design, operating procedure, production method and kind of product being processed. It also depends on the water management and water conservation program being applied. Dairy wastewater might be separated into three significant classes [20]:

i. Water used in the heat exchanger for heating and cooling purpose are generally less polluted and can be reused or discharged safely.
ii. Wastewaters generated by the cleaning process of various dairy plant equipment like containers, vessels, reactor, pipelines, filters, tanks and other related equipment.
iii. Sewage wastewater generated by sanitary use and application.

IOP Conf. Series: Materials Science and Engineering 1146 (2021) 012006 doi:10.1088/1757-899X/1146/1/012006

The characterisation of effluent water of some of the Indian dairies are shown in Table 2. Important parameters like pH, TDS, BOD and COD of the wastewater at the inlet of effluent treatment plant is taken from the data available in open literature.

Table 2. Dairy wastewater characterisation.

Sr. No	Parameter	Location (city/ state)	Sample collection Source	pH	TDS (ppm)	TSS (ppm)	COD (ppm)	BOD (ppm)
1	Mother Dairy [21]	Delhi	Combine waste water	7.5	1500	700	2500	1200
2	Vikas Dairy [22]	Jalgaon	Composite waste water	6.5	1803	729	790	530
3	Sarvottam Dairy [23]	Bhavnagar	ETP inlet	7.8	---	---	2593.3	---
4	Amul Dairy [24]	Anand, Gujarat	Waste water at common storage equalization tank	3.69	3720	---	7110	---
5	Vijaya Dairy [25]	Hyderabad	Raw effluent	7.8	1100	398	880	372
6	Shivamrut Dairy (Jan) [26]	Vijaynagar, Maharashtra	Dairy waste water (whey sample)	10.0	1442	252	1499	649
7	Madavaram Dairy [27]	Chennai, Tamilnadu	Dairy effluent	5.15	1128	356	631	216
8	Purabi Dairy [28]	Guwahati, Assam	Raw dairy waste water	5.64	722	254	1462	758
9	Dairy at Dehradun [29]	Dehradun, Uttarakhand	Dairy outlet waste water	7.31	---	455	1676	770
10	Banas Dairy [30]	Palanpur, Gujarat	Sample collected at collection tank	9.93	1858	1766	2049	1366
11	Lucknow Dairy [31]	Lucknow, Uttar Pradesh	untreated wastewater (influent)	6.2	900	75	---	---
	Min Value			3.69	722	75	631	216
	Max Value			9.93	3720	1766	7110	1366
	Average			6.81	2221	920.5	3870.5	791

3. Expected outcomes:

Based on the mathematical equations (1) to (6), the sample calculation can be made for the design of medium scale ED plant to produce potable water. Feed water salinity taken as 1100 mg/L and product water quality can be comparable to RO water as 100 mg/L. ED system design calculation is done for water capacity of 500 LPH with 70% recovery. The calculated input and output parameters are shown in Table 3.

Table 3. Sample calculation for ground water quality for medium scale ED plant.

Symbol	Description	Value	Unit
	INPUT DATA		
C^{fd}	Concentration at inlet of dilute stream	1100	mg/L
C^{d}	Concentration at exit of dilute stream	100	mg/L
R	Recovery ratio	0.70	
$\rho_A+\rho_C$	Total area resistance of membrane	0.0007	$\Omega\ m^2$
Q^{d}	Product water flow rate	500	L/h
w	Width of cell pair	0.1	m

L_{st}	Length of flow path per stack	0.42	m
ΔP	Pressure drop in stack	350	kPa
η_p	Efficiency of pump	0.6	
	OUTPUT DATA		
C^{Δ}	Concentration difference of dilute stream	1000	mg/L
C^{c}	Concentration at exit of concentrate stream	4100	mg/L
C^{fc}	Concentration at inlet of concentrate stream	3100	mg/L
i_{prac}	Practical limiting current density	7.98	A/m^2
A_{prac}	Area required for stack	12.19	m^2
I_{st}	Current passing through stack	6.97	A
U_{st}	Potential(Voltage) drop across stack	11.02	V
P_{des}	Power required for desalination	76.82	W
E_{des}	Specific energy consumption for desalination	0.12	$kW\ h/m^3$

Expected results for the acid and base production by using BMED stack can be calculated by using equations (7) to (9). If one molar salt solution given as a feed water and reduced to the half molar salt solution then generated acid and base concentration achieved would be the 0.5 molar strength. The generated output data of current requirement, membrane area and voltage drop are shown in the Table 4.

Table 4. Sample calculation for BMED stack.

Symbol		Value	Unit
	INPUT DATA		
C_s^f	Concentration of salt at inlet	1	eq/L
C_s^p	Concentration of salt at outlet	0.5	eq/L
C_a^f	Concentration of acid at inlet	0.5	eq/L
C_a^p	Concentration of acid at outlet	1	eq/L
C_b^f	Concentration of base at inlet	0.5	eq/L
C_b^p	Concentration of base at outlet	1	eq/L
Q_p	Volume flow rate of product	2.78E-05	m^3/s
F	Faraday constant	9.65E+07	A s/keq
ξ	Current utilization	0.7	
Δ	Thickness of individual cell	1.85E-03	m
i	Current density	1000	A/m^2
$\Lambda_{s\ (NaCl)}$	Equivalent conductivity of NaCl	6.32	$S\ m^2/keq$
$\Lambda_{a\ (HCl)}$	Equivalent conductivity of HCl	21.30	$S\ m^2/keq$
$\Lambda_{b\ (NaOH)}$	Equivalent conductivity of NaOH	12.43	$S\ m^2/keq$
$r^{bm}=(r^{tr}+r^{bml})$	Area resistance of bipolar membrane	1.00E-03	$\Omega\ m^2$
$r^{am}+r^{cm}$	Anion and cation exchange membrane's area resistance	1.60E-03	$\Omega\ m^2$
	OUTPUT DATA		
I	Current required for given capacity plant	1914.68	A

A	Membrane area required	1.91	m^2
ΔU_{cell}	Voltage drop across cell unit	4.15	V

4. Conclusion:

Looking at the present scenario of desalination industry and the clean water demand, it is clear that the existing water desalination processes are less efficient in terms of amount of clean water generated from the given input water feed. Electrodialysis process has the flexibility to adjust recovery of the system as well as the degree of desalination required. This unique feature of electrodialysis can provide the new dimensions to the existing brackish water desalination in terms of more productivity and low running cost for the lower feed salinity. Expected results from this study is to demonstrate the electrodialysis process as an emerging desalination technique in the brackish ground water desalination particularly for the Gujarat region. The long term testing of proposed electrodialysis process would increase confidence level in wide spread use of the technology. State of the art pilot scale level configuration of the electrodialysis process at academic institution would open up the gate of collaboration and hybridisation of technologies to improve overall desalination performance. Direct utilisation of DC electrical potential in the electrodialysis and diminishing cost of photovoltaic (PV) solar panel technology would open a door to the potential application in the rural and farthest area from the electricity grid connectivity.

In most of the dairy industries in India generating wastewater with total dissolved solids more than 1000 mg/L will lead to the stringent norms of water disposal. Process of BMED could become a desirable solution with generation of acid and base solution which can be also reused at the same place in the clean in place process. This would create circular economy at the same place with aim of reducing waste.

Acknowledgement:
This project has received funding from the European Union's Horizon 2020 research and innovation programme under grant agreement No 820906 and the Department of Biotechnology, Government of India, Project no: BT/IN/EU-WR/40/AM/2018. We are thankful to the Pandit Deendayal Energy University, Gandhinagar, India for providing necessary infrastructure and facilities to work.

References

[1] Ahmed F E, Hashaikeh R and Hilal N 2020 Hybrid technologies: The future of energy efficient desalination–A review *Desalination* **495** 114659
[2] Gonzalez A, Grágeda M and Ushak S 2017 Assessment of pilot-scale water purification module with electrodialysis technology and solar energy *Appl. Energy* **206** 1643–52
[3] Gurreri L, Cipollina A, Tamburini A and Micale G Elsevier, 2020. Electrodialysis for wastewater treatment—Part I: Fundamentals and municipal effluents in *Current Trends and Future Developments on (Bio-) Membranes* 141–92
[4] Herrero-Gonzalez M, Diaz-Guridi P, Dominguez-Ramos A, Irabien A and Ibañez R 2020 Highly concentrated HCl and NaOH from brines using electrodialysis with bipolar membranes *Sep. Purif. Technol.* 116785
[5] Wang Y, Wang A, Zhang X and Xu T 2011 Simulation of electrodialysis with bipolar membranes: estimation of process performance and energy consumption *Ind. Eng. Chem. Res.* **50** 13911–21
[6] https://www.lenntech.com/processes/electro-membrane-processes.htm lenntech B.V.
[7] Valero F, Barceló A and Arbós R IntechOpen, 2011. Electrodialysis technology-theory and applications in *Desalination, trends and technologies*
[8] Murray P American Water Works Association/Colorado, 1995. *Electrodialysis and*

Electrodialysis Reversal-Manual of Water Supply Practices, M38
[9] Association A W W Membrane Residuals Management Subcommittee (2004) Current Perspectives on Residuals Management for Desalting Membranes *J. Am. Water Work. Assoc.* **96** 73–87
[10] Nicot J-P and Chowdhury A H 2005 Disposal of brackish water concentrate into depleted oil and gas fields: a Texas study *Desalination* **181** 61–74
[11] Watson I C, Morin O J and Henthorne L 2003 Desalting handbook for planners *Desalin. Water Purif. Res. Dev. Progr. Rep.* 5–9
[12] Strathmann H 2010 Electrodialysis, a mature technology with a multitude of new applications *Desalination* **264** 268–88
[13] Hanrahan C, Karimi L, Ghassemi A and Sharbat A 2016 High-recovery electrodialysis reversal for the desalination of inland brackish waters *Desalin. Water Treat.* **57** 11029–39
[14] Turek M and Dydo P 2003 Electrodialysis reversal of calcium sulphate and calcium carbonate supersaturated solution *Desalination* **158** 91–4
[15] Lee H-J, Sarfert F, Strathmann H and Moon S-H 2002 Designing of an electrodialysis desalination plant *Desalination* **142** 267–86
[16] Strathmann H 2004 Operating principle of electrodialysis and related processes *Ion-exchange Membr. Sep. Process.* **9** 147–225
[17] Pourcelly G 2002 Electrodialysis with bipolar membranes: principles, optimization, and applications *Russ. J. Electrochem.* **38** 919–26
[18] Strathmann H 2004. Applications of Ion-Exchange Membrane Separation Processes in 287–330 doi:10.1016/S0927-5193(04)80037-8
[19] Tanaka Y 2007 Bipolar membrane electrodialysis *Membr. Sci. Technol.* **12** 405–36
[20] Britz T J, Van Schalkwyk C and Hung Y-T 2006 Treatment of dairy processing wastewaters *Waste Treat. food Process. Ind.* 1–28
[21] Mehrotra R and Trivedi A 2016 Study on Characterisation of Indian Dairy Wastewater *Int. J. Eng. Appl. Sci. Technol.* **1** 77–88
[22] Patil S A, Ahire V V and Hussain M H 2014 Dairy wastewater-a case study 30–4
[23] Chokshi K, Pancha I, Ghosh A and Mishra S 2016 Microalgal biomass generation by phycoremediation of dairy industry wastewater: An integrated approach towards sustainable biofuel production *Bioresour. Technol.* doi:10.1016/j.biortech.2016.09.070
[24] Kumar A K, Sharma S, Patel A, Dixit G and Shah E 2019 Comprehensive evaluation of microalgal based dairy effluent treatment process for clean water generation and other value added products *Int. J. Phytoremediation* **21** 519–30
[25] K P *et al.* 2014 Performance assessment and hydrodynamic analysis of a submerged membrane bioreactor for treating dairy industrial effluent *J. Hazard. Mater.* **274** 300–13
[26] Shivsharan V S, Kulkarni S W and Wani M 2013 Physicochemical characterization of dairy effluents. *Int. J. Life Sci. Pharma Res.* **2** 182–91
[27] Kotteswari M, Murugesan S and and Ranjith Kumar R 2012 Phycoremediation of Dairy Effluent by using the Microalgae Nostoc sp *Int. J. Environ. Res. Dev.* **2** 35–43
[28] Kumar R V, Goswami L, Pakshirajan K and Pugazhenthi G 2016 Dairy wastewater treatment using a novel low cost tubular ceramic membrane and membrane fouling mechanism using pore blocking models *J. Water Process Eng.* **13** 168–75
[29] Verma R and Suthar S 2018 Performance assessment of horizontal and vertical surface flow constructed wetland system in wastewater treatment using multivariate principal component analysis *Ecol. Eng.* **116** 121–6
[30] Chavda P, Rana A and Deshpande A 2015 Performance evaluation of effluent treatment plant of pharmaceutical industry *J. Ind. Pollut. Control* **31** 87–91
[31] Kothari R, Pathak V V., Kumar V and Singh D P 2012 Experimental study for growth potential of unicellular alga Chlorella pyrenoidosa on dairy waste water: An integrated approach for treatment and biofuel production *Bioresour. Technol.* **116** 466–70

Desalination of brackish water using cascade Rankine cycle based reverse osmosis system

Milan Raninga[1], Anurag Mudgal[2*], Vivek Patel[3] and Jatin Patel[4]

[1] Junior Research Fellow, LC-RED Project, Department of Mechanical Engineering, School of Technology, Pandit Deendayal Energy University, Gandhinagar, India
[2*, 3, & 4] Department of Mechanical Engineering, School of Technology, Pandit Deendayal Energy University, Gandhinagar, India
Email: milan.raninga@sot.pdpu.ac.in; anurag.mudgal@sot.pdpu.ac.in[2*]

Abstract. The desalination of brackish ground water using cascade Rankine cycle is proposed. A pair of a Rankine cycle like steam Rankine cycle (SRC) and organic Rankine cycle (ORC) as a waste heat recovery. The single stage steam turbine for the SRC unit while the scroll expander for ORC unit is selected. Simulation of cascade RO system performance is considered using R245fa as a working fluid for ORC unit. The saturated steam from solar Scheffler disc will expand into steam turbine, where the reject heat from steam turbine will utilize for evaporation of ORC working fluid. The high-pressure RO pumps integrated with SRC and ORC turbines to provide net driving pressure to the RO module. This type of system is well suitable for desalination of brackish water due to moderate working temperature & pressure. Result shows that the pair of Rankine cycle will increase the overall (cascade) efficiency of the system. The basic input parameters are optimised with Taguchi approach. The performance of the system shows a good agreement with variation of mass flow rate (m_{SRC}) of the steam in which the permeate flow rate from RO will increase along with the cycle efficiencies.

1. Introduction

Water scarcity does not occur in arid areas but also, it unbalanced the demand & supply of drinking water. The pollution and exploitation of aquifers may affect quality of natural water resources. The population growth, agriculture, and industries trends towards the increasing demand of water led to increase the per capita consumption of water. The desalination of water is one of the promising techniques towards water shortage in countries. It becomes an alternative solution otherwise the water is not accessible for irrigation, industrial and municipal application. [1]. There are various techniques for water desalination includes thermal desalination like multi-stage flash (MSF), multi-effect distillation (MED), vapour compression (VC), etc and membrane desalination processes like reverse osmosis (RO), reverse electro-dialysis (EDR), etc [2]. Amongst all of them, the reverse osmosis process is one of the key techniques for desalination [1]. According to water salinity, the required pressure to pass through porous membrane of RO which can generate from high-pressure pump. The required energy to drive RO pump can provide through various technologies. The Rankine cycle is one technique to provide required energy by coupling turbine with the RO pump. It increases the power transmission efficiency instead of driving RO pump through an electric source.

The small capacity steam Rankine cycle is the most promising technique to produce the work output from turbine which shows low thermal efficiency. The organic fluid is one which evaporates at low temperature. Therefore, the combination of steam Rankine cycle along with organic Rankine cycle is one technique to produce almost twice of permeate water from a single heat input which may increase the overall efficiency of the system. As comparison of rotary expander with the turbo-expanders, rotary expander can handle the liquid, binary, and vapour phase of the working fluid. The thermal power generation using the solar collector will utilise by the steam Rankine cycle can eliminate the super-heating and operate at moderate temperature and pressure [3]. The direct steam generation (DSG) through the Scheffler disc or solar concentrators can avoid to use of boiler which may reduce the future cost of the system [4]. However, boiler can be used for alternative heating source during cloudy weather or night. The major advantage of solar Scheffler disc for constant heating temperature during pick time and highly efficient for evaporation of water. The demonstration of the Scheffler disc at PDPU, Gandhinagar is used for direct steam generation through the parabolic shaped mirror disc to generate steam at >160 °C temperature. The plant is developed for desalination of water by batch RO system [5].

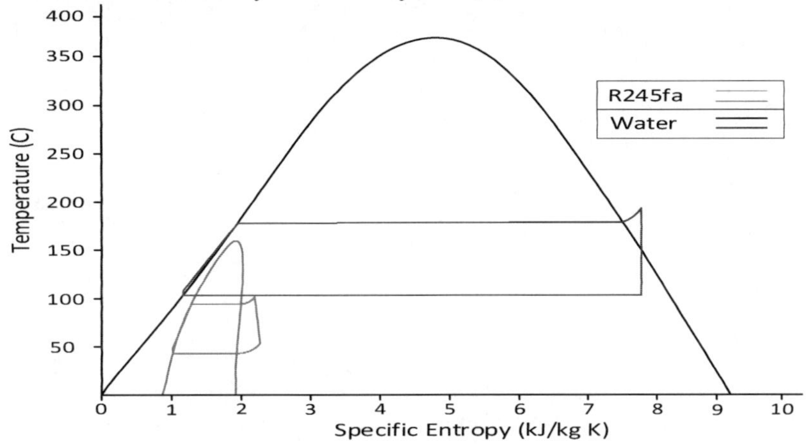

Figure 1 T-S diagram of cascade (pair of steam & organic) Rankine cycle.

A pair of steam Rankine cycle along with organic Rankine cycle is proposed. The T-S diagram of a Rankine cycle for water and R245fa working fluid saturation curve (refer Figure 1) with a basic cycle flow process in which the water shows negative saturation slope while R245fa showing positive saturation slope. The evaporation temperature of steam is higher than R245fa. For the high temperature difference of hot and cold side, the cascading of Rankine cycle is preferred in which the steam Rankine cycle at top cycle while the organic Rankine cycle to be at bottom cycle. It will be more beneficial for single stage small steam turbine and expander showing the pressure ratio of about 2.5 - 8. The low boiling point temperature of the R245fa advantage to evaporate at the saturation temperature of steam during condensation. The bottoming of the organic Rankine cycle allows to design the system at smaller size in terms of power capacity.

In this work, solar powered cascade Rankine cycle-based water desalination process is proposed. A novel design guarantees the continuous water purification at a low consumption of specific energy. The thermodynamic performance of the system is examined by using R245fa with variation of different parameter of the system.

2. System description

The schematic diagram of the cascade-RO system (refer Figure 2) comprises with the steam Rankine cycle & organic Rankine cycle. The turbine shafts are directly integrated with the high-pressure pump of RO unit. The evaporative temperature of organic fluid is lower than the water so that organic Rankine intimate as a bottom cycle in the cascade system. The water is considered as a working fluid of steam Rankine cycle and R245fa for organic Rankine cycle. The generated steam from solar Scheffler disc, or by additional biomass boiler, will be expanded into steam turbine. The rejected heat from the steam turbine will be condensed by an organic fluid of the organic Rankine cycle. This organic fluid will be evaporated up to 100 °C temperature & 11 bar pressure. The super-heated vapour will be expanded into the ORC expander & recirculated by condensing into the ORC condenser. In this manner, the cascading of Rankine cycle will increase the overall efficiency of the system by utilising the rejected heat of steam Rankine cycle. The produced power from the steam turbine & ORC expander will be utilised by the high-pressure RO pump. The bypass line from steam separator can balance the energy (enthalpy) at evaporative condenser to provide constant vapour generation of Organic fluid.

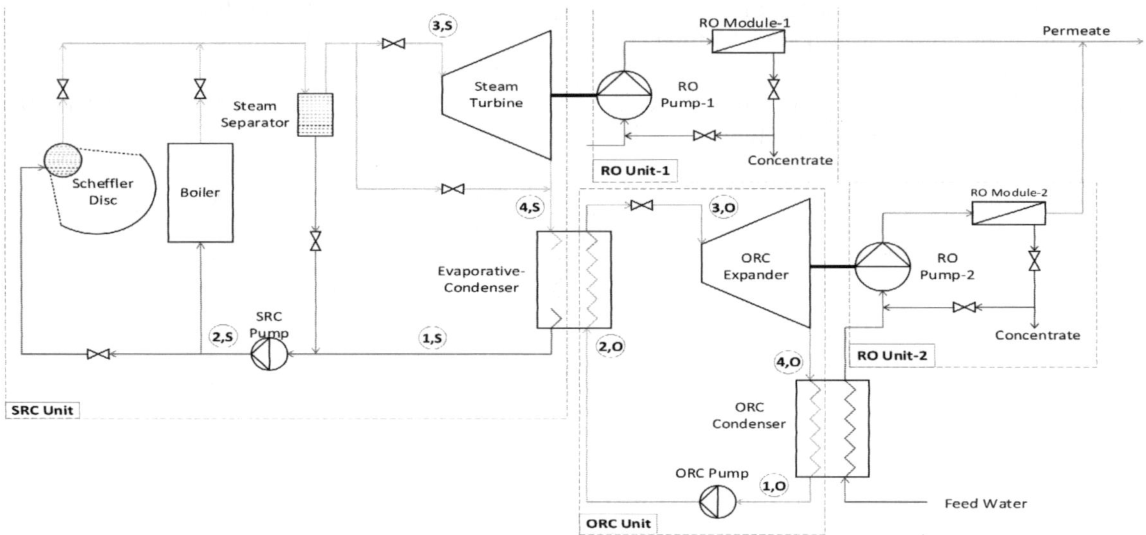

Figure 2 Schematic diagram of cascade RO system.

Reverse osmosis unit is integrated with both Rankine cycles. The recycling of concentrate water is proposed to increase the overall recovery rate of the permeate water. The osmotic pressure of feed water can estimate by considering impurities in feed water. The required feed pressure or net driving pressure of feed water can calculate based on the osmotic pressure to allow salt rejection from water by passing it into the RO module. This net driving pressure can be generated by utilising turbine energy (shaft power) coupled with the high-pressure RO pump. Because of the net driving pressure more than osmotic pressure, the permeate water will separate out and concentrate will be rejected from the RO module. Apart from concentrate water, fractional amount of water will be recycled with the feed water.

In RO unit-2, the feed water from ground or surface will be passed into the ORC condenser as a cooling liquid. It can eliminate the cost of cooling tower and due to this, pump work of circulating water will also reduce.

2.1. Mathematical model

The SRC, ORC along with Cascade efficiencies are defined by:

$$\eta_{SRC} = \left(\frac{(h_{3S} - h_{4S}) - (h_{2S} - h_{1S})}{(h_{3S} - h_{2S})}\right) * 100 \quad (1)$$

$$\eta_{ORC} = \left(\frac{(h_{3O} - h_{4O}) - (h_{2O} - h_{1O})}{(h_{3O} - h_{2O})}\right) * 100 \quad (2)$$

$$\eta_{Cascade} = \left(\frac{(h_{3S} - h_{4S}) - (h_{3O} - h_{4O})}{(h_{3S} - h_{2S})}\right) * 100 \quad (3)$$

The performance formula of Scheffler disc type solar collector is:

$$A_{col} = \frac{Heat_{input}}{\eta_{col} \cdot G_b} \tag{4}$$

Where η_{col} is collector efficiency is calculated by [6]:

$$\eta_{col} = \eta_o - \left(0.0000045.(T_{co} - T_a)\right) - 0.039\left(\frac{T_{co} - T_a}{G_b}\right) - 0.0003\left(\frac{T_{co} - T_a}{G_b}\right)^2 \tag{5}$$

The enthalpy balance of evaporative condenser is based on:

$$m_{SRC}(h_{4S} - h_{1S}) = m_{ORC}(h_{3O} - h_{2O}) \tag{6}$$

The net flow rate by RO pump is defined by:

$$work_{pump-RO} = work_{Turbine} = \frac{Q_{net} * H_S * SG_{water}}{367 * \eta_{pump}} \tag{7}$$

Where, H_S is the suction head of water and SG_{water} is the specific gravity of feed water. the salt rejection from RO module is defined by [7]:

$$X_{concentrate} = \frac{(Q_{net} * X_{net}) - (Q_{permeate} * X_{permeate})}{Q_{concentrate}} \tag{8}$$

Where, $Q_{permeate}$ is the permeate flow can calculated by:

$$Q_{permeate} = Q_{net} * RR_{membrane} \tag{9}$$

Where $RR_{membrane}$ is the recovery rate of RO membrane.

3. Assumptions

Some assumptions for cascade Rankine cycle RO system are proposed for mathematical model may listed as:

- The work produce by the turbine and expander will be consumed by the RO pump, i.e., the mechanical efficiency of $\eta = 100\%$ [8].
- The evaporative-condenser and ORC condenser is well-insulated compact heat exchangers having zero heat loss to surrounding i.e. $\varepsilon = 1$ [9].
- Isentropic efficiency of steam turbine, expander, & pump assumed to be $\eta_{isentropic} = 75\%$.
- The inlet temperature of the cooling water into the condenser is 25 °C whereas the flow rate and the outlet temperature of the cooling water can be adjusted according to achieve 48.5 °C ORC fluid temperature during condensation.
- The ambient temperature for the system assumed to be 25 °C.
- A constant solar radiation assumed to be 850 W/m^2 during sunshine hour.
- A constant salinity of feed water is assumed to be 1000 ppm.

4. Result and discussion

Both Rankine cycles are working under the subcritical cycle where the hottest temperature of the cycle is the evaporation temperature of steam at turbine inlet & coldest temperature is the condensation temperature of the organic fluid. The thermodynamic design of the cascade RO is examined based on the mathematical model. The output parameters of the system are determined using the engineering equation solver software [10]. The inlet variables of the cascade cycle are listed in Table 1.

Table 1 Specific input parameters.

Term	Unit	Value
Evaporation temperature of SRC, (T_{3S})	°C	190
Evaporation pressure of SRC, (P_{3S})	bar	11
Condensation temperature of SRC, (T_{4S})	°C	133
Condensation pressure of SRC, (P_{4S})	bar	2.9
Mass flow rate of SRC, (m_S)	kg/hr	19.67

Evaporation temperature of ORC, (T_{3O})	°C	100
Evaporation pressure of ORC, (P_{3O})	bar	11
Condensation temperature of ORC, (T_{4O})	°C	67
Condensation pressure of ORC, (P_{4O})	bar	3.4
Mass flow rate of ORC, (m_O)	kg/hr	216
Reference Solar radiation, (G_b)	W/m^2	850

4.1. Thermodynamic parameters of optimum cascade power conversion condition for RO

Based on the mathematical model of cascade RO, the optimum output results are listed in Table 2. The cycle efficiencies depending on the temperature of the steam at turbine inlet. Based on feed water salinity of 1000 ppm into the RO, the required specific energy consumption, i.e., power required to drive RO pump is determined. The output results for RO permeate are shown in Table 2. The area of collector is also calculated based on equation (4) & (5). The system may dynamically unbalance due to the variation of solar irradiation during sunshine hours can interrupt the permeate water from RO.

Table 2 Thermodynamic parameter results.

Term	Unit	Value
SRC efficiency (η_{SRC})	%	7.74
ORC efficiency (η_{ORC})	%	7.23
Cascade Efficiency ($\eta_{Cascade}$)	%	14.07
Heat Input ($Heat_{input,S}$)	kW	13.66
Pump Work SRC ($Work_{pump-S}$)	kW	0.01
Pump Work ORC ($Work_{pump-O}$)	kW	0.5
Turbine Power ($Power_{Turbine}$)	kW	0.99
Expander Power ($Power_{expander}$)	kW	0.98
SRC RO Permeate ($Permeate_{S-RO}$)	m^3/hr	0.98
ORC RO Permeate ($Permeate_{O-RO}$)	m^3/hr	0.98
Area of collector (A_{col})	m^2	34

The Taguchi method is examined for optimize input parameters to achieve maximum output from the system [11]. There are 3 different input parameters along with 4 levels are preferred for the orthogonal array (refer Table 3) as evaporation temperature (190, 200, 210, & 220 C), mass flow rate (0.0054, 0.0055, 0.0056, & 0.0057 kg/sec), & evaporation pressure (9, 10, 11 & 12 bar) are considered. According to the array, there are 16 experiment to be examined by arranging these parameters and levels in different sequences (refer Table 4).

Table 3 Number of variable (parameters) along with levels (values).

			Number of Levels			
			1	2	3	4
Number of parameters	1	Evaporation Temperature, T_{3S} (°C)	190	200	210	220
	2	Mass flow rate, m_{SRC} (kg/hr)	19.44	19.8	20.16	20.52
	3	Evaporation pressure, P_{3S} (bar)	9	10	11	12

Table 4 Experiment sequence based of orthogonal array.

Experiment No.	Evaporation Temperature, T_{3S} (°C)	Mass flow rate, m_{SRC} (kg/hr)	Evaporation Pressure, P_{3S} (bar)
1	190	19.44	9
2	190	19.80	12
3	190	20.16	10
4	190	20.52	11

5	200	19.44	10
6	200	19.80	11
7	200	20.16	12
8	200	20.52	9
9	210	19.44	11
10	210	19.80	10
11	210	20.16	9
12	210	20.52	12
13	220	19.44	12
14	220	19.80	9
15	220	20.16	11
16	220	20.52	10

Table 5 Examined output parameters.

Exp. No.	Heat In (kW)	$W_{turbine}$ (kW)	W_{exp} (kW)	Q_{S-RO} (m^3/hr)	Q_{O-RO} (m^3/hr)	η_{SRC} (%)	η_{ORC} (%)	$\eta_{Cascade}$ (%)
1	13.52	0.84	0.98	0.84	0.98	6.18	7.37	13.10
2	13.66	1.05	0.98	1.05	0.98	7.68	7.37	14.49
3	13.98	0.94	1.01	0.94	1.01	6.73	7.36	13.60
4	14.2	1.03	1.02	1.03	1.02	7.23	7.36	14.05
5	13.62	0.92	0.99	0.92	0.98	6.76	7.37	13.64
6	13.84	1.01	0.99	1.01	1.00	7.25	7.37	14.09
7	14.05	1.08	1.00	1.08	1.00	7.70	7.37	14.50
8	14.41	0.90	1.04	0.89	1.04	6.22	7.34	13.11
9	13.71	1.00	0.99	1.00	0.98	7.29	7.37	14.13
10	14.00	0.96	1.01	0.95	1.01	6.81	7.36	13.67
11	14.28	0.90	1.03	0.89	1.03	6.26	7.35	13.16
12	14.58	1.14	1.04	1.13	1.03	7.78	7.35	14.56
13	**13.82**	**1.08**	**0.99**	**1.07**	**0.99**	**7.78**	**7.37**	**14.58**
14	14.15	0.90	1.02	0.89	1.02	6.32	7.36	13.21
15	14.36	1.06	1.03	1.05	1.02	7.34	7.35	14.16
16	14.64	1.01	1.05	1.00	1.05	6.86	7.34	13.69

Based on the sequence of experiments, output parameters are examined (refer Table 5). The heat input of 13.52 to 14.58 kW will be required to achieve cascade efficiency of 13.1 to 14.58%. The work output of 0.838 to 1.1377 kW will be achieved along with 0.836 to 1.134 m^3/hr of permeate water from the RO. The lowest heat input of 13.52 kW will be achieved in experiment 1. While the experiment 12 shows highest work output of the steam turbine & ORC expander. The experiment 13 show significant output parameter with highest system efficiencies & balanced heat input and turbine works.

Based on the optimized result, with a variation of mass flow rate of steam with constant input parameters (such as heat input & evaporation pressure of steam P_{3S}) shows improvement in performance in which evaporation temperature of ORC T_{3O} become increase along with work output from both SRC & ORC cycle. Due to this, the performance of both cycles advantage to production of permeate water from RO.

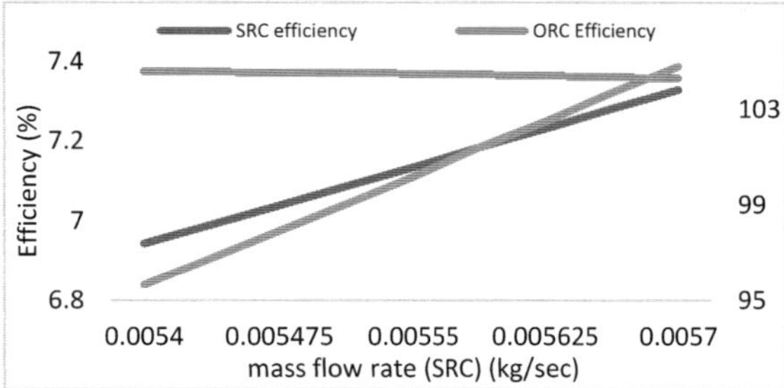

Figure 3 System efficiencies with variation of mass flow rate of steam.

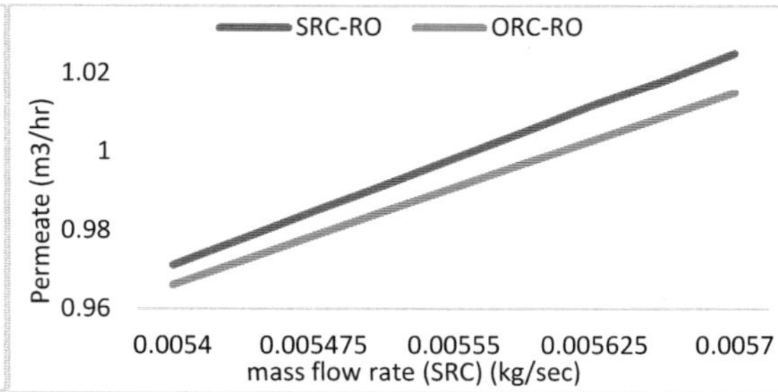

Figure 4 Permeate water with variation of mass flow rate of steam.

As shown in graph (refer Figure 3), the super-heating of R245fa from its saturation temperature of 93°C shows an increase in SRC efficiency η_{SRC} along with variation of mass flow rate (m_{SRC}) of steam. The ORC efficiency η_{ORC} is reduced due to heat reject by ORC condenser will be increased. However, the ORC work output from expander become improve with variation of mass flow rate (m_{SRC}) of steam along with high-pressure pump shows an increase in net driving pressure of feed water, resulting high permeate water (refer Figure 4) from both RO systems is achieved.

5. Conclusion

The proposed cascade Rankine cycle driven RO system comprises two Rankine cycle along with individual RO systems. The good applicability in RO system with a pair of cycle makes the utilisation of solar energy at low or medium heat source temperature. The Scheffler disc-based cascade cycle has advantage to lower temperature and pressure in solar field. The thermodynamic analysis is performed with different input parameters and values. The mathematical model is simulated using the EES software. The investigation performs based on Taguchi approach to optimise the system for suitable output. Based on the analysis, the conclusion remarks:

- By using EES software, the mathematical model is investigated with variation of mass flow rate of steam. The cycle integrates with the RO shows significant performance.
- The literature shows R245fa is suitable working fluid for ORC. The critical temperature is less than the condensing temperature of steam at turbine exit.
- The Taguchi approach for optimising the output parameters is applied. The experiment 13 shows a good agreement of system efficiency along with work output from turbine.
- The variation of mass flow rate of steam m_{SRC} will improve the overall efficiency and performance of the system along with permeate water form RO. The super-heating of the R245fa will increase the work output from the expander but, the ORC efficiency will be reduced because of the increase in heat rejects during condensation of working fluid at expander exit.

It is clear for this analysis that a pair of Rankine cycle can increase in overall performance of the system. However, water need wet expansion turbine because of negative slope of water in saturation curve. Otherwise, the super-heating of steam is required, which increase size of the solar collector. The super-heating of steam required highly stable piping and equipment to sustain which impact on capital cost of the system. The normal boiling point of R245fa is 15 °C will advantage to evaporate at the condensing temperature of steam at turbine exit. The R245fa working fluid is the best alternative to replace with chlorine based working fluid in nowadays. Overall, the solar energy-based cascade RO shown good agreement towards the desalination of brackish water.

6. Future Work

Thermo-economic analysis of the system will be investigated by considering the capital cost of turbine, solar collector, RO pump, RO module, etc along with different working fluids. The experimental investigation will be examined at PDEU, Gandhinagar.

Acknowledgements

This study is sponsored by the Department of Science & Technology, India under the project "Low-cost renewable energy driven (LC-RED) water treatment solutions centre", Grant No.: DST/TM/WTI/WIC/2K17/124. Authors are thankful to Pandit Deendayal Energy University, Gandhinagar, India to provide infrastructure facilities for research work.

References

[1] Fritzmann C, Löwenberg J, Wintgens T, Melin T. State-of-the-art of reverse osmosis desalination. Desalination. 2007;216(1–3):1–76.
[2] Nafey AS, Fath HES, Mabrouk AA. A new visual package for design and simulation of desalination processes. Desalination. 2006;194(1–3):281–96.
[3] Li J, Li P, Gao G, Pei G, Su Y, Ji J. Thermodynamic and economic investigation of a screw expander-based direct steam generation solar cascade Rankine cycle system using water as thermal storage fluid. Appl Energy [Internet]. 2017;195:137–51. Available from: http://dx.doi.org/10.1016/j.apenergy.2017.03.033
[4] Li J, Li P, Pei G, Alvi JZ, Ji J. Analysis of a novel solar electricity generation system using cascade Rankine cycle and steam screw expander. Appl Energy. 2016;165:627–38.
[5] Mudgal A, Davies PA. A cost-effective steam-driven RO plant for brackish groundwater. Desalination [Internet]. 2016;385:167–77. Available from: http://dx.doi.org/10.1016/j.desal.2016.02.022
[6] Blanco et al. J. Technical Comparison of different solar-assisted heat supply systems for a multi-effect seawater distillation unit. Sol World Congr. 2003;(May 2014):14–9.
[7] El-Dessouky HT, Ettouney HM. Chapter 7 - Reverse Osmosis. In: El-Dessouky HT, Ettouney HMBT-F of SWD, editors. Amsterdam: Elsevier Science B.V.; 2002. p. 409–37.
[8] Nafey AS, Sharaf MA. Combined solar organic Rankine cycle with reverse osmosis desalination process: Energy, exergy, and cost evaluations. Renew Energy [Internet]. 2010;35(11):2571–80. Available from: http://dx.doi.org/10.1016/j.renene.2010.03.034
[9] Liu H, Shao Y, Li J. A biomass-fired micro-scale CHP system with organic Rankine cycle (ORC) - Thermodynamic modelling studies. Biomass and Bioenergy. 2011;35(9):3985–94.
[10] SA. K. Engineering equation solver (EES). Madison USA: Klein SA.;
[11] Ivens GP. Chemical process dynamics. Chem Eng J. 1983;27(2):120–1.

IOP Conf. Series: Materials Science and Engineering 1146 (2021) 012008 doi:10.1088/1757-899X/1146/1/012008

Effectiveness of RSM based Central Composite Design for optimization of in-situ biodiesel production process from castor seeds

K Thakkar[1,3], S S Kachhawaha[1,3], P Kodgire[2,3], and M Keshav[1]

[1]Department of Mechanical Engineering, Pandit Deendayal Petroleum University, Gandhinagar, India

[2]Department of Chemical Engineering, Pandit Deendayal Petroleum University, Gandhinagar, India

[3]Centre of Biofuel and Bioenergy Studies (CBBS), Pandit Deendayal Petroleum University, Gandhinagar, India

Corresponding author: kartik.tphd16@sot.pdpu.ac.in, surendra.singh@sot.pdpu.ac.in

Abstract. In the present experimental study, biodiesel is made by reactive extraction (in-situ method) of castor seed. The hybrid (MW+US) apparatus was used to intensify in-situ process. Potassium Hydroxide (KOH) as catalyst and methanol as a reagent are used for the transesterification process. Response surface methodology has been applied to design the experiments. A central composite design (CCD) was adopted to examine the influence of various process parameters (MeOH:seeds ratio (v/w), catalyst loading, reaction temperature and, time) on the biodiesel yield. The optimum conditions for maximum biodiesel yield (93.2%) obtained for castor seed, are 7.47 methanol to oil ratio (v/w), 1.7% KOH catalyst loading, 317 K reaction temperature, and 1788 s reaction time. The regression equation obtained for the model having a coefficient of correlation (R^2), and adjusted coefficient of correlation (R^2_{adj}) are 0.974 and 0.951 respectively shows the goodness of fit for the model.

1. Introduction

Biodiesel is a sustainable alternative to petro-diesel derived from edible/non-edible plant oils, animal fats, and used frying oil including triglycerides. [1]. Vegetable oil feedstocks are widely available from various plant sources, and the glycerides present in the oils are suitable for the transesterification reaction used to obtain fatty acid methyl esters [2].

Biodiesel is a mixture of mono-alkyl esters of various types of fatty acids. Thus, biodiesel is made by converting fatty acid into the respective esters. In transesterification reaction, triglyceride present in vegetable oil is reacted with alcohol and get converted into a mixture of mono-alkyl esters. Conventionally, oil or animal fat is extracted from raw feedstock and the reaction is carried out between extracted oil and methanol. In other words, this method comprises three stages viz. oil-extraction, purification (degumming, deacidification, dewaxing, dehydration, dephosphorization, etc.), and transesterification. The requirement of these processing stages (except transesterification) constitutes over 70% of the total production cost of biodiesel [3]. Therefore, the reactive extraction process (development of in-situ extraction, esterification, and transesterification) has the potential to decrease the processing cost.

The application of ultrasound for intensification of biodiesel synthesis has been reported in many studies in the last decade. It has been reported that ultrasound and cavitation have beneficial effects such as higher yield, faster kinetics, reduction in the number of process steps, etc.[4]. The ultrasound system is specified by frequency (kHz) as well as its intensity (W/cm2). Microwave irradiation, due to the direct delivery of energy to the mixture is a well-established technique to accelerate and enhance chemical reactions [5]. MW enhances the reaction speed and makes the separation process less complex compared to conventional heating [6].

PI techniques (Ultrasound and microwave) have limitations which include attenuation effect in ultrasound and low penetration depth and uncontrolled heating in the microwave [7]. These limitations can be minimized by the hybrid approach and the combined effect of microwave (heat transfer barrier removal) and ultrasound (mass transfer barrier removal) lead to synergistic effects [8].

Hence, the main objective of this research was to incorporate microwave and ultra-sound energy intensification simultaneously to the in-situ process of biodiesel production from castor seeds and to find favorable process conditions for maximum biodiesel yield. RSM based on the CCD procedure is applied to analyze the effects of each factor and their interaction.

2. Methodology

2.1. Materials

The freshly harvested castor seeds from the agricultural field of Gandhinagar, Gujarat were used in the current study. Methyl alcohol (CH_3OH) (≥99%, grade: analytical reagent) and the homogeneous catalyst KOH (≥85%) were procured from Merck and were used without purification. Chemicals required for gas chromatography (GC) which are n-Heptane, (purity ≥99.5%, grade: GC), and internal standard Methyl heptadecanoate (purity ≥99.5%, grade: GC) were purchased from Sigma-Aldrich dealer of Ahmedabad, India.

2.2. Identification of factors and their levels

In this study, RSM is being used. CCD has been adopted to study the effect of individuals and the interaction of various process parameters. As discussed above, RSM-CCD requires a 5-factor level, which is –α, -1, 0, 1, α. Here alpha (α) is the distance between the axial points from the center in a CCD. After identifying the appropriate strategy of the experiments, it is required to decide various factors and their level from the literature study. For example low level, intermediate and high level. For this purpose, '0th' experiments were conducted. From the literature survey, values of different levels can be estimated. In the present study, a set of 31 experiments was designed based on the CCD method which consists of 5 levels for each factor as shown in table 1.

Table 1. Factors and levels in DOE.

Levels		Methanol:seeds ratio (v/w) (A)	% Catalyst (%wt oil) (B)	Temperature (K) (C)	Time (s) (D)
Axial Point	-α	1.6	0.5	313	600
Lower Value	-1	3.2	1.0	318	1200
Intermediate Value	0	4.8	1.5	323	1800
Upper Value	1	6.4	2.0	328	2400
Axial Point	α	8.0	2.5	333	3000

2.3. Experimentation

Castor seeds in the sample size of fifty grams were ground in a household-type domestic grinder for half a minute to achieve 1.3 mm of average particle size. The slurry of seeds was preheated at the required temperature after mixing with methyl alcohol in a pre-defined ratio. Potassium methoxide (catalyst (KOH) + required amount of methyl alcohol) was added to the glass reactor of size 500 ml for the transesterification reaction to begin. The constant Ultrasonication of frequency 40 kHz and 50 W nominal power was applied along with microwave (power 0-50W) to maintain the temperature throughout the reaction. After the completion of the reaction, products with slurry were passed through Whatman filter paper (Grade 42) to separate solid medium using vacuum filtration. The excess methanol was recovered in a rotary evaporator. The resulting mixture consists of FAME (Fatty Acid Methyl Esters), glycerol, and un-reacted triglycerides were analyzed using GC-FID, and % FAME was calculated [8]. The by-product glycerol was separated gravimetrically in a separating funnel and removed. Catalyst (KOH) present in the product mixture was removed by washing three times with the distilled water. Finally, biodiesel was obtained after removing the moisture by heating up to 80 °C with stirring on a hotplate with a magnetic stirrer.

3. Results and discussion

3.1. Determination of regression equation and analysis of variance

The regression equation to determine yield (Y, %) in terms of coded factors using the least square method is given in Eq. (1).

$$\mathrm{Y}\,(\%) = 70.68 + 7.31A + 0.68B + 2.74\mathrm{C} + 0.35\mathrm{D} + 2.26\mathrm{A}^2 - 4.03B^2 - 2.79C^2 - 3.85D^2 + 1.22AB - 5.32AC - 2.28AD - 0.54BC + 0.3BD - 2.81CD \quad (1)$$

The regression equation consists of variables of the predictors viz. methanol:seeds ratio (v/w) (A), catalyst amount (B, % wt.), reaction temperature (C, °C), reaction time (D, min) respectively. The values of R^2 and R^2_{adj} were calculated to be 97.4% and 95.1% respectively which are considered satisfactory. Predicted yield along with experimentally calculated yield of biodiesel (%) obtained through regression equation (Eq. (1)) is given in table 2. It can be observed from table 2 that, maximum experimental biodiesel yield of 91.33% is obtained for experiment no. 28 in which methanol to seed ratio (v/w) is 8.0, catalyst amount is 1.5% (w/w), the reaction temperature is 323 K, and reaction time 1800 s. The predicted biodiesel yield for the same reaction conditions is 94.34%. The minimum experimental biodiesel yield that has been obtained out of performed experiments as given in table 2 is 41.63% for the experimental run 20. The reaction conditions for the observed minimum biodiesel yield are 3.2 methanol to seed ratio (v/w), 1% catalyst amount (w/w), 318 K reaction temperature, and 1200 s reaction time. The reaction conditions corresponding to the minimum and maximum biodiesel yield are specified by the bold letters in table 2.

It is easy to follow from table 2 that, the chosen reaction parameters have significant effects on the biodiesel yield. However, the significance of each reaction parameter can be studied using analysis of variance (ANOVA) as shown in table 3. The ANOVA for the quadratic regression model is carried out using a 95% confidence interval and with a significance level of 0.05. The P-value < 0.05 demonstrates high significance with a 95% of the confidence interval in predicting the yield values and the suitability of the developed regression model as given in table 3. The "lack of fit" parameter is not significant with a p-value higher than 0.05 as can be observed from table 3 which indicates the goodness of fit of the model. The plot of actual yield vs. predicted yield (figure 1) also justifies the goodness of fit of the model.

Table 2. Experimental and predicted values of the yield.

No.	A	B	C	D	Experimental Yield (%)	Predicted Yield (%)
	Methanol:seeds ratio (v/w)	Catalyst amount (wt %)	Reaction Temp. (K)	Reaction time (s)		
1	3.2	2	318	2400	53.51	52.66
2	4.8	1.5	323	3000	54.98	55.99
3	3.2	2	328	2400	62.72	62.06
4	4.8	1.5	323	1800	70.25	70.68
5	6.4	2	328	1200	74.92	72.80
6	6.4	1	328	1200	71.14	70.68
7	6.4	2	328	2400	65.37	63.93
8	4.8	1.5	323	1800	75.38	70.68
9	6.4	1	318	1200	69.4	69.13
10	6.4	1	318	2400	71.38	70.30
11	4.8	1.5	323	1800	70.24	70.68
12	4.8	0.5	323	1800	54.2	53.21
13	1.6	1.5	323	1800	65.89	65.12
14	6.4	2	318	1200	74.81	73.42
15	4.8	1.5	333	1800	63.77	65.00
16	4.8	1.5	323	600	53.34	54.57
17	3.2	1	328	1200	64.24	64.59
18	4.8	1.5	323	1800	73.93	70.68
19	6.4	1	328	2400	60.83	60.61
20	**3.2**	**1**	**318**	**1200**	**41.63**	**41.77**
21	4.8	1.5	323	1800	69.02	70.68
22	3.2	1	328	2400	63.54	63.63
23	4.8	1.5	323	1800	67.92	70.68
24	4.8	1.5	313	1800	53.02	54.04
25	3.2	2	328	1200	62.04	61.81
26	6.4	2	318	2400	77.06	75.79
27	3.2	2	318	1200	41.88	41.17
28	**8.0**	**1.5**	**323**	**1800**	**91.33**	**94.34**
29	4.8	1.5	323	1800	68.01	70.68
30	4.8	2.5	323	1800	52.69	55.92
31	3.2	1	318	2400	50.87	52.06

IOP Conf. Series: Materials Science and Engineering 1146 (2021) 012008 doi:10.1088/1757-899X/1146/1/012008

Table 3. Analysis of Variance (ANOVA).

Source	DF	Adj SS	Adj MS	F-Value	P-Value	Remark
Regression	14	3391.2	242.23	42.29	0	S
A	1	1281.18	1281.18	223.7	0	S
B	1	11.06	11.06	1.93	0.184	
C	1	180.15	180.15	31.46	0	S
D	1	3.01	3.01	0.53	0.479	
A^2	1	146.5	146.5	25.58	0	S
B^2	1	463.87	463.87	80.99	0	S
C^2	1	222.6	222.6	38.87	0	S
D^2	1	423.73	423.73	73.99	0	S
AB	1	23.92	23.92	4.18	0.058	
AC	1	452.07	452.07	78.93	0	S
AD	1	83.18	83.18	14.52	0.002	S
BC	1	4.73	4.73	0.83	0.377	
BD	1	1.44	1.44	0.25	0.623	
CD	1	126.41	126.41	22.07	0	S
Error	16	91.63	5.73			
Lack-of-Fit	10	41.12	4.11	0.49	0.849	
Pure Error	6	50.51	8.42			
Total	30	3482.84				

[‘S’ represents significant terms in regression equation]

Figure 2 shows the standardized effect in the Pareto chart and it is also noticeable that linear terms B and D are insignificant while B^2 and D^2 are significant indicating dominant quadratic effect over linear effect which is also observable from the ANOVA (table 3) and regression equation coefficients. The most significant term affecting the conversion of biodiesel is methanol to seeds ratio (v/w) (A), while interaction terms BC and BD are insignificant as shown in figure 2.

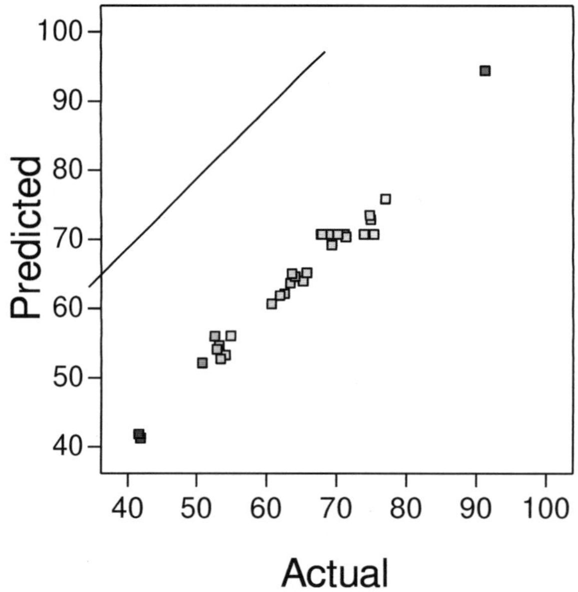

Figure 1. Comparison of predicted vs. actual yield.

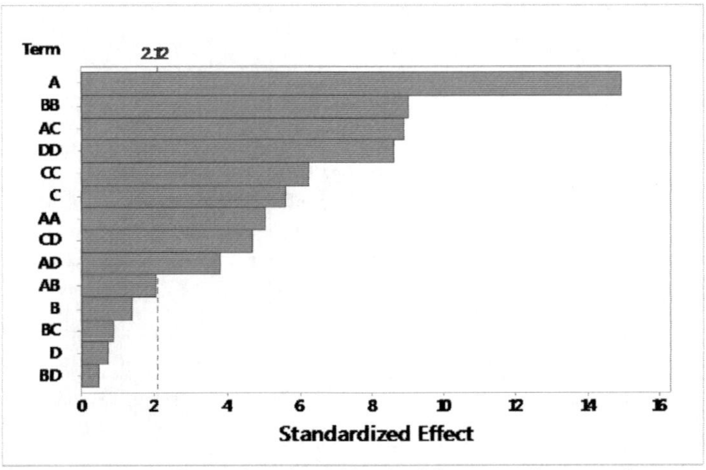

Figure 2. Pareto chart showing the significance of regression terms.

3.2. Main and interaction effects of reaction parameters

Figure 3 shows the main effects of the four factors on biodiesel yield as predicted with a 95% confidence interval. From the trend of methanol:seeds (v/w) ratio (A), it can be followed that there is an increase in biodiesel yield with an increase in A up to upper level (coded level-1 and un-coded level- 8.0), however, increase in A above 7.47 (v/w) is not economically feasible. It can also be observed that variation in catalyst loading (B) has the curvilinear effect on biodiesel yield (Y) and maximum biodiesel yield is predicted around mid-level which is also similar for other factors (C and D). With an increase in temperature (C) there is an increase in biodiesel yield up-to around 318 K, beyond 318 K there is a slow decrease in biodiesel yield with an increase in temperature. Similarly, the effect of reaction time (D) on conversion (%) is also observed to be significant with its steeper slope up to midlevel. The conversion increases significantly with an increase in reaction time (D) up to midlevel, however, there is a reduction in conversion when reaction time is increased beyond the mid-level of 1800 s.

The interaction effect of methanol:seeds ratio (v/w) and reaction temperature on biodiesel yield can be observed in figure 4. At higher methanol:seeds ratio (v/w) with a decrease in reaction temperature, an increase in the biodiesel yield is observed. Moreover, maximum biodiesel yield is obtained when the volume of methanol per unit weight of castor seeds is higher (more than the intermediate level range) due to increased oil extraction. An increase in temperature above 313 K can lead to partial evaporation of methanol which causes less solvent contact for biomass resulting in decreased oil extraction even for higher MeOH:seeds ratio (v/w) as depicted in figure 4. Similarly, figure 5 shows the interaction effect of MeOH:seeds ratio (v/w) and reaction time. It is easy to conclude from figure 5 that, if the MeOH:seeds ratio (v/w) is sufficiently high (above 6.0 v/w), a relatively higher biodiesel yield can be achieved in less reaction time.

The yield was found to be enhanced with an increase in reaction time (D) and reaction temperature (C) up to a mid-level limit. It can be concluded that increasing reaction temperature towards the boiling point of methanol inhibits the reaction. However, as shown in figure 6 that there is less penalty on biodiesel yield as a result of higher temperature if the reaction is carried out for a comparatively shorter time. Exposure to a higher temperature for a longer duration of time can lead to methanol evaporation causing a reduction in yield.

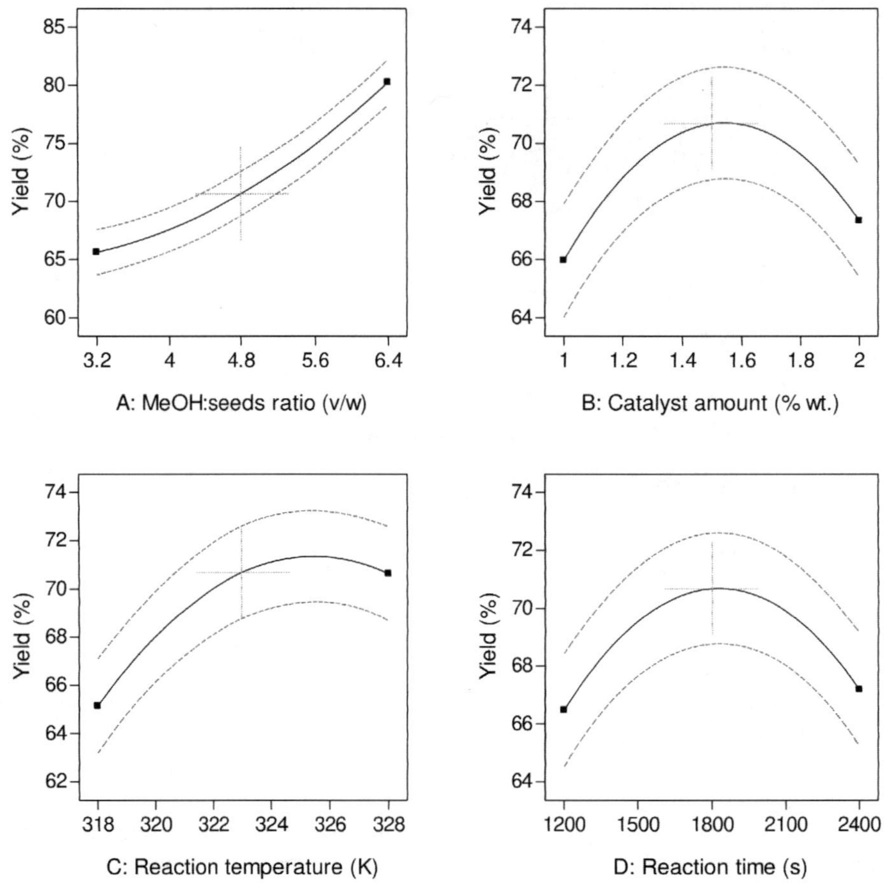

Figure 3. Main effects of factors.

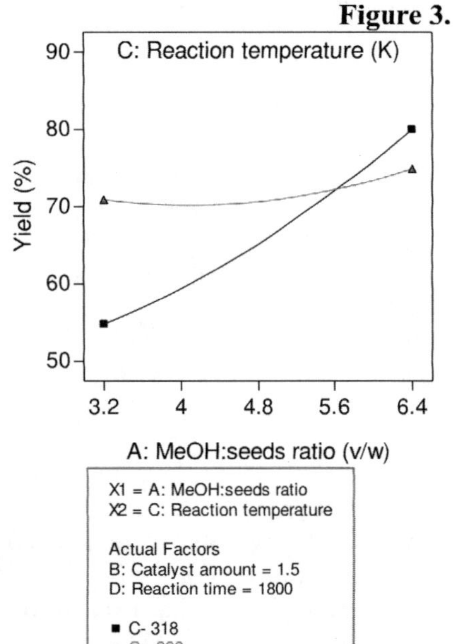

Figure 4. Interaction effects of MeOH:seeds ratio (v/w) and reaction temperature.

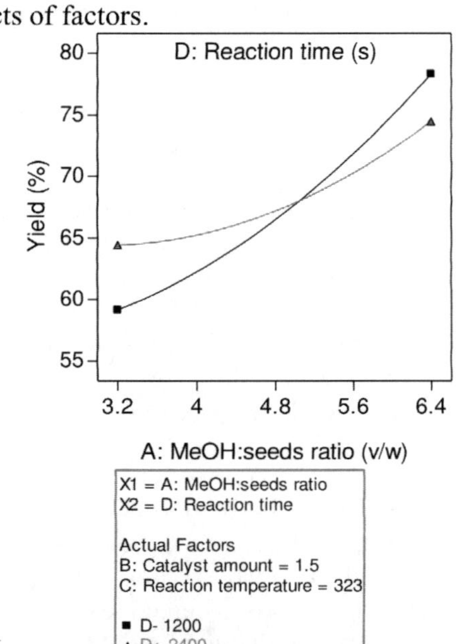

Figure 5. Interaction effects of MeOH:seeds ratio (v/w) and reaction time.

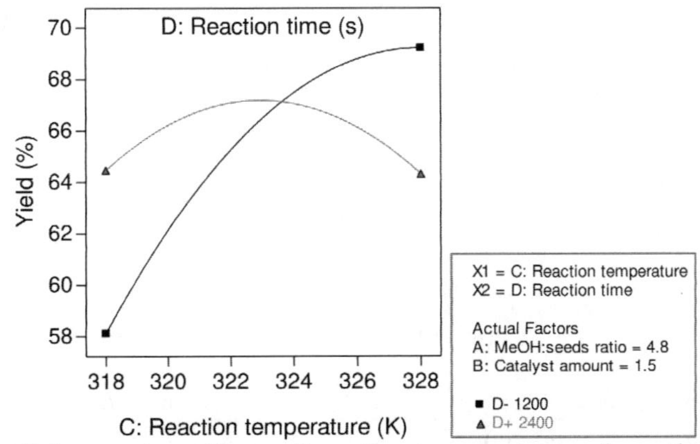

Figure 6. Interaction effects of reaction temperature and reaction time.

3.3. Optimization of reaction conditions

Process optimization for maximum yield was done using State Ease Design Expert (version 10) using desirability function approach maximum desirability value of 0.883 was achieved as given in figure 7). Maximum biodiesel yield of 93.2% yield was predicted for factor level combinations of 7.47 MeOH:seeds ratio (v/w) (coded value=1.67), 1.7% (%wt.) catalyst loading (coded value=0.41), 317 K temperature (coded value= -1.13) and 1788 s of reaction time (coded value=-0.02). Maximum experimental biodiesel yield of 93.5± 0.76 % was obtained at the optimum reaction conditions which is similar to the predicted biodiesel yield calculated using the regression model.

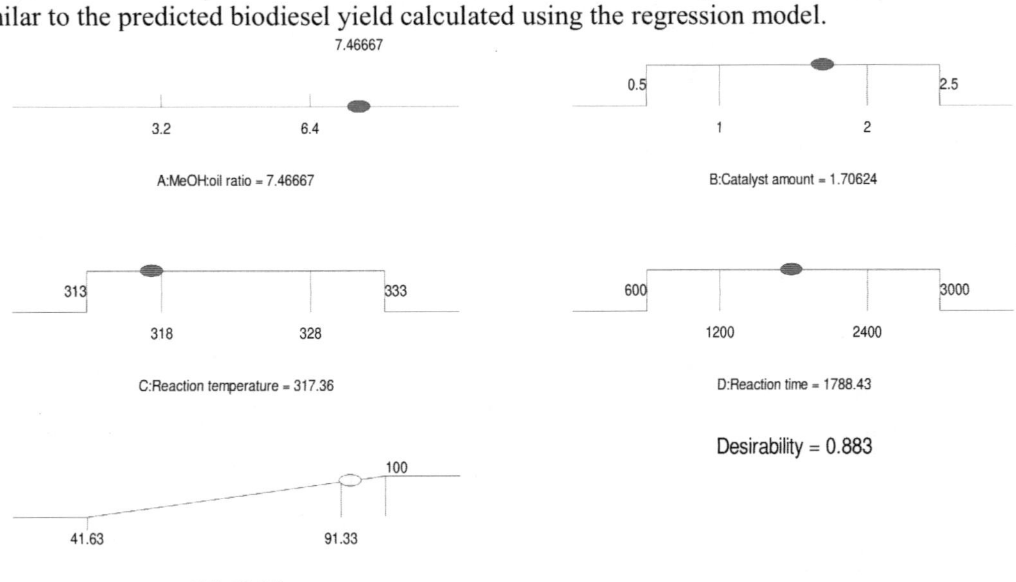

Figure 7. Optimized reaction conditions using the desirability function approach.

3.4. Prediction of biodiesel yield using an extreme learning machine

In the experiment, four parameters have been taken which affect the biodiesel yield. They are MeOH:seeds ratio (v/w) (A), catalyst amount (%wt. oil) (B), reaction temperature (K) (C) and reaction time (s) (D). By using the CCD DoE technique, 31 experimental combinations have been generated and second-order RSM has been fitted as given in Eq. (1).

Further, using the given experimental results, an Extreme learning machine (ELM) has been fitted. ELM is single-layered feedforward neural network (SLFN) with activation function of hidden layer defined by radial basis network and parameters of the hidden layer initialized by random distribution. The output function of the ELM network by L^{th} layer is given by

$$f_L = \sum_{i=1}^{L} \beta_i G(w_i, b_i, x) \ , x \in R^n \quad (3)$$

Where w_i and b_i are the learning parameters of the hidden layer and β_i is the parameter relating the output of hidden nodes with output function. The function G (w_i, b_i, x) is the output of the ith hidden node for the input value x. This function is the radial basis function (RBF) network where the weights are initialized randomly. Here, in the current study, the uniform distribution has been used for initializing the weights of the hidden layer of SLFN. The unknown coefficients, β_i are estimated by using Moore-Penrose pseudoinverse of the design matrix formed by initialized values of hidden node parameters and given values of input variables. It is given by

$$\beta_i = (D^T D)^{-1} D^T f_L \quad (4)$$

Here, 208 hidden nodes are chosen, and uniform distribution of parameters associated with them.

Table 4. Experimental results and predicted values by different data-driven models.

Run No.	Experimental Yield%	RSM	ELM
1	53.51	52.66	53.51
2	54.98	55.99	54.98
3	62.72	62.06	62.72
4	70.25	70.68	70.68
5	74.92	72.80	74.92
6	71.14	70.68	71.14
7	65.37	63.93	65.37
8	75.38	70.68	70.68
9	69.4	69.13	69.4
10	71.38	70.30	71.38
11	70.24	70.68	70.68
12	54.20	53.21	54.20
13	65.89	65.12	65.89
14	74.81	73.42	74.81
15	63.77	65.00	63.77
16	53.34	54.57	53.34
17	64.24	64.59	64.24
18	73.93	70.68	70.68
19	60.83	60.61	60.83
20	41.63	41.77	41.63
21	69.02	70.68	70.68
22	63.54	63.63	63.54
23	67.92	70.68	70.68
24	53.02	54.04	53.02
25	62.04	61.81	62.04
26	77.06	75.79	77.06
27	41.88	41.17	41.88
28	91.33	94.34	91.33
29	68.01	70.68	70.68
30	52.69	55.92	52.69
31	50.87	52.06	50.87

Table 4 shows the experimental yield and its comparison with the predicted yield obtained through RSM and ELM. It also compares the yield predicted by RSM, Improved RSM, and ELM. Table 5 compares the performance of the fitted model using various statistical parameters. Both the models are fitting the experimental yield data with less value of error. The mean average percentage error (MAPE) is smaller in ELM compared to RSM. However, all other indicators are comparatively better for RSM than ELM.

Table 5. Comparison of different fitted data-driven models.

Statistical Parameters ↓ Fitted models→	**RSM**	**ELM**
R^2	0.9740	0.9852
MSE	1.6300	2.9571
RMSE	1.2768	1.7196
SEP	1.9896	2.6798
Adj-R^2	0.9510	-
Correlation coefficient, R	0.9927	0.9867
MAPE	0.0814	0.0319

4. Conclusions

The current study successfully determined optimum reaction conditions for the in-situ transesterification reaction of castor seeds intensified by hybrid energy (ultrasound and microwave) using RSM coupled with CCD. Moreover, the RSM model is also compared with the ELM, and both models are found to be accurate in predicting biodiesel yield. However, higher R^2 and R^2_{adj} values of the RSM model show its fitness for the present application of predicting biodiesel yield. The optimum conditions obtained through RSM corresponding to maximum biodiesel yield of 93.5± 0.76 %, are found to be: MeOH:seeds ratio (v/w) 7.47, catalyst amount (KOH) 1.7%, reaction time 1788 s, and reaction temperature 317 K.

References

[1] Kerschbaum S, Rinke G (2004) Measurement of the temperature dependent viscosity of biodiesel fuels. *Fuel* **83**:287–291

[2] Go AW, Sutanto S, Ong LK, et al (2016) Developments in in-situ (trans) esterification for biodiesel production: A critical review. *Renew Sustain Energy Rev* **60**:284–305

[3] Syers JK, Wood D, Thongbai P (2007) Feasibility of non-edible oil seed crops for biofuel production. In: *International Technical Workshop*

[4] Shivamathi CS, Moorthy IG, Kumar RV, et al (2019) Optimization of ultrasound assisted extraction of pectin from custard apple peel: Potential and new source. *Carbohydr Polym* **225**:115240

[5] Gude VG, Patil P, Martinez-Guerra E, et al (2013) Microwave energy potential for biodiesel production. *Sustain Chem Process* **1**:5

[6] Yuan H, Yang BL, Zhu GL (2009) Synthesis of biodiesel using microwave absorption catalysts. *Energy & Fuels* **23**:548–552

[7] Sharma A, Kodgire P, Kachhwaha S., et al (2018) Application of Microwave Energy for Biodiesel Production using Waste Cooking Oil. *Mater Today Proc* **5**:23064–23075 . doi: 10.1016/j.matpr.2018.11.036

[8] Thakkar K, Shah K, Kodgire P, Kachhwaha SS (2019) In-situ reactive extraction of castor seeds for biodiesel production using the coordinated ultrasound–microwave irradiation: Process optimization and kinetic modeling. *Ultrason Sonochem* **50**:6–14

IOP Conf. Series: Materials Science and Engineering 1146 (2021) 012009 doi:10.1088/1757-899X/1146/1/012009

Experimental Investigation of Three Bladed Inclined Savonius Hydrokinetic Turbine by using Deflector Plate

B Jeeva[1*], S Jai Sandeep[2], N Ramsundram[3], M Prasanth[4] and B Praveen[5]

[1*] Assistant Professor, Department of Mechanical Engineering, Kumaraguru College of Technology, Coimbatore, India. Email ID: jeevmech@gmail.com
[2] PG Scholar, Department of Mechanical Engineering, Kumaraguru College of Technology, India. Email ID: jaisandeep03@gmail.com
[3] Associate Professor, Department of Civil Engineering, Kumaraguru College of Technology, Coimbatore, India. Email ID: ramsundram.civil@kct.ac.in
[4] UG Student, Department of Mechanical Engineering, Kumaraguru College of Technology, Coimbatore, India. Email ID: prasanth.18me@kct.ac.in
[5] UG Student, Department of Mechanical Engineering, Kumaraguru College of Technology, Coimbatore, India. Email ID: praveen.18me@kct.ac.in

Abstract This paper discusses the development of experimental test rig for an inclined Savonius hydrokinetic turbine with improvement in coefficient of power by using deflector plates. The hydrokinetic turbine was designed for different water flow velocities ranging from 1.37 m/s to 2.28 m/s in an open channel with an area of 0.0891 m^2. The hydrokinetic turbine with tip speed ratio (λ) =0.8, aspect ratio (AR) = 0.72 was designed with 222 rpm as the maximum speed. To capture maximum energy from the water, the deflector plate was used to increase the velocity at the advancing side of the blade. The experiment was conducted with a deflector plate and without deflector plate conditions. The extreme coefficient of power obtained was 0.24 for a returning blade angle of 60° with deflector as compared to returning blade angle of 21° and 107°. The coefficient of power without deflector and an inclined shaft was obtained as 0.11. The experiment was performed with deflector plates with various configurations, and the torque was measured for different flow velocities.
Keywords: hydrokinetic turbine, coefficient of power, tip speed ratio.

1. Introduction

The hydrokinetic turbine extracts power from the moving energy of water without a potential head. However, a hydropower turbine requires the head of water to produce electrical energy. In a hydrokinetic turbine, the flow stream of water is from upstream to downstream. Generally, the hydrokinetic turbine is not expensive and requires minimum maintenance *(Kailash et al., 2011),* and the Savonius turbine construction is straightforward (*Wnehenubun et al*., 2015).

A vertical axis Savonius turbine is operable at a shallow head and with remarkable self-begin potential *(Sarma et al., 2014).* The major driving force for the operation of Savonius turbines is the drag force *(Kailash, et al., 2012).* The hydrokinetic turbines are characterised as horizontal axis turbine or axial turbines and a vertical axis turbine or cross-flow turbines *(Kailash et al., 2011).* Generally used vertical axis turbines are the Savonius turbine and Darrieus turbine. *Kamoji et al., (2009)* investigated experimentally on the modified Savonius turbine and deduced an extreme coefficient of power as 0.32.

The three-bladed Savonius turbine resulted in the most exceptional performance at the higher tip speed ratios *(Wnehenubun et al., 2015).* Various efficiency techniques to measure the performance of hydrokinetic turbines of different researchers had been reviewed by *Verma and Saini, (2015). Sahim et al., (2014)* investigated the consequence of deflector position on Darrieus turbine and Darrieus-Savonius turbine blade using NACA 0015 in a water canal. *Wahyudi et al., (2013)* studied three types of tandem blade Savonius rotor namely, overlap, symmetrical and convergence. The CFD simulation predicted that the Savonius convergence type rotor had better performance than symmetrical and overlaps design rotors. *Wahyudi et al., (2015)* developed a crossflow vertical axis turbine with Savonius type rotor, moving deflector blade and Savonius diffuser blade and concluded that use of deflector results in a positive moment compared to without deflector condition. *Wahyudi* and *Adiwidodo, (2017)* have performed the CFD simulations for a tandem blade with radial and tangential moving deflector for a Savonius configuration and validated with the experiments. The investigation revealed that the tandem blade with a tangential moving deflector had higher efficiency than the tandem blade with a radial moving deflector for a Savonius configuration.

Although, many experimental works were performed to increase the coefficient of power of a vertical Savonius hydrokinetic turbine. Variation of various deflector positions and inclination of the turbine shaft was not studied. In the present work, experimentation was performed for three-bladed curved type blades profile with various deflector angles to increase the velocity of the water. A comparative study of the coefficient of power of an inclined hydrokinetic turbine with and without deflector is investigated.

2. Experimentation Methodology

The experiment was performed in the open channel. figure 1 shows the representation of the experimental setup with the hydrokinetic turbine, which is positioned in the open channel with a cross-sectional area of 330 mm × 240 mm. The water in the open channel is provided from a water tank with a 1m head. The height of the turbine is 160 mm, the diameter of the blade is 110 mm, the diameter of the turbine is 240 mm, and the thickness of the blade is 1.5 mm. The swept area of the blade is 0.0384 m^2. Fig. 1 also shows the representation of the experiment set up with the torque measurement by the spring balance method, figure. 2 shows the representation of the inclined turbine arrangement with a 15° inclination to the vertical shaft and 1.7 m from the tank discharge. Due to the inclination of the turbine blade, the drag force has increased. Thus, the higher the drag force, the more power is generated. The velocity of water was measured using the Pitot tube.

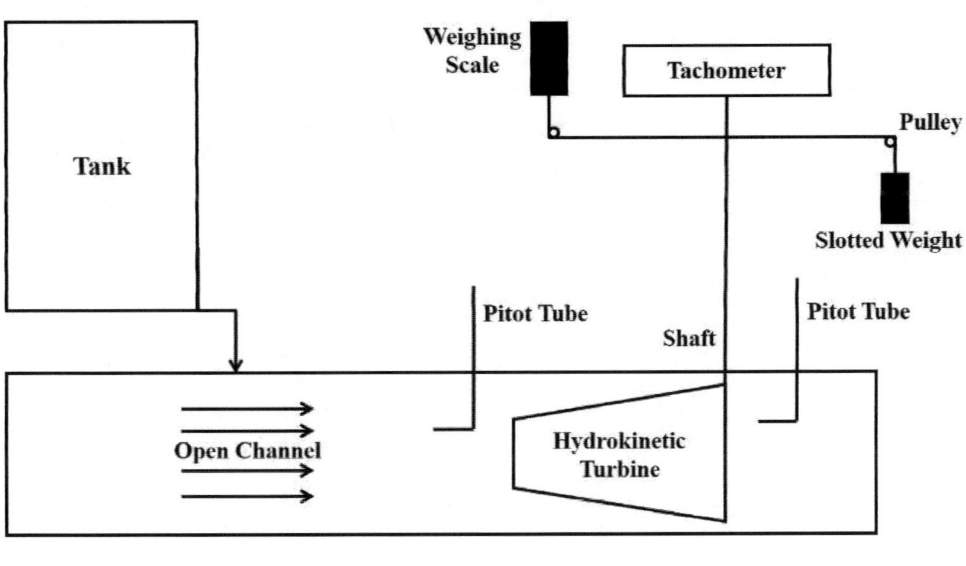

IOP Conf. Series: Materials Science and Engineering 1146 (2021) 012009 doi:10.1088/1757-899X/1146/1/012009

Figure 1. Representation of the experimental setup.

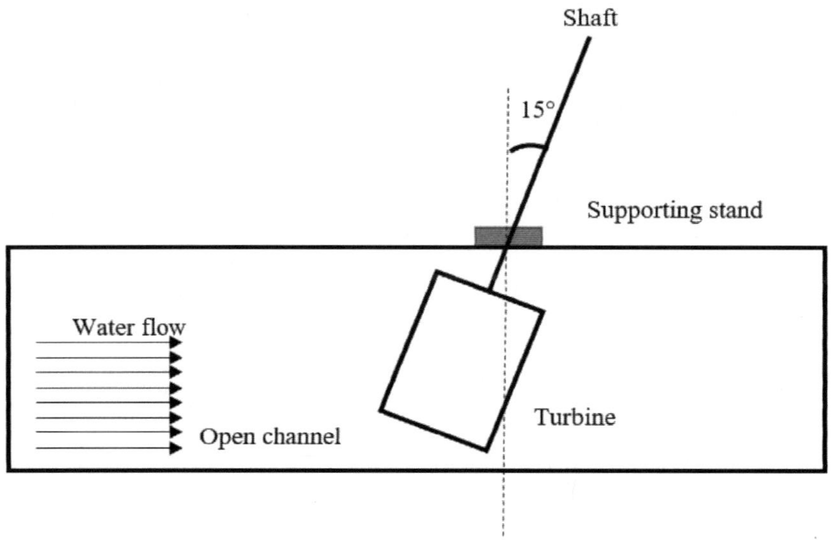

Figure 2. Representation of the turbine inclined at 15° angle.

When the water strikes the blades, the turbine blade rotates with speed. The weights were added in the spring balance set up from 100, 200, 300, 400 and 500 grams respectively to the setup to measure the torque for various conditions. A nylon thread of 1 mm was used for loading the weights.

The deflector plate was used to maximize power extraction. Due to the deflector plate, the water flow speed has increased. There are two deflector plates which are placed in the open channel, namely returning blade, and advancing blade. Advancing blade (α) angles are 35° and 15° and returning blade (β) angles are 107°, 60°, and 21°. Fig. 3 indicates the representation diagram of the deflector plates in the open channel. Advancing blade angle of 35° was useful during the experiment; hence, $\alpha = 35°$ was set for the rest of the experiment. Fig. 4, Fig. 5 shows the arrangements of the deflector plate at 107° and 60° respectively. Fig. 6, Fig. 7 shows the arrangement of the deflector plate at $\beta = 20°$, $\alpha = 35°$ and $\beta = 20°$, $\alpha = 15°$ respectively used in the experiment. Fig. 8 shows the pictorial view of the hydrokinetic turbine system.

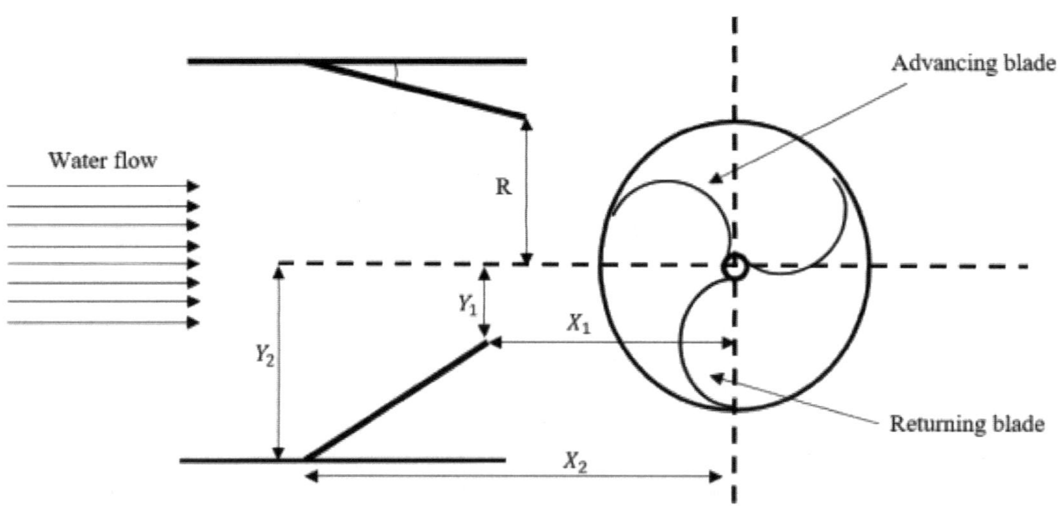

Figure 3. Representation of the hydrokinetic turbine with space position parameter for deflector plate.

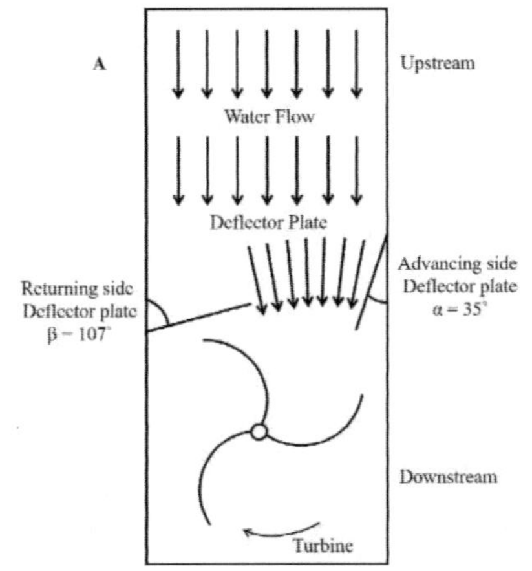

Figure 4. Deflector Plate at β = 107°.

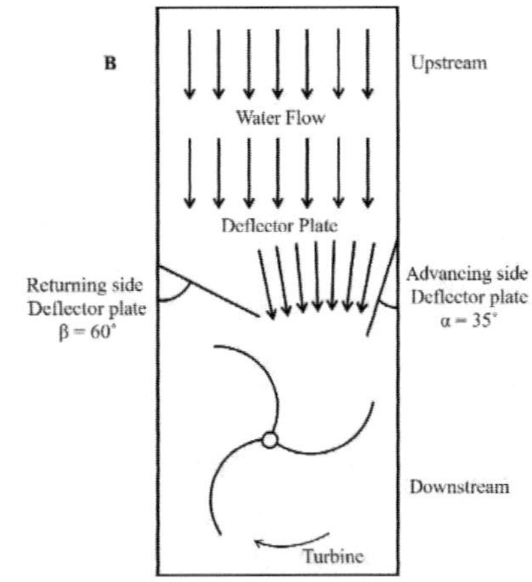

Figure 5. Deflector Plate at β = 60°.

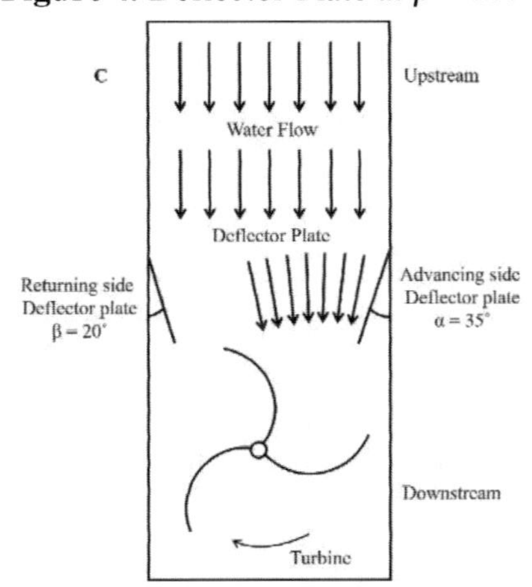

Figure 6. Deflector Plate at β = 20° and α = 35°.

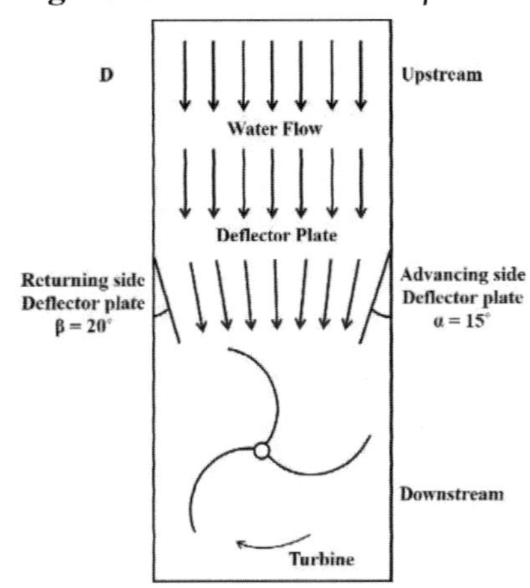

Figure 7. Deflector Plate at β = 20° and α = 15°.

The specification of the deflector plate configurations is itemized in Table 1. The optimum parameter for the deflector plate in the experimental study was $X_1/R = 1.254$, $Y_1/R = 0.545$, $X_2/R = 2.618$ and $\beta = 60°$ by fixing $Y_2 = 164$ and $Y_2/R = 1.490$ fixed. For $\beta = 60°$, a high-water velocity of 2.28 m/s was observed at the upstream of the turbine.

Figure 8. A pictorial view of the experimental setup.

Table 1. Geometrical specifications of the deflector plate with configuration number.

Sl. No.	X_1 (mm)	$\frac{X_1}{R}$	X_2 (mm)	$\frac{X_2}{R}$	Y_1 (mm)	$\frac{Y_1}{R}$	β
1	207	1.881	145	1.318	30	0.272	107°
2	138	1.254	288	2.618	60	0.545	60°
3	120	1.090	275	2.5	100	0.909	20°

2.1 Experimentation Parameters

The performance of the Savonius hydrokinetic turbine can be validated in the method of a coefficient of power (C_p) and the coefficient of torque (C_T) with a comparison with tip speed ratio TSR (λ). The TSR is related to speed of the turbine and diameter of the rotor. The ratio between speed of the blade tip and water velocity through the blade is called the TSR. TSR (λ) can be written as,

$$TSR = \lambda = \frac{V_{blade}}{V} = \frac{\omega R}{v} \tag{1}$$

Where ω is the speed of the rotating turbine in rad/s, R is the radius of turbine blade in m and v is the water flow velocity in m/s.

In the experimentation, the spring balance method was used to determine the actual torque and estimated as,

$$T = \frac{(S-L)\,(r_{shaft}+r_{rope})g}{1000}\ Nm \tag{2}$$

Where, S is the weight balance reading in grams, L is the slotted weight in grams, r_{shaft} is the radius of the shaft in m, r_{rope} is the radius of the rope in m and g is the acceleration due to gravity.

The coefficient of torque (C_T) is the relation between the torque generated by turbine (T) and the theoretical torque (T_w) from water velocity as,

$$C_T = \frac{T}{T_w} = \frac{4T}{\rho A_s d V^2} \tag{3}$$

Where, ρ is the density of water 1000 kg/m^3, T is the actual torque developed by the rotor in N-m and as is the swept area in m^2.

The coefficient of power C_p is the relation between the high power captured from the water flow stream (P_m) and the theoretical power available in water velocity (P_t) as,

$$C_p = \frac{P_m}{P_t} = \frac{P_m}{\frac{1}{2}\rho A_s V^3} \tag{4}$$

Where the maximum power captured from the water flow is determined as,

$$P_m = T\omega in\ (Watt) \tag{5}$$

2.2 Error analysis

The uncertainty of the instruments (ψ_o) is determined as,

$$\psi_o = \left[(\Psi_{pt})^2 + (\Psi_{tm})^2\right]^{\frac{1}{2}} \tag{6}$$

Where Ψ_{pt} is the percentage error of the Pitot tube and $\Psi_{tm}$is the percentage error of the digital tachometer. The uncertainty of the instruments is decided from the manufacturer catalog and by repeating the experiments. The uncertainty of the instruments used in experimentation is ±1.885%.

3. Results and Discussion

The experimentation was performed with various velocity conditions by varying the deflector angles. Fig. 9 shows the speed vs. load for different returning blade angles as compared to without deflector plate. It was observed that as the load increased the speed reduced.

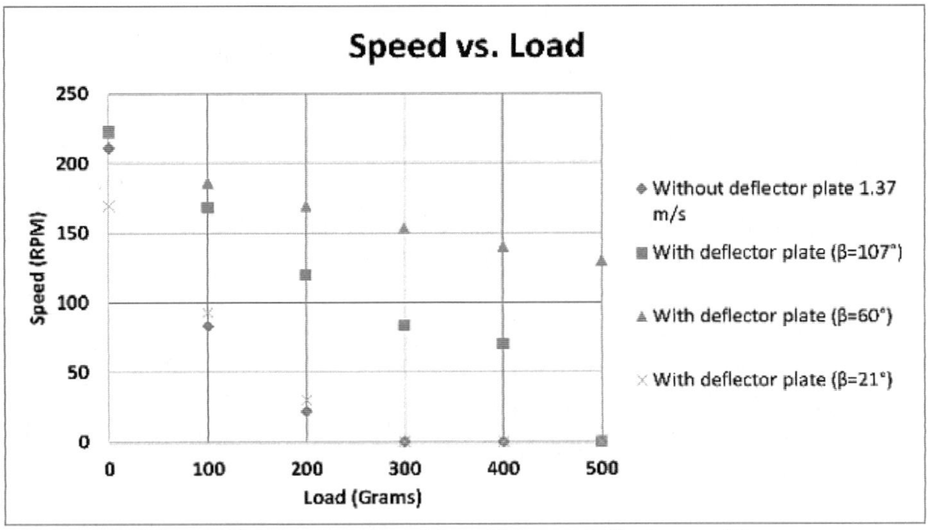

Figure 9. Speed vs. Load for different deflector positions.

Figure 10 shows the hydrokinetic turbine's torque vs. load for different deflector positions as compared to without deflector plates. The torque increased when there was an increase in the load. Further, the torque reduced from the maximum torque value due to the reduction in the speed of the turbine at higher loads. This trend was observed for all conditions except for the deflector plate angle of 60°. The maximum torque could not be determined for a deflector plate of 60° due to limitations in the experiments.

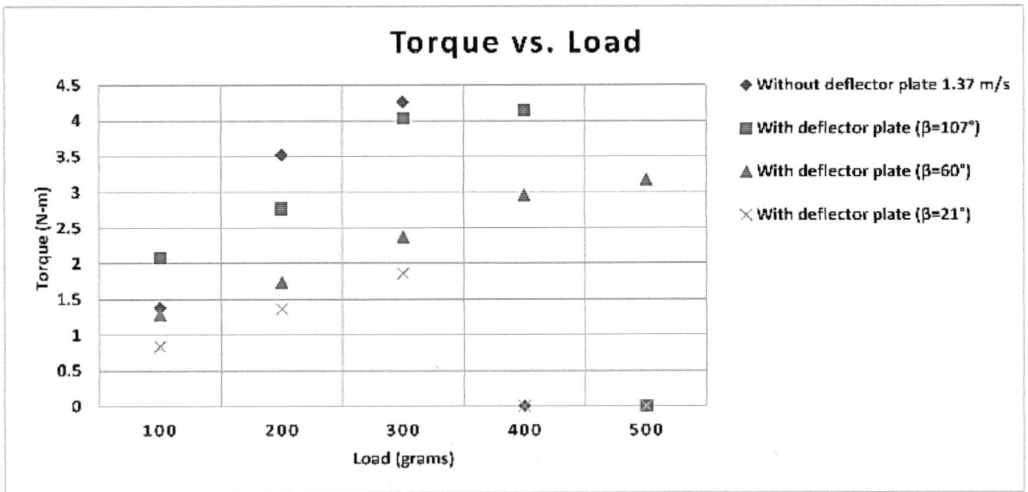

Figure 10. Torque vs. Load for different deflector conditions.

Figure 11 shows the comparison of the coefficient of power vs. tip speed ratio for different deflector plate conditions. The highest coefficient of power obtained in the experimentation was 0.24 for the deflector plate with $\beta = 60°$ and a tip speed ratio of 0.80. The C_p value had a substantial increase for the returning blade angle of β=60° as compared to without deflector plate condition.

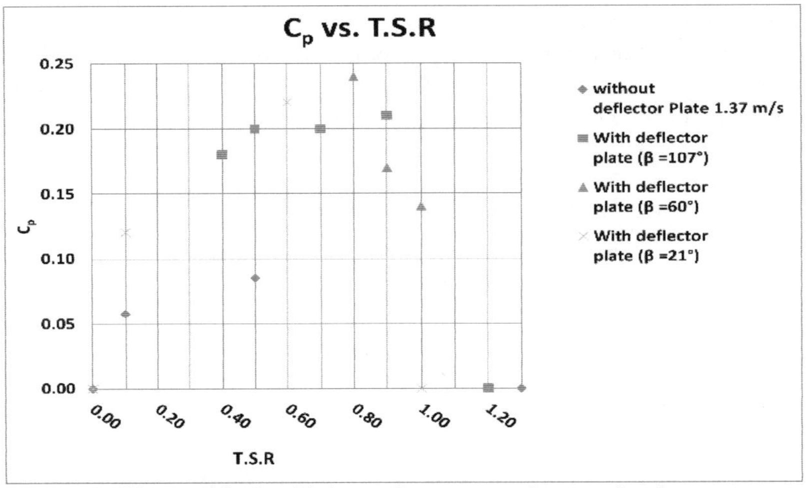

Figure 11. Coefficient of power vs. tip speed ratio.

Figure 12 shows the coefficient of torque vs. tip speed ratio for various deflector plate conditions. The coefficient of torque for β=107° was highest for the tip speed ratio of 0.4 and 0.5 respectively as compared to the coefficient of torque for β= 60°.

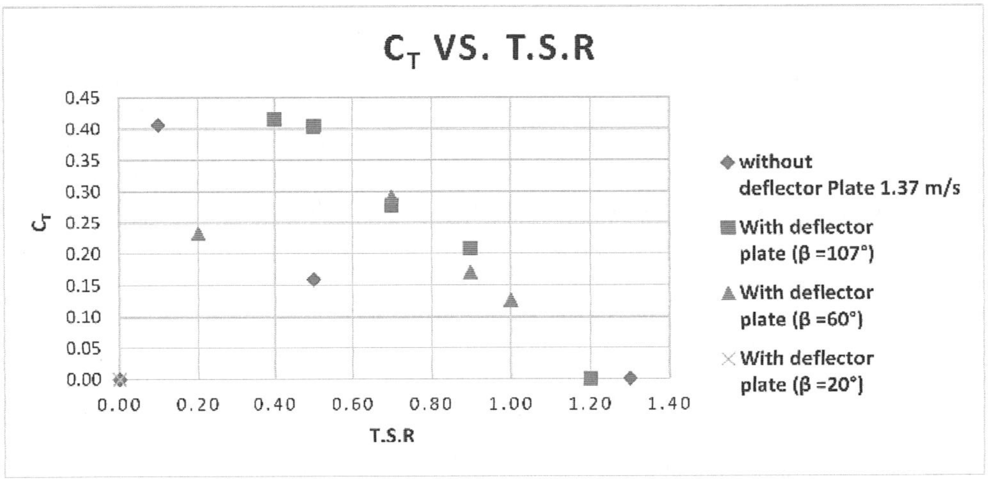

Figure 12. Coefficient of torque vs. tip speed ratio.

Although the coefficient of torque values is higher for the least values of tip speed ratio. For the returning blade angle of β= 60°, the coefficient of power was the highest. The velocity of water measured by the Pitot tube at the upstream of the hydrokinetic turbine for different configurations is listed in table 2.

Table 2 Velocity of Water for Different Configurations.

Sl. No.	The advancing angle of the turbine blade	Water velocity measured at the upstream of the turbine (m/s)
1	Without deflector (β = 0°)	1.37
2	β = 21°	1.7
3	β = 60°	2.28

4	β = 107°	2.0

Comparing the previous work of *Sarma et al*., (2014) and *Kailash et al*., (2012) the C_p was estimated at 0.24 for the aspect ratio of 0.72. The results of the present work with previous work are compared in table 3.

Table 3 Comparison Table of Present Work with Previous Work.

Authors	Aspect ratio	Velocity in m/s	TSR	C_T	C_p
Sarma et al., (2014)	1.21	0.65	0.643	---	0.343
Kailash et al., (2012)	0.7	0.45	1.08	0.32	0.35
Present work	0.72	2.28	0.8	0.29	0.24

4. Conclusion

The following points are concluded:

- The power generated from the hydrokinetic turbine was 45.8 W with the highest coefficient of power as 0.24 for the returning blade angle of β=60°.
- For the above design conditions, the hydrokinetic turbine without inclination, the coefficient of power was 0.15 for the returning blade angle of β=60°.
- The inclination of the shaft improves the coefficient of power by 60%, but the only disadvantage is that the loading of forces acts on the side of the bearing of the shaft.

5. References

[1] Kamoji M A, Kedare S B and Prabhu S V 2009 *Experimental investigations on single stage modified Savonius rotor.* Applied Energy 86, 1064-73

[2] Kailash G, Eldho T I and Prabhu S V 2011 *Influence of the deflector plate on the performance of the modified Savonius water turbine* Applied Energy 88, pp 3207-17

[3] Kailash G, Eldho T I and Prabhu S V 2012 *Performance study of the modified Savonius water turbine with two deflector plates* Int. J. of Rotating Machinery 2012, pp 1-12

[4] Sarma N K, Biswas A and Misra R D 2014 *Experimental and CFD analyses of two-bladed savonius water turbines under low-velocity conditions* Proc. ASME 2014 Power Conference, Baltimore, Maryland, USA

[5] Sahim K, Ihtisan K, Santoso D and Sipahutar R 2014 *Experimental study of Darrieus-Savonius water turbine with deflector: Effect of deflector on the performance* Int. J. of Rotating Machinery 2014, pp 1-6

[6] Verma A K and Saini R P 2015 *Efficiency measurement techniques of hydrokinetic turbines: A Review* Proc. International Conference on Hydropower for Sustainable Development, pp 268-285.

[7] Wahyudi B, Soeparman S and Wahyudi S *2013 A simulation study of low and pressure distribution patterns in and around of tandem blade rotor of the savonius (TBS) hydrokinetic turbine model* J. of Clean Energy Technologies 1(4), pp 286-291

[8] Wahyudi B, Soeparman S and Hoeijmakers H W M 2015 *Optimization design of savonius diffuser blade with moving deflector for hydrokinetic crossflow turbine rotor* Energy Procedia 68, pp 244-253

[9] Wahyudi B and Adiwidodo S 2017 *The influence of moving deflector angle to positive torque on the hydrokinetic cross flow Savonius vertical axis turbine* Int. Energy Journal 17, pp 11-22

[10] Wenehenubun F, Saputra A and Sutanto H 2015 *An experimental study on the performance of Savonius wind turbines related to the number of blades* Energy Procedia 6s8, pp 297-304

Experimental investigations and thermal analysis of a natural draft improved biomass cookstove with different air conditions

Kinnari Modi[1] and Darshit S Upadhyay[1]*

[1]Mechanical Engineering Department, Institute of Technology, Nirma University, Ahmedabad-382481, Gujarat, India

*Corresponding Author:darshitupadhyay@yahoo.com; darshit.upadhyay@nirmauni.ac.in

Abstract. Traditional biomass cookstoves (TCS) are very popular among rural areas and street vendors especially in developing and underdeveloped nations due to their conventional process and simple construction. However, this direct combustion technique is not suitable due to poor efficiency and harmful gaseous-particular emission. Improved biomass cookstoves (ICS) have the potential to overcome the health and environmental problems that are common in the case of TCS. In the present study, investigations were carried out on 3.5 kWth ICS based on gasification design with four different air conditions. The opening to the closing ratio of primary to secondary air vents are taken at 50/50, 50/100, 60/100, and 80/100, respectively for the experiments. Performance of TCS and different ICS cases were compared in terms of burning rate, specific fuel consumption, firepower, useful firepower, thermal efficiency, total particulate matter, and gaseous emissions (CO, CO_2, O_2, NO_x, CC, HC). Apart from that, thermal analysis such as mass balance, energy balance, and exergy efficiency was also calculated. 50/100 opening to the closing ratio of primary to secondary air offers the better performance in terms of higher thermal efficiency and lower gaseous and particulate emission is the major conclusion from this study.

Keywords: Traditional biomass cookstove, improved biomass cookstove, performance, emission, water boiling test, efficiency.

1. Introduction

A huge number of people in the world still depend on traditional (or conventional) biomass cookstove (the bricks/stone stove). It is not limited to rural area people, as people in urban areas are also used the traditional stove (especially homeless people and street vendors) in most of the underdeveloped and developing countries [1]. It is due to the fact that these stoves are inexpensive and simplest design. Apart from that, these types of cookstoves are easily operated [2]. The combustion efficiency of these stoves is very less. Moreover, these stoves generate a large amount of smoke, particulate matter, and other harmful gases. Due to the same reason, traditional cookstoves are not suitable for human health as well as environmental aspects [3]. These hazardous effects resulted in respiratory diseases, adverse pregnancy outcomes, lung cancer, cardiovascular disease (CVD), and asthma. Worldwide annual death of household air pollution is estimating at 3.1 million deaths per year [4]. It includes 1.6 million premature deaths, and 0.9 million children die less than five years of age worldwide as per the World Health Organization (WHO) [5]. To overcome these problems, use improved biomass cookstove is essential nowadays.

The improved biomass cookstoves are made of different materials and designs. The gasification-based natural-draft and forced-draft cookstoves are some of the popular options among researchers. These cookstoves are having two types of air supply i.e., primary holes – to generate producer gas (generally $CO + H_2$) and then secondary holes - to combust producer gas (generally $CO_2 + H_2O$). The top-lit updraft micro biomass gasifier-based cookstoves have primary holes and secondary holes at the bottom part and the upper part for air supply, respectively. the holes are partly or fully closed or opened depending on the needed heating rate to be supplied or controlled which is according to the combustion process [6].

The study of airflow variation in primary to secondary ratio is important due to changes in stoichiometric mixture. The energy and exergy analysis of an improved biomass cookstove was carried out with sheesham and eucalyptus (1 kg to 5 kg). The variation of exergy and energy efficiency was found in the range of 25.83%–30.53% and 3.83%–5.01%, respectively [2][6]. carried out a water boiling test for five different airflow rate values. They concluded that 9.07 Nm^3h^{-1} and 24.3 Nm^3h^{-1} airflow rates provide the highest (64 min) and lowest (50 min) operation time, respectively[7]. performed the experimental study on improved biomass cookstove with woody fuel and different insulating materials. They claimed that the stove with glass wool insulation offers lesser CO emission (29.36 ppm) & PM 10 emissions (3.64 mg) and higher thermal efficiency (32.66%) as compared to the stove with Plaster of Paris[8]. performed a comparative experimental study on gasifier cookstove for four different solid fuels. They observed low CO emission for all feedstock for the lower airflow rate. For the literature survey, it is observed that a limited study with different primary to secondary air ratios was carried out on a gasifier-based cookstove.

In this study, experimentation on traditional cookstove and 3.5 kWth improved biomass cookstove was carried out with different primary to secondary air ratios. Cookstove performance was evaluated by burning rate, specific fuel consumption, thermal efficiency, emissions, and particulate matter. Moreover, thermal analyses such as mass & energy balance, exergy efficiency were also carried out.

2. Experimental

2.1. Material

Wood was selected as a primary food for the experimentation purpose due to its abundant and inexpensive nature especially in the rural area. In the world, most of the cookstoves are operated by wood only. Teak wood was collected from the nearby furniture factory. It was cut and sized before utilizing it in a cookstove. Leco AC-350 Bomb Calorimeter [Test method: IS 1350 (part II)-1970] was used to measure the heating value of the wood. Heating value and ultimate analysis were expressed on a dry basis whereas proximate analysis was carried out on a wet basis. Proximate [Test method: IS 1350 (part I)-1984] and ultimate analysis of the wood is mentioned in Table 1.

Table 1. Properties of feedstock.

		Wood
Physical Properties	Bulk Density (kg m^{-3})	413
	Particle Size (mm)	25×25×5
	Heating Value (MJ kg^{-1})	16.82
	Carbon	45.8

Ultimate analysis	Hydrogen	6.3
	Nitrogen	0.4
	Sulphur	0.002
	Oxygen[a]	40.79
Proximate analysis	Fixed Carbon[a]	5.92
	Volatile matter	82.84
	Ash	4.55
	Moisture	6.69

[a]*by Difference*

2.2. Experimental set-up

3.5 kWth top-lit updraft micro biomass gasifier-based cookstove is designed as per the stoichiometric calculation. It is prepared by mild steel 3 mm thick sheet. Grate and ashpit are added to collect char and ash particles after the combustion process of the feedstock. A number of Primary holes (20 mm) & secondary holes (25.4 mm) and total height were kept 8, 12, and 365 mm, respectively. The shape of the combustion chamber is kept cylindrical whereas the shape of the ashpit is kept rectangular. For the experimentation, two aluminum pots (18L capacity) were used to heat the water. CAD model of biomass cookstove is shown in Figure 1.

2.2.1. Methodology. The experiments were performed on the improved biomass cookstove based on the gasification principle (at atmospheric pressure). Before starting the experiments, the setup is cleaned properly, and then after sized biomass is fed to the reactor. The feedstock is heated by the torch and combustion was taking place. The experiments were carried out three times with the same condition to check the repeatability of the results. To check the performance of the cookstove, three major tests (Water boiling for calculating thermal efficiency, emission, and particulate matter) were carried out.

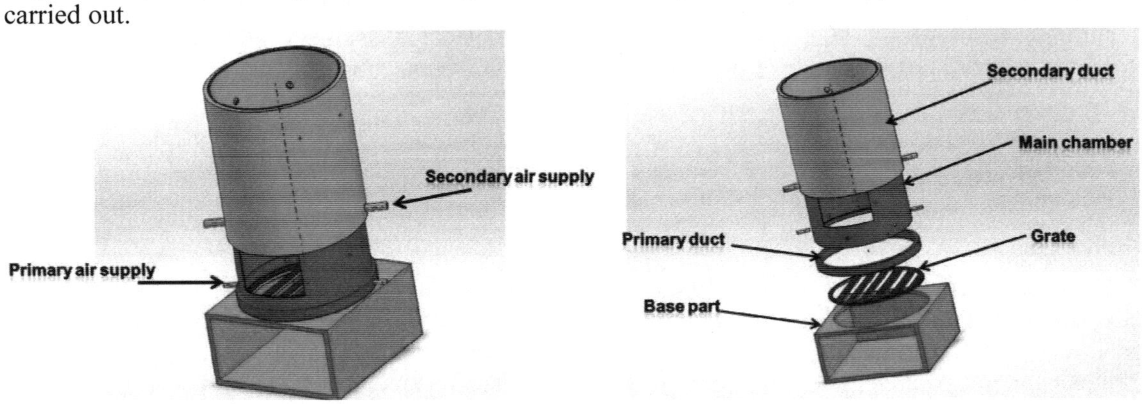

Figure 1. CAD model of ICS.

For the water boiling test, two pots were filled with 18 kg water as per the Bureau of Indian Standards (BIS)[2]. There were two phases in the experiment. In the first phase, an aluminum pot was filled with 18 kg of water and 850 gm of teak wood was filled in the combustion chamber. The initial temperature of the water was measured at the beginning of the experiment along with time. The temperature of the water was constantly monitored at the interval of 5 min until it reaches 95°C with the help of the laboratory thermometer[6]. When 850 gm wood was partially burnt, then again add the remaining wood (350 gm). When the temperature of the water reaches 95°C then pot 1 was replaced by pot 2. The same process was repeated till the utilization of all the fuel in the chamber [2].

An emission test was carried out with five gas analyzer (i3Sys make, EPM 1601 model). Without interfering in any process of combustion in a cookstove, the flue gases were collected in a hood. In order to take a sample of flue gases, this analyzer was fitted to the hood [9]. By this procedure, HC, NO_x, O_2, CC, CO_2 CO, and NO_x were measured relatively in ppm or %.[10]Total Particulate Matter (TPM) was measured as per the guideline of the Ministry of New and Renewable Energy (MNRE), Govt. of India [10][11]. Axiva make glass fiber filters (25 mm diameter, 2.5-micron) were used to measure TPM through the gravimetric method. For suction of the flue gas from the hood to the TPM arrangement, a vacuum pump was used[12]. At the center axis of a duct nozzle, anemometer mounting was placed. Testing of water boiling, gaseous and particulate emission was carried out along with simultaneously during the experimental run. A flow rate of the gas and air was measured with an Amprobe TMA-21HW Hotwire anemometer with a data logger. The temperature of the combustion zone was measured by K (Chromel-Alumel) type thermocouple.

Experiments were carried out on traditional cookstove and improved cookstove with four different primary to secondary air ratios. For creating different conditions, few air vents are kept open and close. The opening to the closing ratio of primary to secondary air vents is maintained at 50/50, 50/100, 60/100, and 80/100, respectively. Example: 80/100 means 80% primary vent open and 100% secondary vent open. The exact mass flow can be found from mass balance Table 2.

2.2.2. Important Parameters and Thermal Analysis. Important parameters of a cookstove performance such as burning rate, firepower, specific fuel consumption (SFC), power output rating are defined as per their definitions available in the literature [2][10]. Mass, energy, and exergy analysis are carried out as per the authors' previous work [13][14][15][16]

Burning rate

$$Burning\ Rate = \frac{Equivalent\ mass\ of\ biomass\ consumed\ in\ kg}{Time\ of\ test\ in\ hour} \tag{1}$$

Firepower

$$Firepower = \frac{Equivalent\ mass\ of\ biomass\ consumed\ in\ kg * CV}{Time\ of\ test\ in\ hour * 3600} \tag{2}$$

Specific fuel Consumption

$$Specific\ Fuel\ Consumption = \frac{Burning\ Rate}{Firepower} \tag{3}$$

Thermal Efficiency

$$? = \frac{(m * cp * \Delta T)water + (\Delta m * \lambda) + (m * cp * \Delta T)pot}{(m * cv)wood} \tag{4}$$

Useful firepower

$$Useful\ firepower = firepower * ? \tag{5}$$

Mass balance

Mass balance was carried out by considering the output mass from a cookstove and input mass to a cookstove. The total input mass of a cookstove consists of air (primary and secondary), fuel while the total output mass of a cookstove consists of gas, char, PM, water, and ash. In the mass balance, the inconsistency finds by mass balance closure. It is defined by the ratio of the total mass output to the input mass. It should close to one proposed the mass balance is accurate.

Mass balance is done by the following equation.

$$M\ fuel + M\ air - p + M\ air - s = M\ gas + M\ char + M\ PM + M\ ash + M\ water \tag{6}$$

M fuel was calculated by weighing machine, Mair and Mgas were converted from respected volumes measured by hot-wire anemometer. M char and M ash were calculated by weighing machine. In fact, both were collected from the same ashpit. However, based on the ash content available in the fuel, ash and char content could be calculated separately. M PM is a mass flow rate of total particulate matter. Mwater was calculated by psychometric chart by considering DBT and WBT observation[15].

Energy balance

Energy balance was carried out by considering output energy from flue gas (utilized or unutilized) and input energy especially from feedstock [17]. The energy balance equation is expressed as:

$$E\ fuel\ + E\ air = E\ gas\ + E\ char\ + E\ pot\ + E\ ash\ + E\ water\ + E\ losses \quad (7)$$

The energy available from fuel and air was calculated by:

$$E\ fuel = m\ fuel * CV\ fuel \quad (8)$$

$$Eair = mair\ *\ Cp, air\ (Tair\ -\ Tref) \quad (9)$$

Similarly, energy from char, water, gas, pot and ash were calculated.

Exergy analysis

Exergy efficiency (or second-law efficiency) computes the effectiveness of a system relative to its performance in reversible conditions. It can also be described as the ratio of the useful work output of the system to the reversible work output for work-consuming systems. It is normally ambient temp and atmospheric pressure [2].

Exergy Efficiency (φ) is expressed as,

$$Exergy\ efficiency\ (\varphi) = \frac{Exergy\ output}{Exergy\ input} \quad (10)$$

Exergy input is calculated based on the following equation:

$$Exi = mwd\ *\ cv\ (1 - \frac{Ta}{Tfuel})\square c \quad (11)$$

Exergy output is calculated based on the following equation:

$$Exo = mw * cpw * (Tfw - Tiw) * \left(1 - \frac{Ta}{Tfw}\right) + mpot * cp\ pot * (Tfp - Tip) * (1 - \frac{Ta}{Tfp}) \quad (12)$$

3. Results and Discussion

3.1. Effects on Useful firepower and firepower

Figure 2 represents the effect of firepower and useful firepower with different airflow. It was observed that TCS has maximum firepower among selected cases, however, the useful firepower of TCS is very less. It is because most of the energy liberated from TCS was not utilized properly and went to the atmosphere. Firepower is depending on the burning rate and rate of oxygen involved in the reaction. Due to the same 50/50 ratio has lower firepower and 80/100 ratio with ICS has higher firepower, comparatively. Useful firepower is depending on efficiency as well as firepower. Due to the same, it is observed that 50/100 ratio has the highest useful firepower compared to other cases due to higher thermal efficiency.

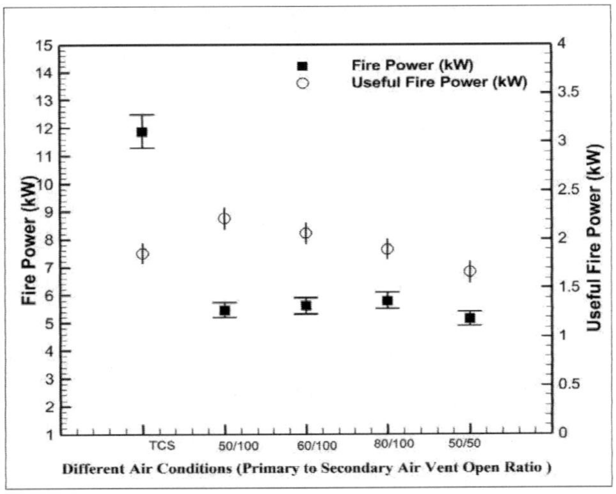

Figure 2. Effects on useful firepower and firepower.

3.2. Effects on burning rate and specific fuel consumption

SFC and burning rate are expressed in Figure 3. More air is involved in the reaction when vents are open to large extents. This process leads to a higher rate of combustion. Due to the same, TCS has a higher burning rate (excess air) whereas ICS with a 50/50 ratio has a lower burning rate. SFC is depending on the thermal efficiency of the system. Therefore, 50/100 ratio has a lower SFC compared to other cases. It is obvious fact that the SFC of TCS is extremely high due to the uncontrolled combustion process.

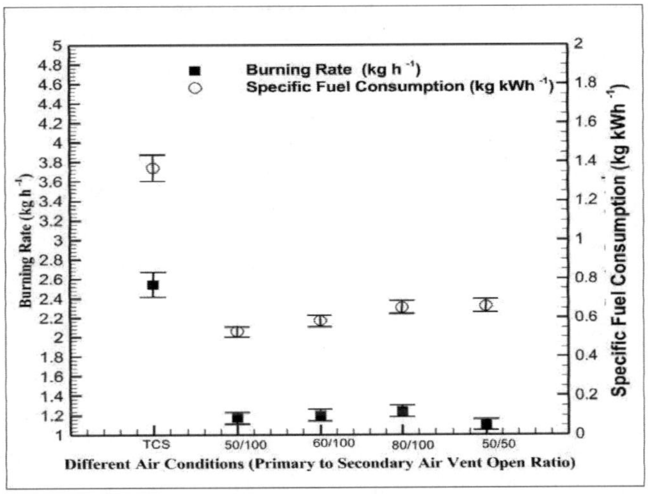

Figure 3. Effects on Burning rate and specific fuel consumption.

3.3. Effects on Thermal Efficiency and Exergy Efficiency

Figure 4 presents the effects on thermal efficiency and exergy efficiency of the biomass cookstove. Thermal efficiency and exergy efficiency were observed in the range of 32.34% - 40.54% and 7.79% - 7.95%, respectively for ICS. For TCS, thermal efficiency was observed only 15.64% and exergy efficiency was observed only 3.93%. It was observed that good quality producer gas may be generated for 50/100 ratio leads to a higher temperature at the secondary air inlet.

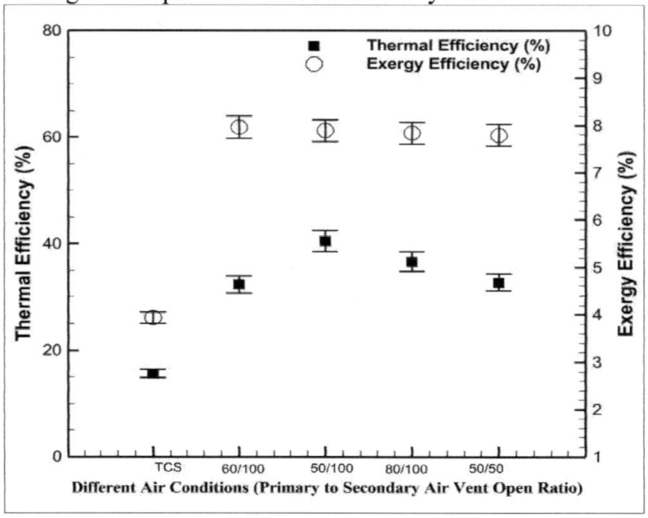

Figure 4. Effects on Thermal efficiency and exergy efficiency.

3.4. Effects on PM and Combustion temperature

Effects on PM and combustion temperature at different airflow are illustrated in Figure 5 Combustion temperatures of TCS and ICS (all cases) were observed in the range between 415ºC to 596ºC. Temperature with TCS was found lower due to improper combustion process. Due to improper combustion, a higher amount of HC and CO was observed in the flue gas for TCS. Due to the same, combustion efficiency was found lower for TCS as compared to ICS. 50/100 and 60/100 ratios offered good combustion temperature comparatively. Particulate emission was found lower as an increment of temperature. It is due to the fact that the vent of the primary holes was placed above the ashpit. Due to the same, fewer particles were carried along with producer gas in the cookstove. Moreover, at a higher temperature, cracking of coarse particles may have resulted in fine particles.

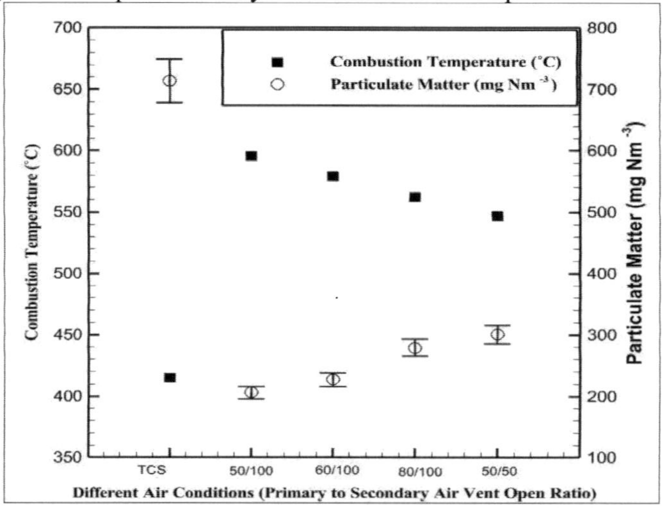

Figure 5. Effects on PM and combustion temperature.

3.5. Effects on Emission gas

Gaseous emissions such as CO, HC, O_2, CC, NO_x, CO_2 are shown in Figure 6 Emissions have a direct relation to combustion characteristics. from figure 6 it is observed that TCS has higher CO and HC contents. These data reveal that combustion products of TCS have the potential to generate more energy. However, due to the unavailability of a sophisticated system, the same components were thrown into the atmosphere. ICS with 50/100 ratio has minimum O_2 and CO_2 compounds. It is because producer gas was combusted properly with secondary air in the combustion chamber.

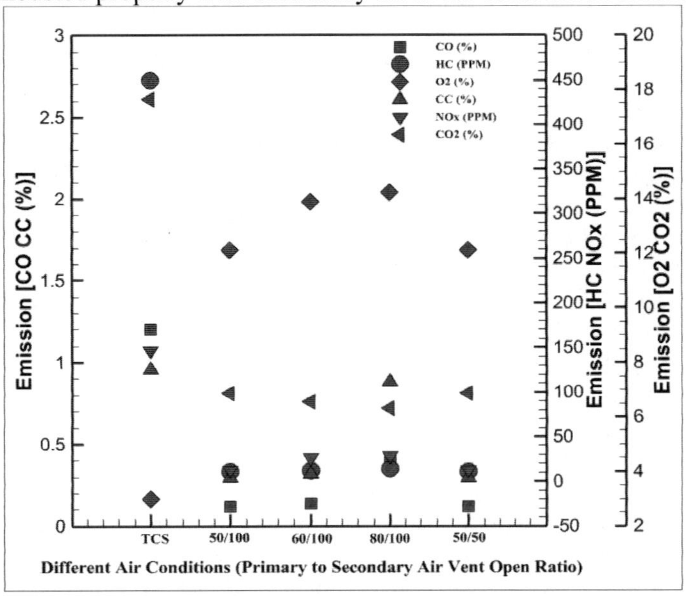

Figure 6. Effects on gaseous emission.

3.6. Mass balance

Table 2 shows a mass balance of different airflow for biomass cookstove. MBC was found in the range of 0.96 to 0.99. 50/50 ratio with ICS has a lower mass of gas. It is due to the lower secondary air providence compared to other cases.

Table 2. Mass balance of different air flow conditions.

Different Airflow (%)	Input Masses ($kg\ h^{-1}$)			Output Masses ($kg\ h^{-1}$)					MBC (%)
	Mfuel	Mair-p	Mair-s	Mgas	Mchar	Mash	MPM	Mwater	
50/50	1.041	1.039	2.01	3.80	0.015	0.062	0.000022	0.085	0.968
50/100	1.178	0.831	4.02	5.69	0.020	0.054	0.000013	0.094	0.971
60/100	1.200	1.039	4.02	5.94	0.018	0.060	0.00001	0.113	0.980
80/100	1.241	1.662	4.02	6.62	0.021	0.064	0.00002	0.144	0.990

3.7. Energy balance

Energy balance was carried out and shown in Table 3 EBC was found in the range of 0.83-0.90. The reason behind lower EBC may be due to the following reasons: 1. No heat loss was considered (pot and reactor), 2. Combustion efficiency was not considered in Efuel, and 3. Unaccounted losses.

Table 3. Energy balance of different air flow conditions.

Different Airflow (%)	Input Masses (kJ h^{-1})		Output Masses (kJ h^{-1})				EBC (%)
	Efuel	Eair	Egas	Echar	Eash	Ewater+pot	
50/50	20184	30.64	4772.04	267.77	2.71	12530.41	0.86
50/100	21462.3	48.75	4245.42	334.72	2.53	14704.09	0.83
60/100	20184	50.84	4614.62	301.24	2.62	13419.37	0.90
80/100	20184	57.10	5040.57	351.45	2.71	12616.32	0.88

4. Conclusions

Experiments were carried out with wood feedstock on TCS and ICS with different air conditions. The opening to the closing ratio of primary to secondary air vents was taken at 50/50, 50/100, 60/100, and 80/100, respectively. Following are the major conclusions from this study.

1. TCS has poor thermal efficiency (15.64%) and higher gaseous and particulate emission (>550 mg Nm-3). Even the combustion temperature of TCS is very less compared to ICS in all cases.
2. Burning rate and SFC are found in the range of 1.1 kg h^{-1} to 1.24 kg h^{-1} and 0.52 kg kWh^{-1} to 0.66 kg kWh^{-1} for different cases of ICS.
3. Firepower is found higher for 80/100 ratio (5.79 kW) whereas useful firepower is found higher for 50/100 ratio (2.21 kW).
4. 50/100 ratio has obtained maximum combustion temperature (596 ºC) and minimum PM as compared to other selected cases.
5. The thermal efficiency of ICS with 50/100 ratio is found maximum (40.54%). In fact, all ICS air conditions offer good thermal efficiencies in the range of 32.34% to 40.54%.
6. ICS with 50/100 ratio offers better output (better thermal efficiency and lower gaseous – particulate emission) compared to other selected conditions. It is due to the generation of good quality producer gas and achieved better stoichiometry during the combustion process.

Acknowledgment

The experimental setup is developed as a part of Idea Lab project (Grant No.: IDEA-2017-ME-04) funded by the Institute of Technology, Nirma University. Authors' express their gratitude towards Nirma University Management to provide all necessary support.

Abbreviations

m	Mass
Cp	Specific heat
ΔT	Difference between initial and final temperature
Δm	Mass of water evaporation
λ	Latent heat of water evaporation
η	Thermal efficiency
η_c	Combustion efficiency
T_{ref}	Reference temperature
M_{wd}	Mass of wood
M_w	Mass of water
M_{pot}	Mass of pot
Cp_w	Specific heat of water
Cp_{pot}	Specific heat of pot
T_a	Ambient temperature
T_{fuel}	Temperature of burning fuel

T_{fw}	Final temperature of water
T_{iw}	Initial temperature of water
T_{fp}	Final temperature of pot
T_{ip}	Initial temperature of pot
M air – p	Mass of primary air
M air – S	Mass of secondary air

References

[1] De C, Lumbreras J, Pérez N, Ealo M and Kane M 2018 Indoor air pollution from biomass cookstoves in rural Senegal Energy for Sustainable Development Indoor air pollution from biomass cookstoves in rural Senegal *Energy Sustain. Dev.* **43** 224–34

[2] Jain T and Sheth P N 2019 Design of energy utilization test for a biomass cook stove: Formulation of an optimum air flow recipe *Energy* **166** 1097–105

[3] Memon S A, Jaiswal M S, Jain Y, Acharya V and Upadhyay D S 2020 A comprehensive review and a systematic approach to enhance the performance of improved cookstove (ICS) *J. Therm. Anal. Calorim.*

[4] Haggerty L, Reischl U, Handy R G, Sleeth D K and Adams K 2020 The thermodynamics of indoor air pollution : A pilot study emulating traditional Kenyan homesteads *Sustain. Cities Soc.* **53** 101926

[5] Mehetre S A, Panwar N L, Sharma D and Kumar H 2017 Improved biomass cookstoves for sustainable development : A review *Renew. Sustain. Energy Rev.* **73** 672–87

[6] Pal K, Jha M K, Gera P and Tyagi S K 2018 with varying quantities of different biomass feedstocks Energy and exergy analysis of a natural-draft improved biomass cookstove with varying quantities of different biomass feedstocks

[7] Bhatta S, Pratap D, Gakkhar N and Rajput J P S 2021 A Comparative Experimental Investigation of Improved Biomass Cookstoves for Higher Efficiency with Lower Emissions 961–71

[8] Kirch T, Medwell P R, Birzer C H and van Eyk P J 2020 Feedstock Dependence of Emissions from a Reverse-Downdraft Gasifier Cookstove *Energy Sustain. Dev.* **56** 42–50

[9] Deng L, Torres-rojas D, Burford M, Lehmann J, Fisher E M and Whitlow T H 2018 Fuel sensitivity of biomass cookstove performance *Appl. Energy* **215** 13–20

[10] Govt. of India 2013 *Portable Solid Biomass Cookstove*

[11] Govt. of India 2000 *Qualifying, Testing and Performance Evaluation of Biomass Gasifiers and Gasifier-Engine Systems, Gasifier Action Research*

[12] Upadhyay D S, Makwana H V. and Patel R N 2019 Performance evaluation of 10 kWe pilot scale downdraft gasifier with different feedstock *J. Energy Inst.* **92** 913–22

[13] Upadhyay D S D S, Khosla A, Chaudhari A, Patel R N R N, Chaudhary A and Patel R N R N 2019 Effect of catalyst to lignite ratio on the performance of a pilot scale fixed bed gasifier *Energy* **189**

[14] Upadhyay D S, Sakhiya A K V, Panchal K R and Patel R N 2018 Thermodynamic Analysis of Lignite Gasification in the Downdraft Gasifier *J. Energy Environ. Sustain.* **5** 58–64

[15] Kumar A, Chaturvedi A, Barasara V, Panchal K, Upadhyay D S, Sakhiya A, Chaturvedi A, Barasara V, Panchal K, Upadhyay D S and Patel R N 2019 Elemental and Heat Loss Analysis of Different Feedstock in Downdraft Gasifier *J. Energy Environ. Sustain.* **7** 10–6

[16] Upadhyay D S, Panchal K R, Sakhiya A K V. and Patel R N 2020 Air-Steam gasification of lignite in a fixed bed gasifier: Influence of steam to lignite ratio on performance of downdraft gasifier *Energy* **211**

[17] Park S R, Pandey A K, Tyagi V V and Tyagi S K 2014 Energy and exergy analysis of typical renewable energy systems *Renew. Sustain. Energy Rev.* **30** 105–23

IOP Conf. Series: Materials Science and Engineering 1146 (2021) 012011 doi:10.1088/1757-899X/1146/1/012011

Exploiting the challenges of copper to austenitic stainless steel bimetallic joining by gas tungsten arc welding: A fluid flow perspective

Raghavendra S. Darji[1,*], Gaurang R. Joshi[2, 3], Ankit D. Oza[4], Vishvesh J. Badheka[5]

[1*]Research Scholar, Department of Mechanical Engineering, School of Technology, Pandit Deendayal Petroleum University, Gandhinagar, Gujarat, 382007, India
[2]Assistant Professor, Department of Mechanical Engineering, Marwadi University, Rajkot, Gujarat, 360003, India
[3]Post-doctoral Fellow, Fusion blanket division, Institute for Plasma Research, Bhat, Ahmedabad, Gujarat, 382428, India
[4]Assistant Professor, Institute of Advanced Research, Gandhinagar, Gujarat, 382426, India
[5]Professor, Department of Mechanical Engineering, School of Technology, Pandit Deendayal Petroleum University, Gandhinagar, Gujarat, 382007, India

Corresponding author's E-Mai;: meetraghav123@gmail.com

Abstract. Miscible nature of the material is vital to join dissimilar materials together. The wide range of properties within the single structure requires merging of multiple materials. Hence, sustainable welding technology development to join distinct material with extreme immiscibility such as Cu and SS is emerging as a recent research interest. Because of its requirement in nuclear reactors' critical components, plan wave linac structures and ultra-high vacuum systems. Gas tungsten arc welding is a well-established commercial process. Hence, a study on optimum weld condition to get better quality of Cu to SS bimetallic joint is advantageous and desirable. However, complex flow nature of the molten Cu and SS is a big challenge. Hence, authors have reviewed and discussed the fluid flow of Cu and SS during gas tungsten arc welding. Which, concludes that the flow characteristics of Cu and SS must be studied through experimental and simulations to set up the favourable weld conditions while joining it with gas tungsten arc welding. The mentioned area of interdisciplinary research and not explored yet about the joint's mechanical and microstructural properties. The said study can potentially predict what will be the final weld properties well before the metallurgical testing. Moreover, the article enlightens the interdisciplinary research aspect of Cu to SS bimetallic joining.

IOP Conf. Series: Materials Science and Engineering 1146 (2021) 012011 doi:10.1088/1757-899X/1146/1/012011

1. Introduction

Diversified material amalgamation imparts multiple mechanical and metallurgical properties within the single structural component. Hence, dissimilar material joining is a thought-provoking demand, especially in cryogenics, vacuum and nuclear systems. Among these, copper (Cu)/austenitic stainless steel (SS) bimetallic joining is of great importance due to their distinct properties across the joint line. SS brings strength, articulated Ductile to Brittle Transitional (DBT) behaviour, ductility, low thermal conductivity and corrosion resistance while Cu inhales excellent thermal conductivity, better corrosion resistance and inherent DBT nature. Thus, Cu/SS joints can also be crucial and necessary for the applications above. Therefore, it is worth to study the mechanism behind Cu/SS joining.
Interestingly, metallurgy (solubility, multiple elements, liquid metal characteristics and solidification pattern) of Cu and SS are incompatible. Even though sufficient efforts were made previously, to explore Cu/SS's joining behaviour by various fusion welding processes. Moreover, individual metallurgy of Cu and SS is a dominating factor while joining of both materials. Cu is metal while SS is an alloy. Nonetheless, both have the same crystal structures, i.e. Face Centered Cubic (FCC), but Cu has an inherent FCC crystal structure while SS secured it through addition of Ni into Fe. Which largely contribute to DBT characteristics. In addition to that, Cu's corrosion resistance is natural whereas corrosion resistance of SS is due to Cr addition, which form a film of Cr_2O_3 layer on the surface. In line with this, Cr and Ni improve the mechanical properties of SS. The ternary phase diagram of Fe, Cr and Ni (see Fig. 1) which form the SS reflect it. The melting points of Fe (1539 ^{0}C), Cr (1907 ^{0}C) and Ni (1484 ^{0}C) individually are higher than the Cu (1084 ^{0}C). Thus, it is obvious that Fe, Cr and Ni have distinct liquid metal flow than Cu. Fe – Cr – Ni alloy's (SS) thermophysical properties, such as surface tension, viscosity, thermal conductivity, and heat capacity, are of great interest in heat and mass transport processes, i.e. welding [1]. The estimation, measurement or calculation of the viscosity for these alloys are complex [1]. However, it has been exploited and reported (see Fig. 2) [1]. Wherein, pure Ni has the lowest viscosity, the addition of Cr and Fe improves the ternary liquid alloy's viscosity. Whereas, Cr/Fe alloy has the highest viscosity, as shown in Fig 2. Comparing the viscosities of individual elements (see Fig. 3) considering the Cu/SS welding, pure Cu has the lowest viscosity than pure Fe and Ni. Hence, the fluidity of Cu predicted higher during fusion welding.

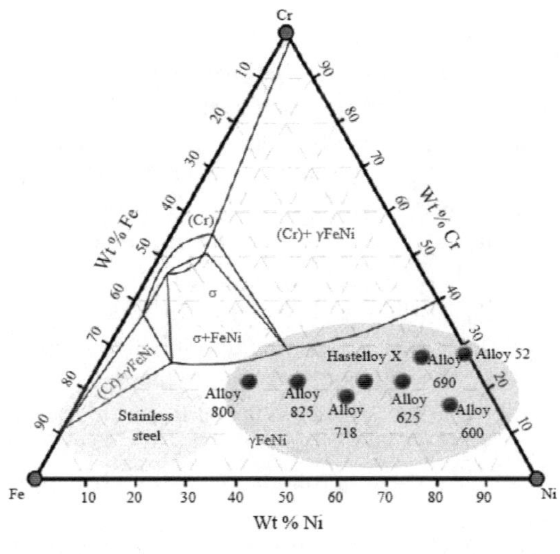

Figure 1: Fe-Ni-Cr ternary phase diagram.

IOP Conf. Series: Materials Science and Engineering 1146 (2021) 012011 doi:10.1088/1757-899X/1146/1/012011

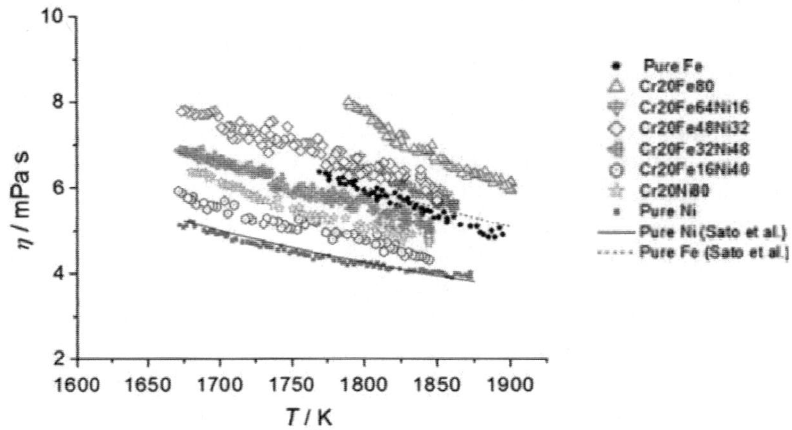

Figure 2: Measured viscosities of various Cr-Fe-Ni alloys as a function of temperature, including pure Fe and Ni [1].

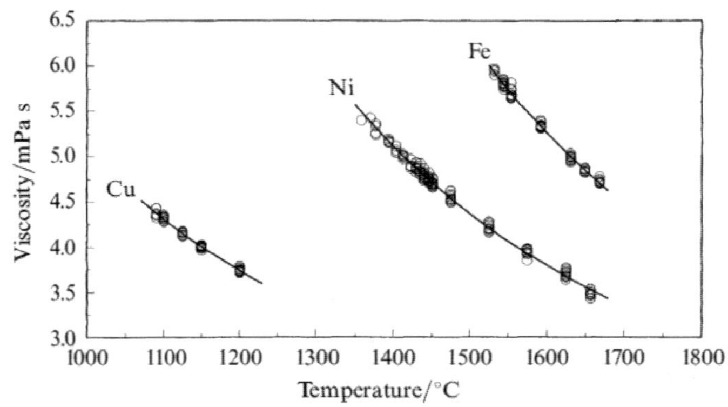

Figure 3: Measured Viscosities of Cu, Ni and Fe by Oscillating Viscometer [2].

Protruding the concept of liquid metal fluidity in welding, the optimum fluidity is required to weld any metal or alloy successfully. Cu and Ni are the two best examples of the metal that can be weld with the metallurgical compatibility but have a vast difference in the fluidity (see Fig. 3); Cu produces the very fluid weld pool on the other side Ni makes the viscous weld pool. In both cases, the chances of getting defects are equal. Cu being erratic liquid metal generates the turbulence which interacts with the atmospheric gases and ending up with the defects such as porosity and crack [3], [4]. Ni being viscous liquid, extra care must be taken while welding pure Ni for the sidewall fusion as Ni weld pool cannot flow easily inside the groove and chances of getting the lack of fusion are very high. The optimum fluid liquid metal provides stability to the weld pool formed during the initial stage of the welding, which can be easy to control and flows inside the intended localized region such a way that sound joint can obtained. In contrast, highly fluid weld pool takes the stability of the process away. The unstable weld pool is prone to produce the defective weld joint due to turbulence caused by the extremely fluid liquid metal under the arc force. Highly viscous weld pool cannot flow easily makes the welding a bit difficult. Still, manipulating the highly viscous metals is possible by implementing the oscillations and changing the welding torch angle.

Similarly, each welding process has its stability during the welding based on their principle (type of shielding, generation and maintain the welding arc, etc.). Gas Tungsten Arc Welding (GTAW) is the most stable welding process being non-consumable process, and the arc is produced using the non-consumable tungsten electrode [5]. A second most crucial factor of the GTAW is the shielding provided through inert gas such as Argon or Helium; thus, the process can be used for any metal and alloy including highly reactive metals [6], [7]. Due to the ingrained stability of the GTAW, it can be an excellent process to implement for Cu to SS dissimilar metal joining [8]. Nonetheless, for successful welding of Cu to SS bimetal joint selection of filler wire becomes very important [9]. The choice of filler wire is the most crucial factor for this type of bimetallic joints. As discussed previously, Cu is a pure metal while SS is an alloy primarily of Fe, Cr and Ni. The alloying elements alter the Fe's characteristics, which must be incorporated while selecting the filler wire. Considering the fluid behaviour in the Cu/SS welding, the complicated occasion can be estimated because of the many elements interacting in the small-localized region in liquid form. Thus, it becomes exciting and challenging to initiate the study in this direction, which is one of the motive behind this article. This article aims to introduce the challenges caused by the fluid flow during the GTAW of Cu to SS and how it affects the microstructure's development and attract the multidisciplinary researcher to investigate the mechanism of fluid flow behind the Cu to SS weld joint. Simulation, Mathematical modelling or temperature distribution can be studied incorporating the fluid flow of individual materials, i.e. Cu and SS, which might be able to produce the results which can predict the parameters which can reduce the formation of metal matrix composited in the Cu to SS bimetallic joint.

2. Fluid flow in the GTAW process

The most influenced parameter on the fluid flow during the GTAW process is the arc force ; under the arc force, weld pool moves in a reverse direction that of the welding (reverse direction that of the tungsten electrode). Along with the arc force, electromagnetic forces and surface tension forces are also affecting the weld pool flow. The liquid metal movement under the electromagnetic and surface tension forces mentioned (see Fig. 4 [10]). It is quite important to observe that the surface tension forces dominate the flow over the electromagnetic forces. In such circumstances, if the liquid metal has low viscosity (i.e. Cu), the weld pool becomes erratic, making the shielding ineffective. Fluid flow velocities of the order of 1 m/s was observed under the GTAW process in normal conditions [11]. As a result, porosities can be observed in the final weld. Camurri et al. [7] reported porosities in Cu weld due to the reason mentioned earlier. While in the SS welding by GTAW, no such evidence was mentioned. SS (6 mm thick) can be welded in a single pass by surface activating elements along with the 150 amps welding current [12] while the same procedure for 6 mm thick Cu welding ended up with the porosities observed by Rana et al. [13] without full penetration using 200 amps welding current. The contradiction in the above two experiments defines the difficulties in the welding of Cu than SS; however, while welding Cu/SS dissimilar joint using GTAW process, the selected parameters remains the same.

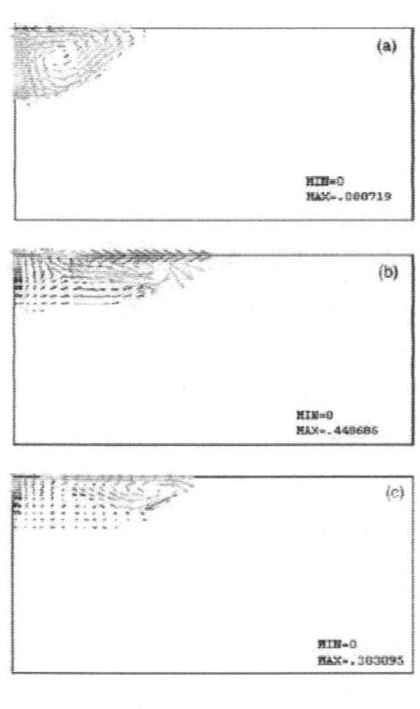

Figure 4: Flow of the fluid inside the weld pool under, (a): electromagnetic force, (b): surface tension forces and (c): combined action of electromagnetic and surface tension forces [10].

3. The flow of Cu and SS during GTAW

The fluidity plays a crucial role in affecting the solidification patterns [9]. Apart from the extreme fluid flow, Cu has excellent thermal conductivity. The weld pool size increases continuously during the welding from start to end reported by Darji et al. [14]. Summing the fluidity and effect of thermal conductivity up, the result would be the relatively larger weld pool having extreme fluid nature, which makes the welding of Cu volatile. Generally, these types of weld pools are prone to react with the atmospheric gases causing defects in the final weld. The surface appearances of the Cu weldments are also not as good as SS [15], [16]. These contradictions are due to the difference in fluidity of Cu and SS as mentioned above (refer fig. 2 and fig. 3).
However, the above discussion was based on the similar metal/alloy joining and not for the Cu/SS joint. Concerning the fluid flow of Cu to SS weld joint, two of the latest studies were considered, executed by Saranarayanan et al. [11], Joshi and Badheka [12]. In both the studies, two different filler wires were selected, as it is a critical parameter for the dissimilar joining by GTAW. Saranarayanan et al. [11] used ER NiCu – 7-filler wire and ER NiCrMo – 3 filler wire was used by Joshi and Badheka [12]. Table 1 shows both the filler wires' chemical compositions. It can be said that both of them are Ni alloys. NiCu-7 is the most compatible filler wire considering the welding of Cu. At the same time, NiCrMo-3 is excellent filler wire considering the welding metallurgy of the SS.

Table 1: Chemical compositions of the ER NiCu-7 and ER NiCrMo-3.

Filler Wire	C	Mn	Fe	P	S	Si	Cu	Ni	Al	Ti	Cr	Nb+Ta	Mo
ER NiCu-7	0.15	4	2.5	0.02	0.015	1.25	Rem.	62 - 69	1.25	1.5 - 3	-	-	-
ER NiCrMo-3	0.10	0.50	5	0.02	0.015	0.50	0.50	58 min.	0.40	0.40	20 - 23	3.15 – 4.15	8 - 10

One has to secure the minimum amount of δ- Fe to avoid the solidification cracking [13]. As both the filler wires were having Ni, the filler wire's fluidity must be incorporated in this discussion. However, both the filler wires' fluidity was relatively low compared to that of the Cu [14]. Embossing these conditions on the actual Cu/SS welding, the scenario would be as if one metal (Cu) is extremely fluid the alloy (SS) is viscous and filler wires were (either ER NiCu-7 or ER NiCrMo-3) viscous too. The amalgamation of elements such as Cu, Fe and filler material within the joint area make a complex scenario. Because of diverse fluid characteristics of mentioned components, significantly lesser viscosity of Cu than filler wire. Cu primarily governs the resultant weld joint's microstructure since it possesses efficient fluid nature in a liquid state. It signifies the importance of the Cu HAZ, Cu fusion line, and weld region. Because of Cu's higher fluidity, it is an exciting aspect as far as Cu/SS joining is concerned. Fig. 5 and Fig. 6 show the Cu's macro interfaces to SS weld joint using the GTAW process. The SS to weld interface observed was very smooth, which describes that the filler wire has fused adequately with the SS base metal during the welding irrespective of welding parameters and filler wire. Both being less fluid, the weld pool towards the SS interface was stable as a result straight, or smooth curved surfaces were received in both the studies performed [9], [17].

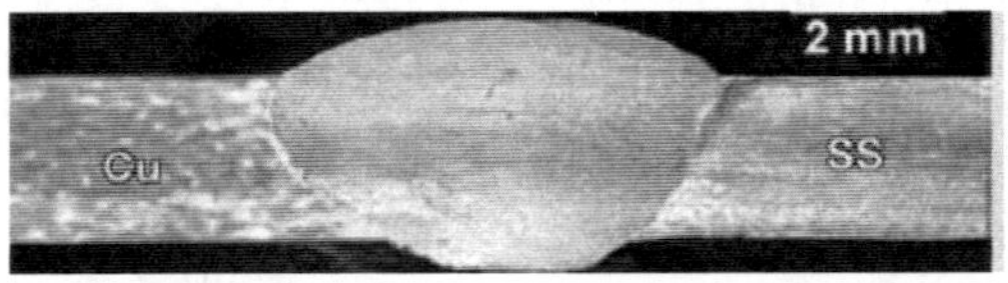

Figure 5: Interfaces of Cu/SS joint by GTAW using ER NiCu-7 filler [18].

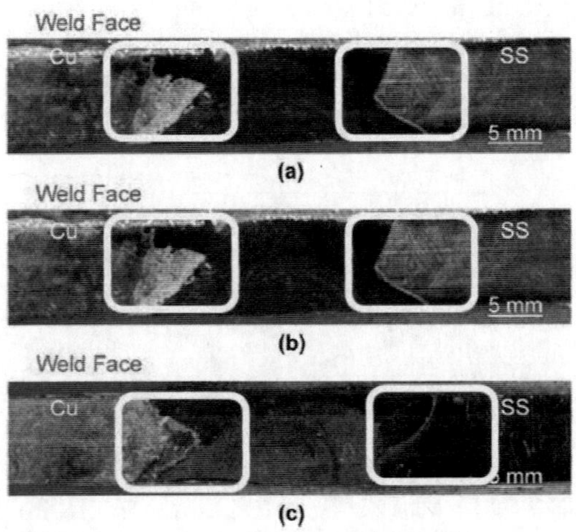

Figure 6: Macro sections of the Cu/SS weld joint performed using GTAW with ER NiCrMo – 3 filler wire at three different welding currents, (a): 125 A, (b): 150 A, and (c): 200 A [17].

However, Fig 6 (C) shows that the weld edge of SS (SS interface) has melted and mixed thoroughly with the weld metal due to very high welding current. The welding current and the compatibility of the filler wire with the SS base metal, as ER NiCrMo-3 is more compatible as compare ER NiCu-7, the filler wires' elements were Cr, Mo and Cu. Out of these, Cr and Mo can be soluble to a more considerable extent in Fe, while Cu is not soluble in Fe it is immiscible in Fe due to the metallurgical incompatibility [17]. Fe having allotropic nature, the solubility of Cu varies in Fe; nevertheless, at room temperature, the solubility of Cu in Fe is significantly less. On the other side, Cr and Mo are the two parameters responsible for the corrosion resistance and improvement in Fe's mechanical properties due to their higher solubility as that of Cu. While Cu being fluid and under the arc force, moved quite volatile during the welding as an outcome, the Cu to weld interface has sharp hills and non-uniform shape. It is important to note that these observations are only at the macroscopic level [9], [17]. The non – uniformity in the Cu's geometry to the weld interface increased significantly at a microscopic level. It can be seen in Fig. 7 and Fig. 8. After looking at the macroscopic results, this was not exciting, but the microscopic results were in line with the macro examination. The reason behind these results was the same as discussed earlier, due to the excess fluidity of Cu. However, these observations are not very exciting. The most exciting outcomes were the fluidity effect on developing the Cu/SS weld joint's microstructure.

The ERNiCu-7 is compatible filler material than ERNiCrMo-3 concerning the Cu welding, as Cu and Ni both, the elements have just 2 % variation in the atomic size and same crustal structure [19]. Hence the solubility of both the components in each other is 100 % in liquid and in solid-state. Despite that, they have significant difference fluidity (see Fig. 3). Cu being higher fluid elements dominate the microstructure formation in the weld metal. Due to the fluidity of the Cu, Metal Matrix Composite (MMC) structure was reported [9]; Fig 8 represents the Cu to SS weld joint's SEM images using compatible filler [6]. On the other hand, ER NiCrMo – 3 filler wire is less compatible considering the Cu base metal as Cr practically having extremely low solubility. In this case, the steel particles with uneven shape can be seen at the microscopic level (see fig. 9), which was very interesting that Cr and Mo have low solubility in Cu but soluble to considerable extent in Fe. These three elements with or without Ni forms a steel and these steel particles inside the weld near to the Cu base metal side were definitely the outcome of the extreme fluidity of the liquid Cu.

When welding the Cu side weld pool remains erratic as discussed previously, the liquid Cu encompasses the steel compositions formed due to the interaction between the filler wire and the base metal due to solidification pattern and excess fluidity of Cu. Ideal microstructure (see Fig. 10) consist of diluted weld metal at the centre of the weld area following by interface region of the weld metal to Cu and SS. In which filler material suppressed within.

Furthermore, the diluted filler material can be found adjacent to both interfaces, i.e. Cu/weld metal and SS/weld metal. Wherein, Cu-Ni solid solution present at the Cu/weld metal interface. It is to identify the elemental effect on joint quality.

(a) **(b)**

Figure 7: Micrographs of the (a): Cu and weld interface, (b): heterogeneity in the microstructure due to interaction between Cu base metal and ER NiCu-7 filler wire [18].

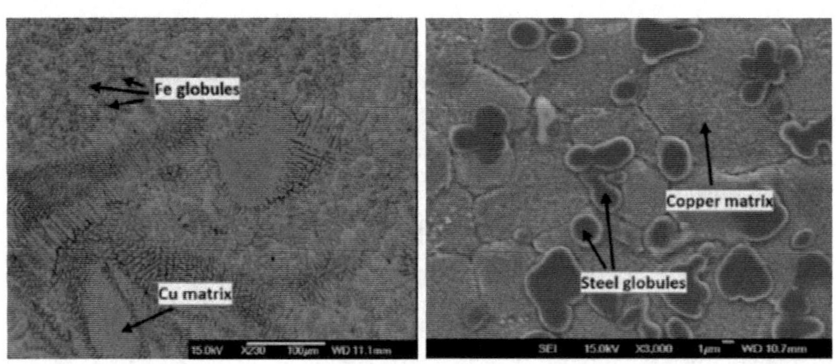

Figure 8: Scanning electron microscopy images of the Cu to SS weld joint using ER NiCu-7 filler wire [9].

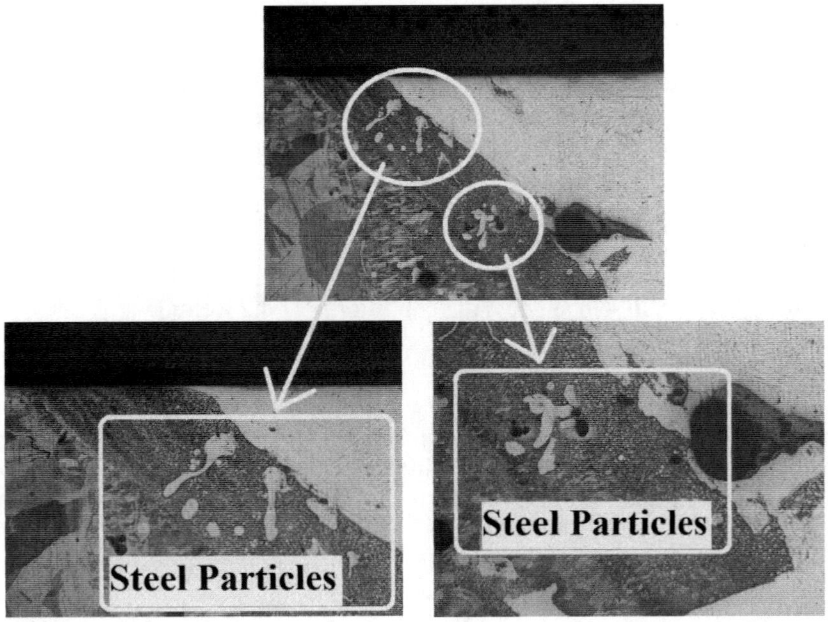

Figure 9: Micrographs of the Cu to weld metal interface using ER NiCrMo – 3.

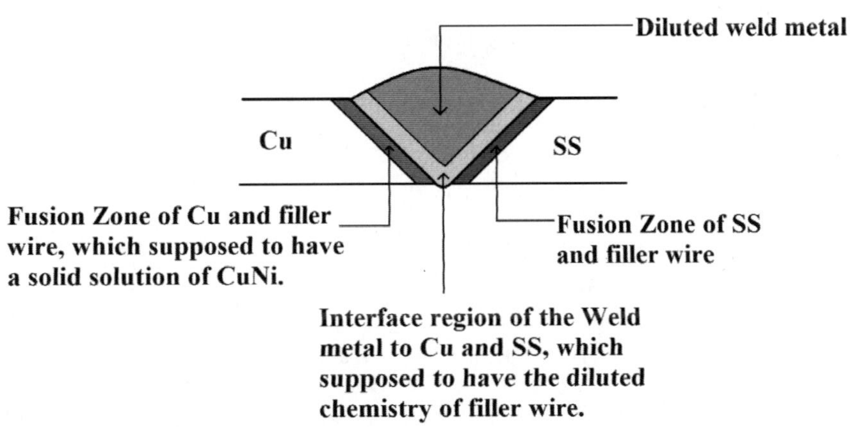

Figure 10: Ideal zones of the Cu/SS dissimilar joint welded by GTAW using compatible filler wire such as ER NiCu-7 or ER NiCrMo-3.

The missing part between the particles and the matrix is coherency, as the steel particles were observed in the Cu matrix towards the weld and Cu interface. The limited mutual solubility of Cu and Fe is one reason for non-coherence between the matrix and particles. Because of the non-coherency between the matrix and particles, the interface remains weaker than the particles and the matrix, which can cause the failure from the Cu weld side or the Cu HAZ side. However, the size of the steel particles observed was in microns; the cumulative effect has to be considered, which degrade the mechanical properties through metallurgical non-uniformities formed caused by Cu's fluidity.

4. Conclusions

1) Fluid flow during the Cu/SS welding is still less researched despite the most crucial parameter.

2) Flow characteristics of Cu and SS play a vital role in the formation of microstructure during GTAW.
3) Excess fluidity of Cu is one of the significant concern to join Cu/SS bimetallic joint. Hence, SS particles entrapped and form Cu metal matrix composite.
4) To control the fluid flow of Cu buttering of Ni on the Cu weld edge or pulsing can be adopted during the welding.
5) Effect of individual elements from Cu to SS can be studied and incorporated during the Cu/SS welding on the fluid flow under the arc pressure.
6) A simulation model can be prepared for understanding the fluid flow for dissimilar Cu/SS bimetallic joint, which will suggest the conditions for the welding.

Acknowledgement

The authors are thankful to the Board for Research in Fusion Science and Technology (BRFST), for sponsoring the research project through Project Number NFP/ MAT/A10/04. We would also like to acknowledge the support of Pandit Deendayal Petroleum University, Gandhinagar, during the conduct of the research project.

References

[1] H. Kobatake and J. Brillo, "Density and viscosity of ternary Cr-Fe-Ni liquid alloys," *J. Mater. Sci.*, vol. 48, no. 19, pp. 6818–6824, 2013, doi: 10.1007/s10853-013-7487-2.

[2] R. F. Brooks, I. Egry, S. Seetharaman, and D. Grant, "Reliable data for high-temperature viscosity and surface tension: Results from a European project," *High Temp. - High Press.*, vol. 33, no. 6, pp. 631–637, 2001, doi: 10.1068/htwu323.

[3] C. Camurri, M. Lopez, R. Fernandez, and V. Osorio, "Copper welding with copper filler," *Weld. Int.*, vol. 10, no. 5, pp. 387–389, 1996, doi: 10.1080/09507119609549014.

[4] L. Yinan, Z. Shanbin, P. Zilong, and G. Feng, "A Study on the Mechanism of Crystal Cracking in GTA Welding of Copper Plates," *Mater. Manuf. Process.*, vol. 31, no. 16, pp. 2143–2151, 2016, doi: 10.1080/10426914.2015.1103861.

[5] E. Spaniol, T. Ungethüm, M. Trautmann, K. Andrusch, M. Hertel, and U. Füssel, "Development of a novel TIG hot-wire process for wire and arc additive manufacturing," *Weld. World*, vol. 64, no. 8, pp. 1329–1340, 2020, doi: 10.1007/s40194-020-00871-w.

[6] B. Baufeld and O. Van Der Biest, "Mechanical properties of Ti-6Al-4V specimens produced by shaped metal deposition," *Sci. Technol. Adv. Mater.*, vol. 10, no. 1, 2009, doi: 10.1088/1468-6996/10/1/015008.

[7] J. Y. Bai, C. L. Fan, S. Lin, C. L. Yang, and B. L. Dong, "Effects of thermal cycles on microstructure evolution of 2219-Al during GTA-additive manufacturing," *Int. J. Adv. Manuf. Technol.*, vol. 87, no. 9–12, pp. 2615–2623, 2016, doi: 10.1007/s00170-016-8633-1.

[8] Y. Poo-arporn *et al.*, "Gas tungsten arc welding of copper to stainless steel for ultra-high vacuum applications," *J. Mater. Process. Technol.*, vol. 277, p. 116490, 2020, doi: 10.1016/j.jmatprotec.2019.116490.

[9] S. RAMACHANDRAN and A. K. LAKSHMINARAYANAN, "An insight into microstructural heterogeneities formation between weld subregions of laser welded copper to stainless steel joints," *Trans. Nonferrous Met. Soc. China (English Ed.*, vol. 30, no. 3, pp. 727–745, 2020, doi: 10.1016/S1003-6326(20)65249-9.

[10] F. Lu, S. Yao, S. Lou, and Y. Li, "Modeling and finite element analysis on GTAW arc and weld pool," *Comput. Mater. Sci.*, vol. 29, no. 3, pp. 371–378, 2004, doi: 10.1016/j.commatsci.2003.10.009.

[11] J. Song, R. Field, A. Clarke, Y. Fu, and M. Kaufman, "ASM Vol 6 WELDING, BRAZING, AND SOLDERING," *Acta Mater.*, vol. 165, pp. 362–372, 2019.

[12] R. Nakhaei, A. Khodabandeh, and H. Najafi, "Effect of active gas on weld shape and microstructure of advanced A-TIG-welded stainless steel," *Acta Metall. Sin. (English Lett.*, vol. 29, no. 3, pp. 295–300, 2016, doi: 10.1007/s40195-016-0381-6.

[13] H. Rana, V. Badheka, P. Patel, V. Patel, W. Li, and J. Andersson, "Augmentation of weld penetration by flux assisted TIG welding and its distinct variants for oxygen free copper," *J. Mater. Res. Technol.*, vol. 10, pp. 138–151, 2021, doi: 10.1016/j.jmrt.2020.12.009.

[14] R. Darji, V. Badheka, K. Mehta, and J. Joshi, "Processing of copper by keyhole gas tungsten arc welding for uniformity of weld bead geometry," *Mater. Manuf. Process.*, vol. 00, no. 00, pp. 1–10, 2020, doi: 10.1080/10426914.2020.1784932.

[15] H. Yin *et al.*, "Jo ur l P of," *Sci. Total Environ.*, p. 141926, 2020, doi: 10.1016/j.jmrt.2020.12.009.

[16] R. H. G. e. Silva, M. B. Schwedersky, and Á. F. da Rosa, "Evaluation of toptig technology applied to robotic orbital welding of 304L pipes," *Int. J. Press. Vessel. Pip.*, vol. 188, no. June, 2020, doi: 10.1016/j.ijpvp.2020.104229.

[17] G. R. Joshi and V. J. Badheka, "Metallographic and Microstructure Analysis of Gas Tungsten Arc-Welded Bimetallic Copper-to-Stainless Steel Joints," *Metallogr. Microstruct. Anal.*, vol. 9, no. 2, pp. 180–193, 2020, doi: 10.1007/s13632-020-00632-7.

[18] R. Saranarayanan, A. K. Lakshminarayanan, and B. Venkatraman, "A combined full-field imaging and metallography approach to assess the local properties of gas tungsten arc welded copper–stainless steel joints," *Arch. Civ. Mech. Eng.*, vol. 19, no. 1, pp. 251–267, 2019, doi: 10.1016/j.acme.2018.08.009.

[19] Davis, J. R.;Copper and Copper Alloys, ASM Speciality Handbook ,2001,32-33.

IOP Conf. Series: Materials Science and Engineering 1146 (2021) 012012 doi:10.1088/1757-899X/1146/1/012012

Fuel Property Investigation: Jatropha Biodiesel, Used Cooking Oil Biodiesel and Fuel-grade Ethanol Blends

Tomesh Kumar Sahu[1*], Pravesh Chandra Shukla[1], Vikas Kumar Sahu[1], Satyajit Gupta[2] and Sumit Sarkar[3]

[1] Department of Mechanical Engineering, Indian Institute of Technology Bhilai, India
[2] Department of Chemistry, Indian Institute of Technology Bhilai, India
[3] Chhattisgarh Biofuel Development Authority, Chhattisgarh, India
[*] Corresponding author's e-mail ID: tomeshs@iitbhilai.ac.in

Abstract. Compression ignition (CI) engines are one of the promising power producing units which provide higher efficiency and durability. Increasing environmental pollution and depleting resources led the research in the field of alternative fuels such as biodiesels, ethanol etc. for utilization in CI engines. In the present study, two different types of biodiesels i.e. Jatropha Biodiesel (JBD), Used Cooking Oil (UCO) Biodiesel and Ethanol were investigated for the fuel properties and compared. Various fuel properties like density, kinematic viscosity, calorific value, oxidation stability, cold weather properties (CFPP, CF, and PP), and flash point temperature were measured and compared for pure form of alternative fuels. Different blends (5%, 10%, 15% and 20%) of JBD, UCO biodiesel and ethanol were also investigated. It has been observed that ethanol showed lower density, lower viscosity and superior cold weather properties compared to the tested fuels/blends. The average viscosity for ethanol was measured to be 1.08 mm2/sec which is ~61.8 % less compared to mineral diesel. Oxidation stability tests were also performed where mineral diesel shows 83.15 min for 10% pressure drop at 140°C and 700 kPa, UCO biodiesel and JBD shows 9.2 min and 8.32 min. 5% ethanol blend showed sharp reduction in flash point compared to mineral diesel, however, it remained almost at the same level with further blending.

1. Introduction

Diesel engines have significant importance in global goods transportation and heavy vehicle operations due to their higher efficiency and durability [1]. Limited reserves, uneven distribution of petroleum reserves among countries and increasing pollution level led to the research activities for alternative fuel options for mineral diesel. Biodiesel is an accepted alternative fuels of mineral diesel being used for a very long time in diesel engines [2]. In their early-stage research, biodiesels were produced and used using edible seeds oils as a first-generation (1G) biodiesels, however, second-generation (2G) biodiesel includes non-edible oil seeds for the production of biodiesel such as Jatropha, Karanja etc. [3,4]. Increasing global emission is a major concern these days and biodiesel are effective in reducing the pollution level. Another option for diesel engines are various alcohols, primarily methanol and ethanol which are relatively easier to produce and used as an alternative of diesel. In India, Jatropha is one of the abundantly available resources for feedstock oil. On the other hand, waste cooking oil (WCO) is an emerging alternative for biodiesel production [5,6]. The usage of vegetable oil is not desirable for cooking when its polar compound concentration increases beyond 25% [7]. Cooking oil with higher concentration of polar compounds can be alternatively used for biodiesel production and known as used cooking oil (UCO). Ethanol is also an important possible alternative fuel with relatively lower calorific values (~27 MJ/kg) [8,9]. Biofuels such as biodiesels and ethanol possess different properties compared to mineral diesel in context of utilization diesel engines. Fuel properties like viscosity, calorific value, density etc. largely affect the combustion and performance behaviour of the diesel engines [10]. Silitonga et al., 2013 [11] have reviewed fuel properties, ignition properties and cold weather properties of jatropha biodiesel (JBD)/mineral diesel blends (10%, 20%, 30%; v/v). They observed that higher proportion of JBD resulted into increased flash point (~18% increase for 30% blend), kinematic viscosity (~38% increase for 30% blend), density and cloud point temperature while calorific value of the blend decreased (~4.3% decrease for 30% blend) [11]. Chen et al., 2012 [12] have observed increase in the oxidation stability by blending the mixture of soapnut biodiesel (SNBD) and JBD (35:65 (w/w)) ratio to mineral diesel (with blends ratios of 1%, 2%, 5%, 7%, 10%, 20%, 40%, 60% and 80%; v/v). They obtained -6°C cold filter plugging point (CFPP), oxidation stability in the range of 9.1-17.3 hours and kinematic viscosity (at 40°C) in the range of 3.03-3.51 mm^2/s upto 40% mixed biodiesel blends in mineral diesel [12]. Chen et al., 2013 [13] have recommended upto 40% blending of JBD (doped with 250 ppm pyrogallol) to counter the problems of oxidation stability [13]. National Renewable Energy Laboratory (NREL), USA has reported [14] that the use of JBD reduced the greenhouse gases (GHGs) emissions by 62% while testing in locomotives. According to their report, 5% biodiesel blend, 20% biodiesel blend and neat biodiesel may replace 4%, 17% and 86% petroleum use respectively [14]. UCO biodiesel's total environment footprint is 40% lesser than first-generation biodiesel and 1/4th (3 times lower) of the mineral diesel's total environmental footprint [15]. Singh et al., 2019 [16] studied the properties of pure diesel and diesel/ethanol blends with co-solvent oleic acid (7% ethanol, 1% oleic acid, 92% diesel; 13% ethanol, 5% oleic acid, 82% diesel; 19% ethanol, 7% oleic acid, 92% diesel; v/v). They concluded that decrease in density and viscosity due to blending leads to retarded injection timing and decreased sulphur content in the blend lower the toxic pollutants in the exhaust [16]. Jimenez et al., 2009 [17] blended ethanol with rapeseed biodiesel at 5%, 10% and 15% and investigated the effect on fuel properties. They suggested that low flash point and hygroscopic nature of ethanol in these blended fuel drawn a major setback for possible practical application [17]. Fernando and Hanna, 2004 [18] experimented two separate blends of ethanol-biodiesel-diesel in certain proportions (3.75%: 25%: 71.25%; 4.00%: 20%: 76.00%; v/v) and found minimum phase transition temperature (PTT) of 11°C in both LSD and ULSD [18]. Further characteristic's influencing the lubrication, ignition and combustion of mineral diesel and its ethanol blends (5%, 10%, 15%; v/v) studied by Jimenez et al., 2011 [19]. Certain properties like Density, viscosity, Pour point, flash point, lower specific energy content, wear scar and filter plugging tendency decreased due to ethanol addition [19].

The objective of the present study is to investigate the physical and chemical properties of alternative fuel and look for their variation by varying percentage of alternative fuel into baseline diesel, to relate them with the desirable specifications of diesel fuel. Second generation (non-edible source) were used for in-house production of biodiesels. Fuel properties are important to understand the combustion chemistry and emission behaviour in IC engines. Calorific value, viscosity, cloud and pour point, density, oxidation stability, cold filter plugging point, the flash point were measured and compared for JBD, UCO biodiesel (UCO), and ethanol (E). These properties were also measured and analysed for their various blend ratios (JB05, JB10, JB15, JB20; U05, U10, U15, U20; E05, E10, E15, E20). Where 5-20 representing the percentage of respective alternative fuel into mineral diesel.

2. Methodology

In the present study, three alternative fuels i.e. JBD, UCO biodiesel and Ethanol, were chosen and investigated for various fuel property parameters of pure fuels and their blends. Methodology of the study adopted the selection and preparation of fuel blends, measurement of various property parameters and their comparison for the utilization suitability in compression ignition (CI) engines.

2.1. *Preparation of Fuels*

UCO was collected from the restaurants of Dhamtari city of Chhattisgarh, and Jatropha seeds were collected from the locally available resources/fields. Further, two separate batches of biodiesel (UCO and JBD) were produced using the transesterification process in a Biodiesel production pilot plant of Chhattisgarh Biofuel Development Authority (CBDA), Raipur. Methanol as a reactant and sodium hydroxide (NaOH) as a catalyst was used for the same. Chemical grade ethanol (99.9%) and mineral diesel were procured from authorized supplier and government authorized diesel outlet respectively. UCO biodiesel, JBD and ethanol were blended with mineral diesel in the range of 5-20% (v/v) blend ratio. Prepared blends include, Ethanol-diesel blends (E5, E10, E15, and E20; v/v), JBD blends (J5, J10, J15, and J20; v/v), UCO biodiesel blends (U5, U10, U15, and U20; v/v), where the numeral values indicate the volume fraction of biofuel in mineral diesel. Figure 1 shows the chemical process of transesterification adopted in the laboratory for biodiesel making.

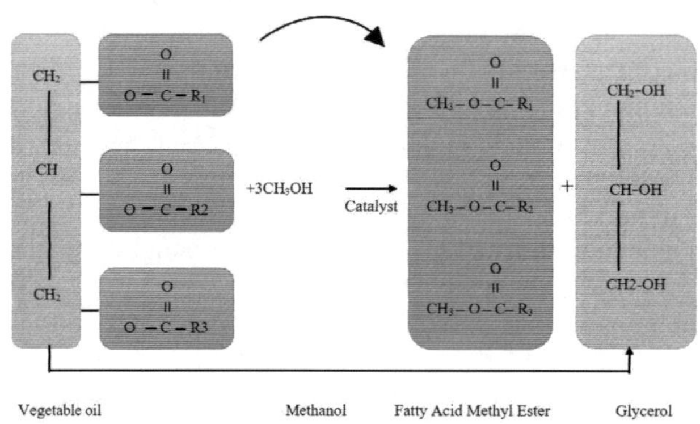

Figure 1: Transesterification process for biodiesel production [20]

2.2. Fuel Properties Measurement

Fuel properties of test fuels were determined as per the international standard of ASTM (Table 1). The knowledge about the fuel property parameters are the essential part for the biofuel utilization in CI engines as varying physico-chemical properties may largely affect the combustion and performance parameters. Measurement of fuel property parameters include density, calorific value, viscosity, flash point, cold filter plugging point (CFPP), cloud point (CP), pour point (PP), and oxidation stability.

2.2.1. Density. The density of fuels were measured according to the criteria specified by ASTM 4052 standard. Density of tested fuels were measured by keeping the room temperature and the device temperature at 25°C. The equipment used for the density measurement was Rudolf Research Analytical [Model: DDM2909]. Small amount of fuel is injected through a syringe in a temperature-controlled U tube to measure the density of fuel. Mechanical oscillation method was used to measure the density. The oscillation frequency *f* was measured, density was calculated using the following expression shown in equation 1 [16, 19]. Each measurement was repeated 3 times to reduce measurement error and average results are used for analysis.

$$\rho=A.\tau^2+B \;\ldots\ldots\ldots\ldots\ldots (1)$$

Where ρ is the density of fuel, τ is the time-period of oscillation ($1/f$) in seconds and A and B are the respective instrument constants.

2.2.2. Kinematic Viscosity. It is defined as a property due to which it offers resistance again motion. It was measured for the biofuels (fuels used in this study) and their blends as per the method specified in standard ASTM D445 at 40°C. Viscometer from OMNITEK B. V. [Model: U-Visc 110] was used with a measuring range of 0.5 to 10000 mm²/sec. The working principle of kinematic viscosity meter include the measurement of the travel time required for the fixed volume of the given fuel through the capillary under the influence of gravity. The equation used for kinematic viscosity calculation is shown in equation 2 [19]. Each measurement was repeated 3 times to reduce measurement error and average results are used for analysis.

$$\upsilon = C.t \quad \text{.................} \quad (2)$$

Where, υ: kinematic viscosity (mm^2/s), C: calibration constant of the viscometer (mm^2/s^2) and t: mean flow time required to pass capillary.

2.2.3. Calorific Value. Calorific values were measured using bomb calorimeter for the chosen alternative fuels (UCO biodiesel, JBD and ethanol). The Bomb calorimeter offers a quick and precise calorific value calculation technique. Measurements of fuel sample calorific values were conducted in compliance with the procedure and parameters defined by the ASTM D240 standard. A small amount of fuel (1 gm) was taken into crucible and oxygen is supplies through the cylinder to fill the bomb till a pressure of 3 MPa. For the calculation of the calorific value of the fuel measured, the original temperature of the water in the calorimeter after stirring and the highest temperature indicated by the thermometer are used. The instrument used for measuring of cv value was from Advance research instruments co. [Model: ARICO IP-3].

2.2.4. CP, PP, and CFPP. For running the engines under cold weather conditions, cold flow characteristics are important. The temperatures of the cloud point shall be calculated as the temperature at which the wax crystal cloud is first appeared under a controlled condition in a liquid during standard testing (ASTM D2500). The temperature below which the fuel cannot be poured due to the forming of gel (ASTM D6892) is known as pour point, and CFPP is the lowest temperature at which the amount of sample fuel (70ml) is passes under standardized filtration unit within a specified time (60 sec) (ASTM D6371). ORBIS B. V. [Model: Air STAR cold block CPPP and Model: Air STAR cold block CFPP] were used for measurement.

2.2.5. Oxidation stability. Oxidation stability test of diesel and biodiesel have been performed to estimate its long term storage capability. Oxidation stability tests are normally performed at relatively higher temperature and also used to add oxygen to speed up the oxidation of the test fuels [21]. ASTM standard D7545 was adopted for performing oxidation stability test for the tested fuels. Test were performed in Anton Par GmbH small scale rapid oxy apparatus [Model: RapidOxy 100].

2.2.6. Flash point. Flash point temperature is important for transportation and storage purposes. It was measured for pure alternative fuels used in this study and their blends as per criteria specified by ASTM D93-A. Pensky martin closed cup auto flash tester by Acute Instruments Pvt. Ltd. [Model: PMCC Auto Flash – 93] was used for measurement. The test involves the use of a sample volume of approximately 75 ml which is then placed in the closed cup of the tester, subsequently firing was performed with 10°C of interval till the flashpoint obtained.

3. Results and Discussion

In the present study, three alternative fuels i.e. JBD, UCO biodiesel and ethanol were chosen for the fuel property parameter measurement. Since, the objective of the current study is to compare the alternative fuel properties for the possible utilization in CI engine, fuel properties were measured for the neat fuels and their blends with mineral diesel. The discussion in this section is subdivided for the property measurement of pure and blends of these tested fuels with mineral diesel.

3.1. Properties of Pure Alternative Fuels

ASTM standards were referred for the methodology adopted for the measurement of fuel properties.

The evaluation of fuel properties is essential to understand their combustion and performance characteristics. Table 1 shows the measured properties of ethanol, UCO biodiesel and JBD along with mineral diesel, it was noted that UCO Biodiesel shows similar value compare to BIS Standard, Viscosity value is ranging on higher side for biodiesel, which may require extra pump power while injecting into engine. Flash point and density values of produced fuel were within limits, and cold weather properties shows superior compare to standard B100.

Table 1: Fuel properties of JBD, UCO biodiesel, ethanol and mineral diesel

Properties	Method	Standard (B100)	Diesel	Ethanol	UCO BD	JBD
Density at 25 °C, (kg/m3)	ASTM D4052	860-900	822	795	863	892.5
Kinematic Viscosity at 40 °C, (mm^2/sec)	ASTM D445	3.5-5	2.83	1.08	5.1 ±0.4	5.6 ±0.3
Calorific value (MJ/kg)	ASTM D240	-	45.6	29.3	39	39.3
Flash point, min (°C)	ASTM D93A	101	56	12[a]	163	160
Cloud point (°C)	ASTM D2500	-	-2	-34	5	6
Pour Point (°C)	ASTM D6892	-	-6	-117.3[b]	-3	-2
Cold filter plugging point (°C)	ASTM D6371	6 °C for winter 18 °C for summer	-6	-	-2	-1
Oxidation Stability at 140 °C, (min)	ASTM D7545	8	83.1 5	-	8.32	9.25

[a] Adapted from *Inchem 2013 Icsc : 0044 Ethanol (Anhydrous) 2–4* [22]
[b] Adapted from *Li D et al., 2005* [23]

3.1.1 Viscosity, Density, and Calorific Value. Figure 2 illustrates the density, viscosity, and calorific value of pure JBD, UCO biodiesel and ethanol. It was observed that both biodiesels (863 kg/m^3 and 892.5 kg/m^3 for UCO biodiesel and JBD respectively) have higher density compared to mineral diesel (822 kg/m^3), however, ethanol density (795 kg/m^3) measured to be the lowest among tested fuels, lower density of ethanol were also observed by Jimenez et al [19]. The effect in density can affect start of ignition and combustion duration, which can adjusted by tuning the injection timing [19, 24]. Reduction in density reduces its sauter mean diameter (SMD) and improve the spray characteristics [19,25]. From Figure 2, it can be observed that the average viscosity for ethanol is 1.08 mm^2/sec, which is ~61.8 % less compared to mineral diesel (2.83 mm^2/sec), however, biodiesels show significantly higher viscosity (5.1 mm^2/sec and 5.6 mm^2/sec for UCO biodiesel and JBD). The lower viscosity of ethanol may cause pump and injector leakage [26,27] while using ethanol in conventional diesel engine.

3.1.2 Cold Weather Properties (CP, PP and CFPP). Figure 3 shows the cold flow properties of Diesel, Biodiesel, and ethanol. It includes CP, PP, and cold filter plugging point (CFPP) temperature for all tested fuels, which are essential for operating an engine in cold countries or in cold climate regions. CP is the maximum absolute temperature at which the first cloud of crystal appears during cooling. CFPP is the minimum temperature below which fuel is unable to pass through fuel filtration device due to gel formation. On the other hand, PP is the minimum temperature below which fluid can't be poured in any system. Figure 3 shows that pure biodiesels possess inferior cold flow properties. CP, PP and CFPP temperatures for biodiesels (JBD and UCO) were found to be higher in comparison to mineral diesel.

Higher CP, PP and CFPP of biodiesels indicates that these fuels may not be suitable for extreme cold weather. Superior cold weather properties of ethanol make it suitable for these purpose (PP and CFPP of ethanol were found to be below -34°C).

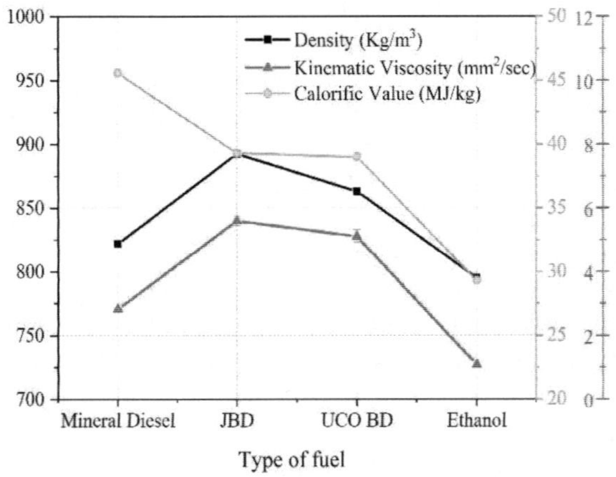

Figure 2: Density, viscosity, calorific value of ethanol, diesel and biodiesel (UCO and JBD).

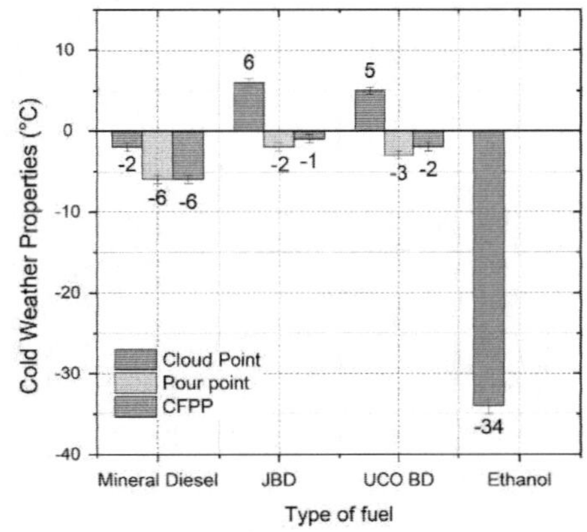

Figure 3: Cold weather properties of tested fuels.

3.1.3 Flash Point and Oxidation Stability. Figure 4 illustrates the flash point temperature and oxidation stability of pure biodiesels and ethanol. Flash point temperature is a very critical parameter since this plays an important role in handling and storage of the fuels. Biodiesels showed a much higher flash point (160°C and 163°C for JBD and UCO biodiesels respectively) in comparison to mineral diesel (56°C), signifies biodiesel handling and storage is safer and special management is thus not required.

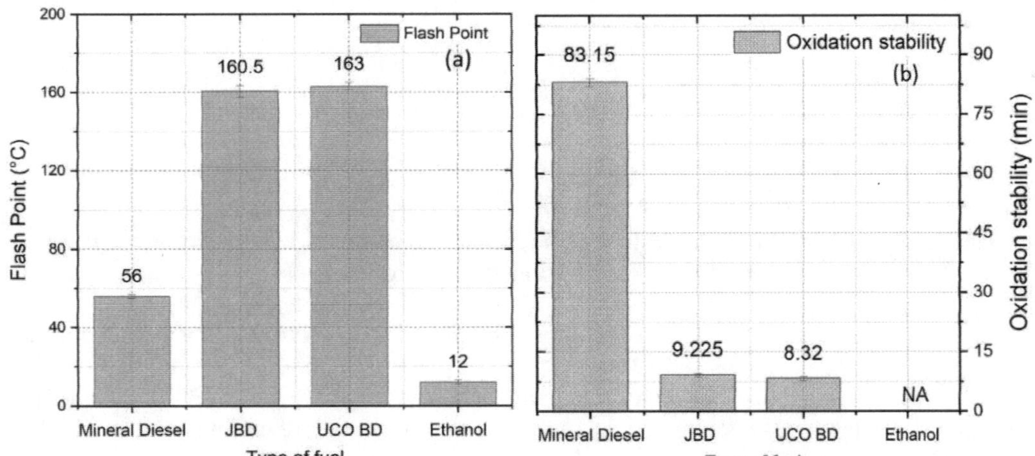

Figure 4: Flash point and oxidation stability of ethanol, diesel and biodiesel (UCO and JBD).

Flash point temperature of ethanol is below room temperature (12°C), and it can catch the fire easily therefore it need special transportation and handling. Apart from the flash points determination, oxidation stability of the pure alternative fuels was also investigated. Oxidation stability tests were also performed where mineral diesel shows 83.15 min for 10% pressure drop at 140°C and 700 kPa, UCO Biodiesel and JBD shows 9.2 min and 8.32 min. Biodiesel oxidation time is 10 times lesser compare to mineral diesel, biodiesel standard oxidation time as per BIS standard is 8 min.

3.2. Properties of Fuel Blends of Alternative Fuels

Due to wide variation in the major properties of the alternative fuels, it becomes difficult to use the pure form of alternative fuels in the compression ignition engines. For example, ethanol shows very high autoignition value (~363°C) which cannot be ignited in conventional CI engine. Similarly, biodiesel shows high viscosity and density and therefore it is difficult to use biodiesel for a longer term in CI engines as it can clog the injector nozzles and fuels lines. To resolve these issues with alternative fuels, blending method has been adopted in which one or more alternative fuels can be blended with mineral diesel to keep the fuel properties closer to mineral diesel. However, it is essential to measure properties of blends for proper functioning of diesel engines because blending two different fuel can alter density, viscosity, stability, cold flow properties, and energy content of the fuel. Ethanol-diesel blends (E5, E10, E15, and E20; v/v), JBD blends (J5, J10, J15, and J20; v/v), UCO biodiesel blends (U5, U10, U15, and U20; v/v) where ethanol, JBD, and UCO biodiesel percentages are 5, 10, 15 and 20% respectively (v/v) into mineral diesel were prepared and tested. Properties of blended fuels were discussed below. Tested fuel properties were analyzed for their variation with blends percentage and alteration with baseline diesel properties.

3.2.1 Viscosity, Density and Calorific Value. Figure 5(a) shows the density variation for various blends of ethanol, UCO biodiesel, and JBD into diesel. E20 showed 0.24% reduction in density compared to mineral diesel while JBD20 and UCO20 showed 1.70% and 1.34% increase in the density respectively. Figure 5(b) shows the variation in viscosity of the blends of three alternative fuels (i.e. JBD, UCO biodiesel and ethanol) with mineral diesel.

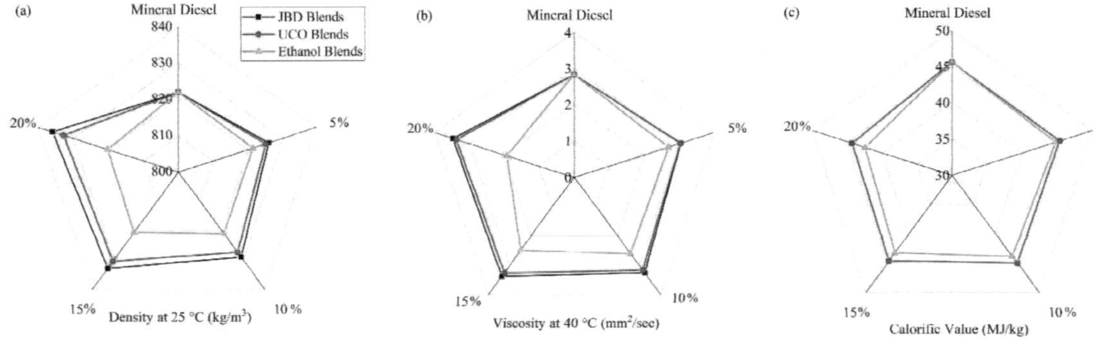

Figure 5: Relative change in fuel properties with blend percentage (a) density variation (b) viscosity variation (c) variation in calorific value.

The increasing fraction of JBD and UCO biodiesel increased the viscosity. It was 2.8 mm^2/sec for mineral diesel and gradually increased to the level of 3.5 mm^2/sec for JBD20 and UCO20. Ethanol showed the opposite trend and its viscosity reduced just below 2 mm^2/sec for E20. It should be noted that significant reduction in viscosity may add a lubrication problem to the engine, which will affect the performance of the engine. Similar to viscosity, density also showing the similar trend for the blends of three alternative fuels tested. Density affects the fuel injection spray behavior such as spray penetration length, spray area, atomization etc. Due to the conical and cylindrical design features of nozzle holes, higher density fuels also prone to create cavitation in the injection, thus reduces the fuel flow rate. Calorific value for a given fuel is also an important property as it influences the engine efficiency and fuel consumption rate. Figure 5(c) illustrates the calorific value of various blends of JBD, UCO biodiesel and ethanol blends. Blends for JBD, UCO biodiesel and ethanol showed reducing trend in calorific values unlike the viscosity and density where it increased for JBD and UCO biodiesel. Reduction in calorific value (~6.8% reduction for E20, ~2.6% reduction for JBD20 and UCO20) was observed with an increase in fraction of alternative fuel in the fuel blend.

3.2.2 Cloud Point (CP), Pour Point (PP), Cold Filter Plugging point (CFPP). Cold weather properties CP, PP, and cold filter plugging point were measured for all blends of ethanol, UCO biodiesel, and JBD is shown in Figure 6. For ethanol-diesel blends, it was observed that, ethanol addition into diesel improves cold weather properties. However, for biodiesel blends, CP appears at a much higher temperature, whereas PP and cold filter plugging point are not much affected, biodiesels possess inferior cold flow properties. From the above result, it can be concluded that ethanol blends can work at extremely lower weather conditions, however, the use of biodiesel blends for the same is not advisable.

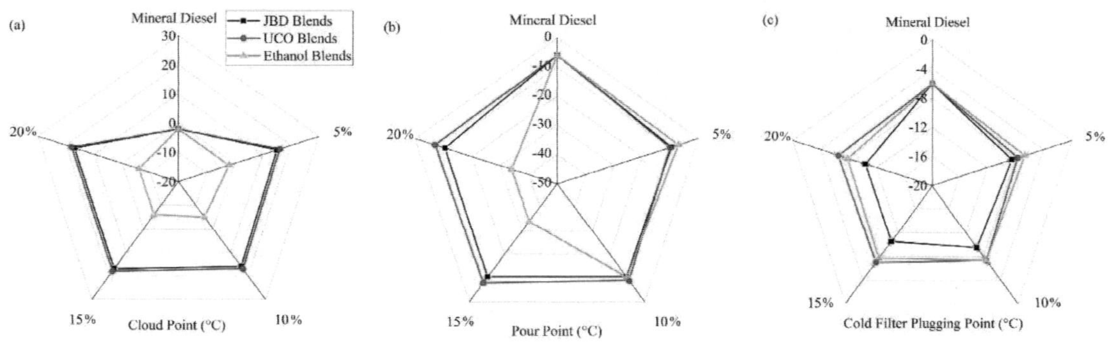

Figure 6: Variation in fuel properties with blend percentage (a) cloud point variation (b) pour point variation (c) variation in CFPP.

3.2.3 Flash point and Oxidation Stability. Figure 7 (a) and (b) illustrate the measured flash point and oxidation stability of blended fuels. Flash point of any fluid is the minimum temperature at which it catches fire when the ignition source is provided.

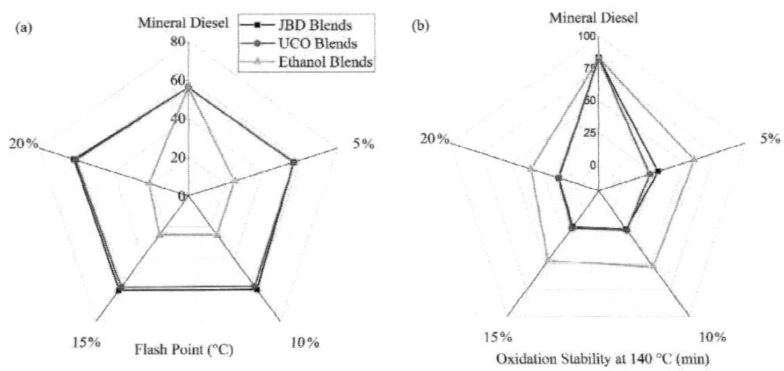

Figure 7: Flash point and oxidation stability variation of blends.

Flash point is important for handling and storage purpose of any fluid. Diesel have flash point of 56°C and biodiesels have flashpoint in the range of 160°C, however, ethanol has flash point at 12°C (which is below room temperature for Indian climate condition). Ethanol needs safe transportation, handling, and storage methods compared to other fuels. Similarly, blending of ethanol into diesel will continue to reduce its flash point temperature shown in Figure 7(a), although biodiesel blending in diesel increases the flash point. The variation in flash point is sharp in the case of ethanol blends, only 5% ethanol reduces flash point from 56°C to 25°C, however beyond 5% of ethanol blend reduction in flash point is negligible. Figure 7(b) shows the oxidation stability test results performed at 140°C for biodiesel blends and ethanol blends. It should be noted that adding ethanol or biodiesel into mineral diesel reduces the oxidation stability of the fuel. However, ethanol shows less reduction compare to biodiesel blends.

4. Conclusions

Physico-chemical properties of trending alternatives (JBD, UCO biodiesel and ethanol) of mineral diesel were investigated and compared in their pure and blended forms. Biodiesel blends exhibit the nearer value for flash point, calorific value and PP while ethanol blends are favourable in terms of viscosity, density and CP. Based on the current study, following inferences may be summarized:

- Biodiesels (863 kg/m^3 and 892.5 kg/m^3 for UCO biodiesel and JBD respectively) have higher density compared to mineral diesel (822 kg/m^3) while ethanol density (795 kg/m^3) measured to be the lowest among tested fuels.
- Measurements showed that biodiesel and ethanol possess lower calorific values compared to mineral diesel (~15-20% lower for biodiesels and 35.74% lower for ethanol). On the other hand, cold weather properties for ethanol showed superior quality.
- Oxidation stability was measured to be 9.225 min and 8.32 min for JBD and UCO biodiesels respectively which was significantly lower compared to mineral diesel (83.15 min).
- Concerning to the blends of biodiesels, it showed increasing trend with higher fraction of biodiesels in blend while ethanol blends with diesel resulted in decreasing values of these properties. Cold whether properties of ethanol blends improved with increasing fraction of ethanol in the blend

Overall, it was observed that biodiesel (pure and blended forms) may be suitable to use in CI engines when it comes to flash point, calorific value and PP. While ethanol (pure and blended forms) may be suitable for CI engines in terms of viscosity, density and CP. To overcome the demerits of these alternative fuels, blended form of these fuels may be recommended for the use in CI engines

References

[1] Agarwal A K, Gupta T, Shukla P C and Dhar A 2015 Particulate emissions from biodiesel fuelled CI engines *Energy Convers. Manag.* **94** 311–30
[2] Agarwal A K, Shukla P C, Patel C, Gupta J G, Sharma N, Prasad R K and Agarwal R A 2016 Unregulated emissions and health risk potential from biodiesel (KB5, KB20) and methanol blend (M5) fuelled transportation diesel engines *Renew. Energy* **98** 283–91
[3] Anand G and Kuchhal P 2018 Experimental and theoretical investigation of viscosity, density and sound velocity of jatropha diesel and pure diesel blends *Biofuels* **9** 1–6
[4] Mathew A and Anand K 2018 Comparison of engine characteristics with biodiesels produced from fresh and waste cooking oils *Biofuels* 1–9
[5] Sahu T K, Sarkar S and Shukla P C 2021 Combustion investigation of waste cooking oil (WCO) with varying compression ratio in a single cylinder CI engine *Fuel* **283** 119262
[6] Sheinbaum-Pardo C, Calderón-Irazoque A and Ramírez-Suárez M 2013 Potential of biodiesel from waste cooking oil in Mexico *Biomass and bioenergy* **56** 230–8
[7] FSSAI 2018 Repurpose Used Cooking Oil (RUCO)
[8] Murali G and Shastri Y 2019 Life-cycle assessment-based comparison of different lignocellulosic ethanol production routes *Biofuels* 1–11
[9] Sahu T K, Gupta S and Shukla P C 2020 Second Generation Bioethanol Production from Organic Waste - Alternative Fuels and Their Utilization Strategies in Internal Combustion Engines ed A P Singh, Y C Sharma, N N Mustafi and A K Agarwal (Singapore: Springer Singapore) pp 49–64
[10] Kannan G R and Anand R 2012 Effect of injection pressure and injection timing on DI diesel engine fuelled with biodiesel from waste cooking oil *Biomass and bioenergy* **46** 343–52
[11] Silitonga A S, Masjuki H H, Mahlia T M I, Ong H C, Chong W T and Boosroh M H 2013 Overview properties of biodiesel diesel blends from edible and non-edible feedstock *Renew. Sustain. Energy Rev.* **22** 346–60
[12] Chen Y-H, Chiang T-H and Chen J-H 2012 An optimum biodiesel combination: Jatropha and soapnut oil biodiesel blends *Fuel* **92** 377–80
[13] Chen L-Y, Chen Y-H, Hung Y-S, Chiang T-H and Tsai C-H 2013 Fuel properties and

combustion characteristics of jatropha oil biodiesel–diesel blends *J. Taiwan Inst. Chem. Eng.* **44** 214–20

[14] Whitaker M and Heath G 2009 *Life-Cycle Assessment of the Use of Jatropha Biodiesel in Indian Locomotives (Revised)* (National Renewable Energy Lab.(NREL), Golden, CO (United States))

[15] Foteinis S, Chatzisymeon E, Litinas A and Tsoutsos T 2020 Used-cooking-oil biodiesel: Life cycle assessment and comparison with first-and third-generation biofuel *Renew. Energy* **153** 588–600

[16] Singh G N and Bharj R S 2019 Study of physical-chemical properties for 2nd generation ethanol-blended diesel fuel in India *Sustain. Chem. Pharm.* **12** 100130

[17] Torres-Jimenez E, Svoljšak-Jerman M, Gregorc A, Lisec I, Dorado M P and Kegl B 2010 Physical and chemical properties of ethanol− biodiesel blends for diesel engines *Energy & fuels* **24** 2002–9

[18] Fernando S and Hanna M 2004 Development of a novel biofuel blend using ethanol− biodiesel−diesel microemulsions: EB-diesel *Energy & Fuels* **18** 1695–703

[19] Torres-Jimenez E, Jerman M S, Gregorc A, Lisec I, Dorado M P and Kegl B 2010 Physical and chemical properties of ethanol–diesel fuel blends *Fuel* **90** 795–802

[20] Banerjee A and Chakraborty R 2009 Parametric sensitivity in transesterification of waste cooking oil for biodiesel production-A review *Resour. Conserv. Recycl.* **53** 490–7

[21] Hartikka T, Kiiski U, Kuronen M and Mikkonen S 2013 *Diesel fuel oxidation stability: a comparative study* (SAE Technical Paper)

[22] INCHEM 2013 ICSC : 0044 ETHANOL (ANHYDROUS) 2–4

[23] Li D, Zhen H, Xingcai L, Wu-gao Z and Jian-Guang Y 2005 Physico-chemical properties of ethanol–diesel blend fuel and its effect on performance and emissions of diesel engines *Renew. energy* **30** 967–76

[24] Kegl B 2006 Experimental investigation of optimal timing of the diesel engine injection pump using biodiesel fuel *Energy & Fuels* **20** 1460–70

[25] Park S H, Kim H J, Suh H K and Lee C S 2009 Atomization and spray characteristics of bioethanol and bioethanol blended gasoline fuel injected through a direct injection gasoline injector *Int. J. heat fluid flow* **30** 1183–92

[26] Hansen A C, Lyne P W L and Zhang Q 1998 Ethanol-diesel blends: a step towards a bio-based fuel for diesel engines *2001 ASAE Annual Meeting* (American Society of Agricultural and Biological Engineers) p 1

[27] Hansen A C, Zhang Q and Lyne P W L 2005 Ethanol-diesel fuel blends - A review *Bioresour. Technol.* **96** 277–85

Acknowledgments
The authors acknowledge the support of Mr. Puranchandra Sahu and Mr. Kanhaiya Yadav for the experimental work. The authors also would like to thank the CBDA, Chhattisgarh Government, for the resource arrangement. The authors would also like to thank DST-SERB for supplying funds for consumables and staff (under the project code: SRG/2019/001664). Some of the key results will also be used in the final report on the submission of the project to DST-SERB.

Heat transfer and fluid flow characteristics within non-circular duct

S K Mahato, S C Rana, R N Barman
National Institute of Technology Durgapur, 713209, West Bengal, India

E-mail id - skm.16me1101@gmail.com

Abstract. Heat transfer and the characteristics of a fluid flow of water flows in the twisted square duct are studied numerically using commercially available software. Numerical investigations of fully-developed laminar fluid flow within duct was performed for water with constant wall temperature boundary conditions, twist ratios of 16.5, 11.5, and 7.5, Reynolds number 100-6000, Prandtl numbers 8 and 10. The numerical results of friction factor and Nusselt number found in this present study were compared with the results obtained from the correlations for the same working fluid such as water. Nusselt number in the twisted square duct is higher than the result found by correlation together with a lower friction factor. Percentage increase in average Nusselt number found for twist ratios 16.5, 11.5, and 7.5 compared to the theoretical values was 25.56, 33.21, and 37.73 The heat transfer coefficients of the square twisted duct are increased with a lower twist ratio and higher Prandtl number.

1. Introduction

Energy saving is the main need of this universe in these recent years. Due to the industrial importance of the compact heat exchanger, most of the research works are available in this area. The twisted duct used in this study is therefore especially relevant to the design of industrial heat exchanger devices. The reason of using twisted duct is that it creates swirl inside the duct which intermixed the flow rapidly and causes to increase the heat transfer more. Based on this key point, two techniques are employed to improve the efficiency of heat exchangers such as passive and active heat transfer enhancement techniques. Passive heat transfer enhancement techniques that employed no external power supply and in opposite active techniques that work on the principle of external power supply. Twisted ducts work on the principle of passive heat transfer technique [1] and found that ducts with larger traced circle diameters offer higher values of Nusselt number and friction factor.

Wang et al. [2] investigated the flow of air within three ducts, namely, the square duct of twisted uniform cross-section, twisted divergent, and convergent, and verified that the duct with a divergent twist improves the heat transfer while convergent section decline.

Yang et al. [3] experimentally examined the water flow performance in the twisted elliptical tubes and tested the impact of parameters obtained by different geometry such as aspect ratio and twist pitch. The tubes examined were found to increase pressure drop and heat transfer compared to straight tubes. Khoshvaght et al. [4] investigated heat transfer in laminar and monitored workflows in twisted mini-

channel with variant cross sections and revealed that the best thermohydraulic performance was obtained in a twisted square mini channel.
An experiment on convection inside a novel alternating elliptical axis tube and heat transfer and flow resistance correlations were performed for Reynolds number of 500-50000. It was depicted that at low flow resistance, the heat transfer rate is higher [5].
Mahato et al. [6] studied that use of nanoparticles with different concentrations into water inside the T_wSD (twisted square duct) enhanced more heat than normal water. Maximum limits of concentration of nanoparticles mixing into the water were also specified to 1%.
The characteristics of heat transfer and friction factor of the twisted oval tube were numerically studied by a realized k-epsilon method with varying aspect ratios and twisted pitch length. They revealed that the Nusselt number and friction factor increases with the increasing aspect ratio as well as pitch length [7].
Yadav et al. [8] studied flow of fluid inside the non-circular duct and investigated the properties of heat transfer and friction factor through it for a laminar flow regime. They used the twisted band inside the channel with several twisting ratios such as Tr = 3.5, 4.5, 5.5, and 6.5. They revealed that the twisted tape insertion into the duct increases the Nusselt number and the friction factor.
Mahato et al. [9] found heat transfer gain using a twisted hexagonal channel and with a twisted square channel. Comparisons between them were also presented. They found that increasing the number of channel sides further improved heat transfer.
Masliyah and Nandakumar [10] simulated a fully developed stable laminar flow with a rotational coordinate system through the twisted square channels. The axial and fluid lines fall to preserve the natural dimension of the registered problem. The temperature was assumed to be constant along the periphery of each wall. They observed that the amplification of heat transfer is greater when the fluid is in rotary motion for the rotation ratio of 2.5 and the Reynolds number range from 100 to 1000.
Bhadouriya et al. [11, 12] tested the airflow through the twisted square channel to define the properties of heat transfer and pressure drop covered by different rotational conditions, the Reynolds and Prandtl numbers are (16.5 and 11.5), (1, 00 to 10,000) and (0.7, and 20). They noticed that Nusselt and friction values of twisted square duct (T_wSD) are higher in comparison to a square straight channel.

The research work declared above defines that variant research has been performed to increase the transfer of heat by incorporating twisted tapes inside the plain duct which proces swirl in it. As twisted tape restrict the flow area inside the plane duct causes to increase the higher pressure drop.This problem was improved by incorporating twisted square duct (T_wSD) instead of twisted tape considering twist ratios like 11.5, 16.5, and 7.5. Very few research works were available in the area of heat transfer enhancement within the T_wSD where water is considered as a working fluid with different Prandtl numbers (Pr). This research seeks to find the characteristics of heat transfer and friction factor in T_wSD by considering water as a normal fluid and compare these results with the Gnielinski correlation.

Some important terms of twisted duct are depicted in Figure 1.

Pitch (P): The distance among two successive points on the pipe length where the position of the cross-section completely matches one another. Along with the pitch distance, the cross-section is rotated by 360^0.
Twist ratio (T_{wr}): The ratio between pitch (P) to hydraulic diameter (d_h = 4*Area / Perimeter) of the twisted square duct defines the T_{wr}.

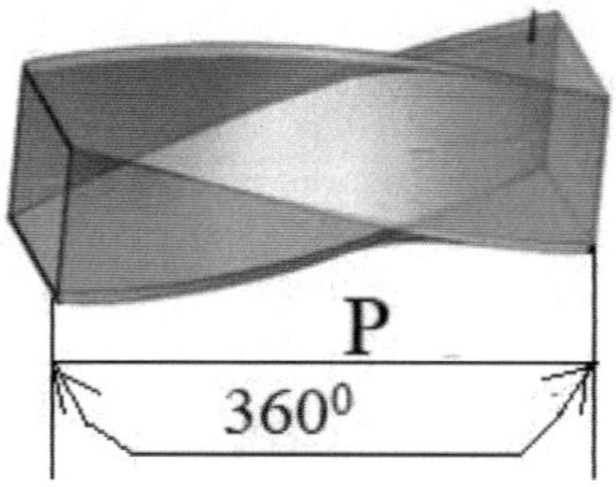

Figure 1. Terminology

2. Methodology

2.1. Geometry of twisted duct

The inlet diameter (d_i), outlet diameter (d_o), and thickness (δ) used for the twisted duct are 18.5 mm, 20.5 mm, and 2 mm. Twist ratios used are 7.5, 11.5, and 16.5, respectively. The inside diameter (D_i) and outside diameter (D_o) of the straight circular cylinder tube is 31 mm and 37 mm depicted in Figure 2.

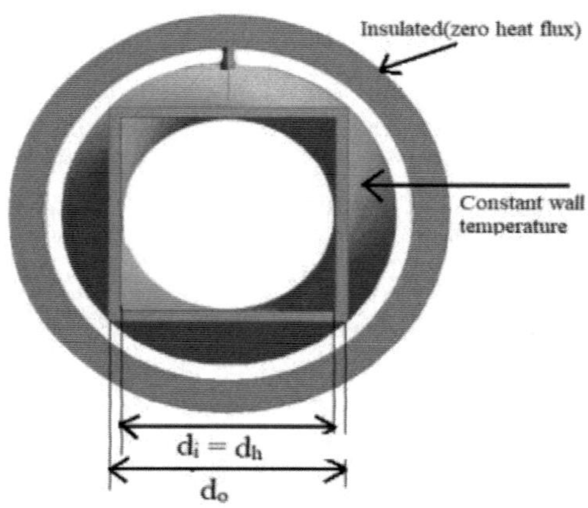

Figure 2. Dimensional properties

3. Numerical methods

3.1. *Governing equations and boundary conditions*. Generally, shell and tube heat exchanger used concentric tubes which are utilized in this study. This tube was cooled by a counter-flow heat technique to provide a greater temperature gradient over the shell and tube surfaces. The flow considered in this study is fully developed hydrodynamically and thermally, periodic, incompressible, and in steady-state. The characteristics of the flow are repeated along the length (L) of the channel, the flow is said to be periodic. Governing equations for such type of flows are the following

$$\text{Continuity: } \nabla \cdot \vec{v} = 0 \tag{1}$$

$$\text{Momentum: } \rho \vec{v} \cdot \vec{\nabla} \vec{v} = -\nabla p + \mu \nabla^2 \vec{v} \tag{2}$$

$$\text{Energy: } \vec{v} \cdot \vec{\nabla} T = \alpha \nabla^2 T \tag{3}$$

Where $\alpha = \dfrac{K}{\rho C_p}$ is thermal diffusivity.

The acceleration term which is treated for the twisted tube, that is, $\rho\vec{v}\cdot\vec{\nabla}\vec{v}$ is not linear, and therefore analytically, it is not easy to solve. This can be solved by the final volume Discretization method which provides an approximate solution.
The twisted wall was kept at a constant temperature at all times and the outer walls of the cylinder were considered adiabatic which is illustrated in Figure 2. On the interior and exterior walls, it is considered a slip limit. A constant success rate ($\dot{m}$= 0.35 kg / s) is used inside the outer cylinder. At the input of each repeated module, a constant flow rate and the periodic boundary condition are chosen for the upper channel. In general, the degree of dependency is 10^{-3}, 10^{-5}, and 10^{-6} considered for continuity, velocity, and energy respectively.

4. Grid independency test

An unstructured mesh is generally used for numerical analysis of the twisted square duct consequently of high complexity and frictionally twisted configuration. A grid-independent study is carried out to make sure the independence of the grid is considered for calculations.

Nusselt numbers were established for several disorderly grids with a constant Reynolds number of 500 in the grid independence study. Grid independence was found for the values of grid which consist of 284647 elements and size of face 1.1419×10^{-3} mm. It is illustrated in Figure 3 that after the grid elements of 284647, there was not much increase in the Nusselt number observed. Therefore, this grid is used for subsequent calculations. Grid independence values are outlined in table 1.

Table 1. Grid parameters

Elements of grid	Nusselt values	Size of the face (mm) $\times 10^{-3}$
194159	6.152	1.3411
217349	6.273	1.2787
243679	6.294	1.2212
284647	6.266	1.1419
305526	6.265	1.0831
346225	6.267	1.0092

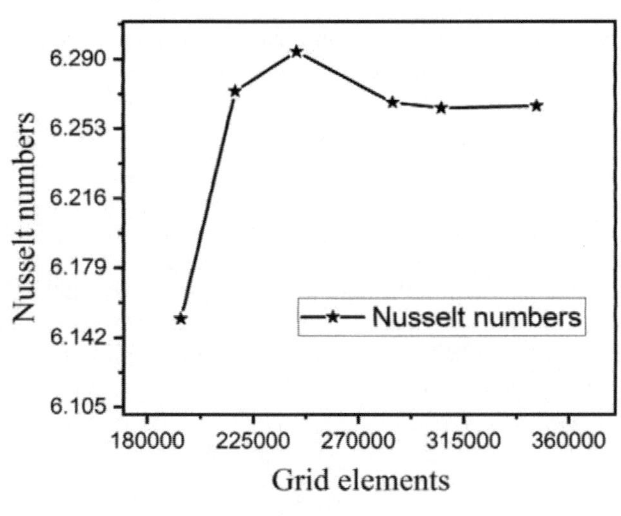

Figure 3. Schematic of grid validation

IOP Conf. Series: Materials Science and Engineering 1146 (2021) 012013 doi:10.1088/1757-899X/1146/1/012013

5. Mathematical evaluation

The three-dimensional simulations for the periodic, stable, and incompressible flow through the model were compiled by commercially available software. Since the geometry of the channel repeats the path several times along the pipe, the use of periodic flow is indicated. As the flow considered is periodic, the numerical simulation may be limited to the repetitive units of all pipelines. Therefore, the modeling length (L), P / 4 is sufficient for modeling a square channel. Based on the numerical results, the laminar flow regime is treated within the Reynolds number range of 1, 00-2,000.

Equation (4) and equation (5) depict the heat transfer rate for hot fluid (Q_h) i.e. water and cold fluid (Q_c) i.e. air where T_{hi}, T_{ho}, C_{ph} and $\dot{m}$ represent the inlet and outlet bulk temperature, specific heat of hot fluid, and mass flow rate. Equation (6) is used for the calculation of the average heat transfer rate ($\overline{Q}$) where A_p and ΔT_{LMTD} stand for heat transfer area and logarithmic mean temperature difference.

$$Q_h = \dot{m}_h C_{ph} (T_{hi}-T_{ho}) \quad (4)$$

$$Q_C = \dot{m}_c C_{pc} (T_{co}-T_{ci}) \quad (5)$$

$$\overline{Q}=(Q_h + Q_c)/2 = UA_P\Delta T_{LMTD} \quad (6)$$

Overall heat transfer coefficient (U) is calculated by equation (7) where R_w acts as the resistance of twist wall.

$$1/(UA_p) = 1/(hA_p)_c + 1/(hA_p)_h + R_w \quad (7)$$

Resistance (R_w) obtained by the wall of the twisted duct is neglected because of the smaller wall thickness (δ = 2 mm) and superior thermal conductivity of the twisted duct. Heat transfer coefficient (h) is studied towards length of twisted duct illustrated in equation (8) where T_b, T_w, and q' denote the mean bulk and wall temperature, and wall heat flux.

$$h = \frac{q'}{T_w - T_b} \quad (8)$$

The Nusselt values are computed by the formula displayed in equation (9)

$$Nu = (h.d_h)/k \quad (9)$$

To evaluate the hydraulic performance of duct, friction factor determined by the equation (10) where τ_w, and q' represents the wall shear stress and heat flux.

$$\bar{f} = 2 \times \frac{\tau_w}{\rho V_m^2} \quad (10)$$

6. Results and discussions

6.1. Validation of results

The numerical analysis of pure water flowing through the channel was verified and the results were obtained with theoretical values given by standard correlations that are widely used in the literature. The numerical values of the Nusselt number were correlated with the theoretical values estimated by Gnielinski equation shown in equation (11) and it was realized that the deviations were within 4.5% Figure 4.

$$Nu = \frac{\xi/2(Re-100)Pr}{1+12.7\left(\frac{\xi}{2}\right)^{0.5}\left(Pr^{\frac{2}{3}}-1\right)} \quad (11)$$

Where $\xi = \frac{1}{(1.58 \ln Re - 3.82)^2}$

The numerical friction factors (f) were compared with the theoretical values obtained using the Blasius equation depicted in equation (12).

$$f = \frac{0.316}{Re^{0.25}} \quad (12)$$

The numerical values of the friction factor determined in reasonably good agreement with the theoretical values obtained from the Blasius equation are shown within a 5% shift from Figure 5.

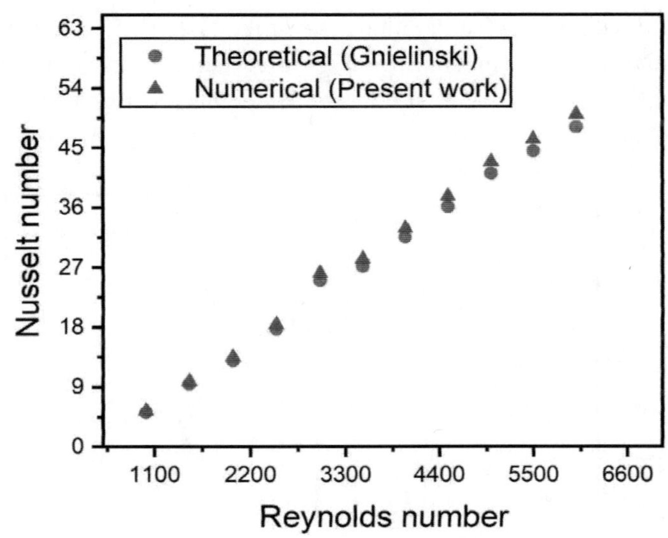

Figure 4. Nusselt validation

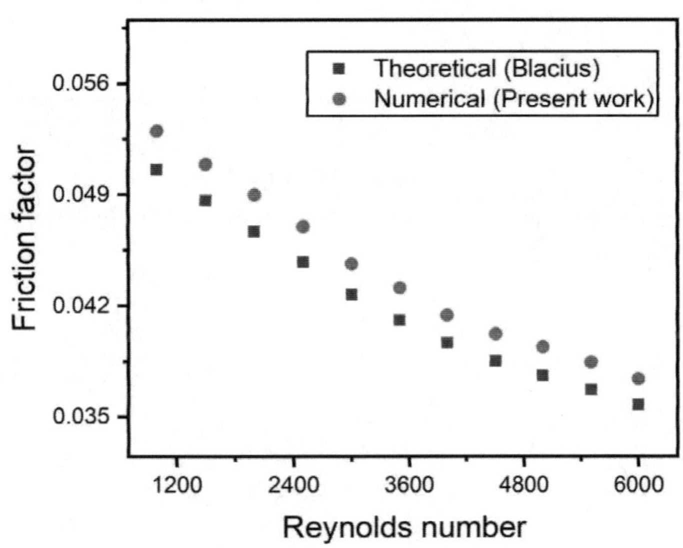

Figure 5. Friction Validation

6.2. Nusselt number and average friction factor

After validating the numerical results, the simulation simulates for the twist ratios of 7.5, 11.5, and 16.5

for Re 100- 6,000, Prandtl numbers of 8, 10. The Nusselt numbers outlined against Reynolds numbers presented in Figures 6, 7, 8, and 9 and Obtained results of T_wSD are plotted against the corresponding theoretical values obtained by Gnielinski correlation using water as working fluid. Result reveals that Nusselt numbers were observed to be superior in T_wSD than the values of Gnielinski correlation. It is also seen that an increase in Pr and at lower T_{wr} , Nusselt number increases.

Figure 6 shows that the T_wSD at Tr of 7.5, Pr number 10 is having a higher Nusselt value compared to the Nusselt number obtained by the Gneilinski equation. Similarly, in Figure 7, and 8 Nusselt numbers are found to be higher in Tr of 16.5 and 11.5 of Pr values of 10 compared to the Nusselt number found in Gnielinski correlation for the same working fluid. Nusselt number found to be maximum at twist ratio 7.5 than the Nusselt value at twist ratios of 11.5, 16.5, and theoretical value of Gnielinski equation which is shown in Figure 9. Figure 10 depicts that a very less friction factor is observed in a T_wSD with twist ratios 16.5, 11.5, and 7.5 compared to the theoretical friction factor values of the Blasius equation.

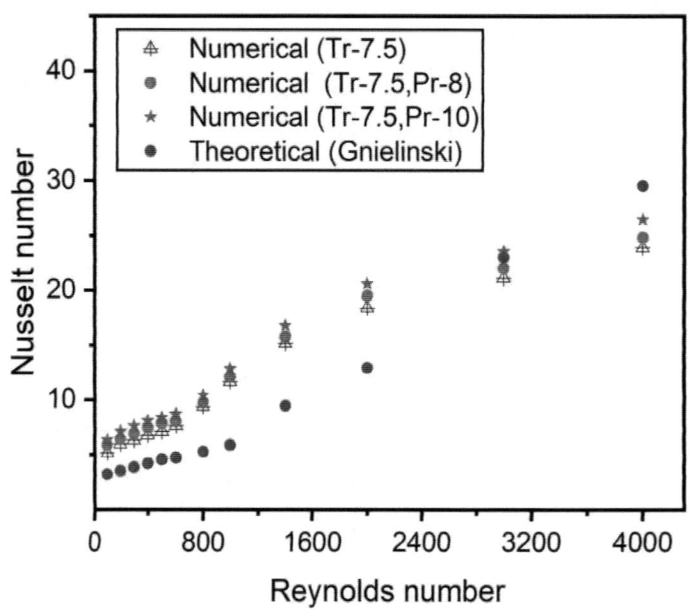

Figure 6. Comparison of numerical Nusselt number of Tr-7.5with Gnielinski equation.

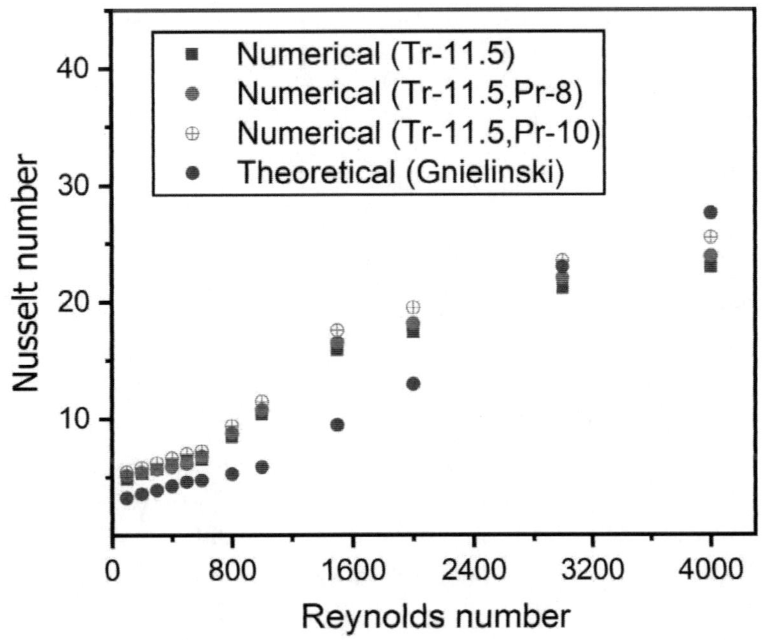

Figure 7. Nusselt number comparison of Tr-11.5 with Gnielinski equation.

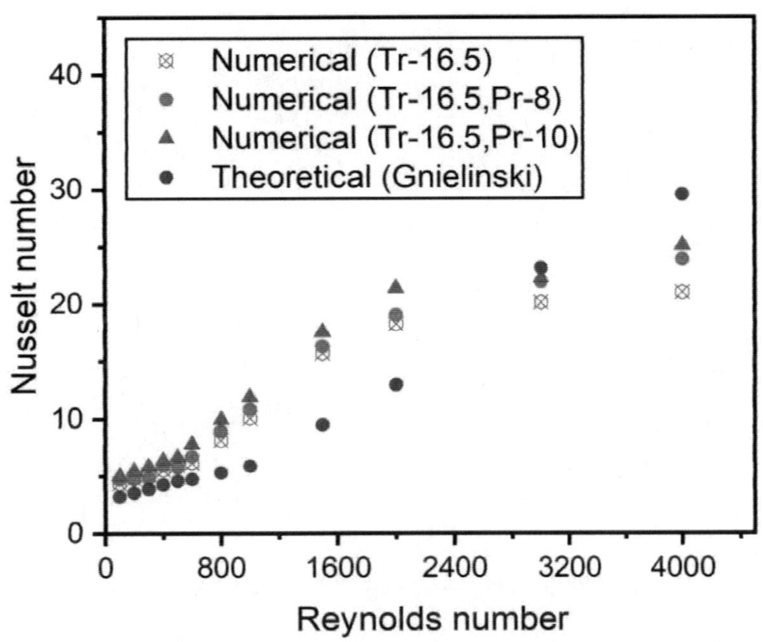

Figure 8. Nusselt number comparison of Tr-16.5with Gnielinski equation.

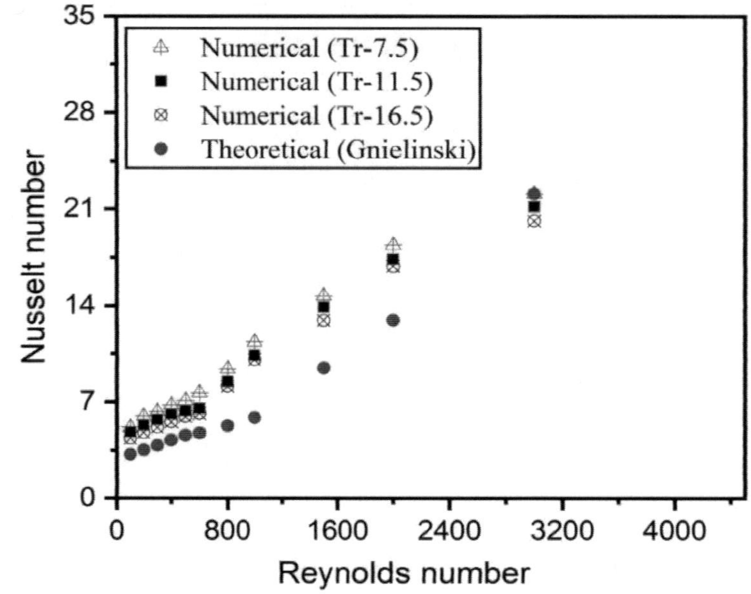

Figure 9. Nusselt number comparison of all twists ratios with Gnielinski equation.

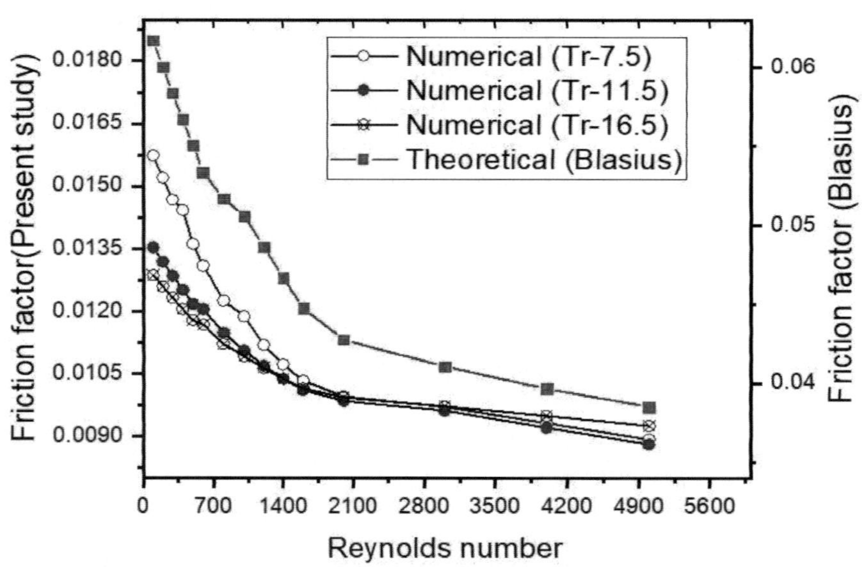

Figure 10. Friction factor comparison.

7. Conclusion

In this study, the thermal-performance of water flow inside the twisted square duct is numerically compared with the correlation value of the Gnielinski and Blasius equation. The twisted square duct with twist ratios 16.5, 11.5, and 7.5 was used in the numerical simulation. The average increase in Nusselt number and friction factor corresponding to different twist ratios for the Reynolds value 100-6000 and Prandtl number of 8,10 were properly defined. It is seen that increase in Prandtl number, Nusselt number increases, and average friction factor decreases.

The percentage increase in average Nusselt number found for twist ratios 16.5, 11.5, and 7.5 compared to the theoretical values of Gnielinski was 25.56, 33.21, and 37.73 and similarly, for friction factor,

percentage decrease was found to be 76.1, 77.08, and 73.45 respectively for twist ratios 16.5, 11.5, and 7.5. It is also found that Nusselt numbers are higher at T_{wr} of 7.5 compared to 11.5, 16.5, and the theoretical value of Gnielinski correlation.
A very less friction factor was observed in T_wSD (twisted square duct) with Tr (twist ratios) of 16.5, 11.5, and 7.5 than the friction values obtained by the Blasius correlation. In the future, various additives like nanoparticles with different concentrations can be used with normal water to increase the heat transfer and to some extent; different geometry can also be checked for this heat transfer enhancement.

References

[1] L. Todd, Some comments on steady, laminar flow through twisted pipes, J. Eng. Math. 11 (1977) 29–48.
[2] L.B. Wang, W.Q. Tao, Q.W. Wang, Y.L. He, Experimental and numerical study of turbulent heat transfer in twisted square ducts, J. Heat Transf. 123 (2001) 868–877.
[3] S. Yang, L. Zhang, H. Xu, Experimental study on convective heat transfer and ow resistance characteristics of water ow in twisted elliptical tubes, Applied Thermal Engineering 31 (2011) 2981-2991.
[4] M. Khoshvaght-Aliabadi, Z. Arani-Lahtari, Forced convection in the twisted mini channel (tmc) with different cross-section shapes: a numerical study, Applied Thermal Engineering 93 (2016) 101-112.
[5] Meng, Ji-an and Liang, Xin-Gang and Chen, Ze-Jing and Li, Zhi-Xin, Experimental study on convective heat transfer in alternating elliptical axis tubes, Experimental Thermal and Fluid Science,29 (2005). 457-465.
[6] Mahato, Shambhu Kumar, and Rana, Subhas Chandra and Barman, Rabindra Nath and Goswami, Sohom, Numerical analysis of heat transfer and fluid ow through the twisted square duct (TSD): Nanouid as working fluid, Journal of Mechanical Science and Technology,33 (2019) 5507-5514.
[7] Tan, Xiang-hui and Zhu, Dong-sheng and Zhou, Guo-yan and Yang, Liu,3D numerical simulation on the shell side heat transfer and pressure drop performances of the twisted oval tube heat exchanger, International Journal of Heat and Mass Transfer,65 (2013)244-253.
[8] R. J. Yadav, S. Kore, V. Raibhole, P. S. Joshi, Development of correlations for friction factor and heat transfer coefficient for square and hex duct with twisted tape inserts in laminar ow, Procedia Engineering 127 (2015) 250-257.
[9] Mahato, Shambhu Kumar and Rana, Subhas Chandra and Barman, Rabindra Nath and Goswami, Sohom, Numerical analysis of heat transfer and fluid flow through the twisted hexagonal and square duct and their comparisons, CHEMICAL ENGINEERING,71,1351-1356(2018).
[10] J.H. Masliyah, K. Nandakumar, Steady laminar flow through twisted pipes – fluid flow in square tubes, ASME J. Heat Transfer 103 (1981) 785–790.
[11] Bhadouriya, Rambir and Agrawal, Amit and Prabhu, SV, Experimental and numerical study of fluid flow and heat transfer in a twisted square duct, International Journal of Heat and Mass Transfer,82, (2015) 143-158.
[12] Bhadouriya, Rambir and Agrawal, Amit and Prabhu, SV, Experimental and numerical study of fluid flow and heat transfer in an annulus of the inner twisted square duct and outer circular pipe, International Journal of Thermal Sciences,94 (2015) 96-10.
[13] V. Gnielinski, New equations for heat and mass transfer in turbulent pipe and channel flow, Int. Chem. Eng. 16 (1976) 359–368.

IOP Conf. Series: Materials Science and Engineering 1146 (2021) 012014 doi:10.1088/1757-899X/1146/1/012014

Investigation of co-pyrolysis characteristics of high-ash Indian coal and rice husk

Mohd. Moiz*, Saurabh Gupta, Santanu De

Combustion and Energy Conversion Systems Laboratory, Department of Mechanical Engineering, Indian Institute of Technology Kanpur, Kanpur, Uttar Pradesh, 208016, India

*Corresponding author: moiz7786@gmail.com

Abstract. This study investigates the non-isothermal pyrolysis characteristics of a low-grade, high ash Indian coal, and rice husk using a thermogravimetric analyzer (TGA). A series of thermo-gravimetric experiments are conducted to examine the effects of coal and biomass blending on their thermal degradation behavior and synergetic effects in the nitrogen atmosphere. The thermo-gravimetric (TG) curves show two distinct tips where the larger tip is obtained for biomass as compared to the coal. The thermogravimetric study does not demonstrate any significant synergetic interaction between the biomass and high ash coal in the blends, and the additive rule describes the behavior of their combined pyrolysis. The activation energy and preexponential factor for coal and biomass are calculated based on the TGA data using the Friedman method. the corresponding calculated mean activation energy values for the biomass and Indian low-rank coal were found to be 203.463 KJ/mol, and 66.97 KJ/mol respectively.

1. Introduction

Due to global warming concerns, renewable energy resources are getting more attention in the present global energy scenario. Fossil fuel-based processes contribute about 80% of the share in primary energy production globally however they pose serious environmental concerns including emissions of harmful pollutants and greenhouse gases. Biomass stock originated from lingo-cellulosic sources is a sustainable renewable energy source that can replace fossil fuels and be converted into energy, power, liquid fuels, and other useful chemicals [1]. Bioenergy is also considered carbon-neutral as CO_2 released during biomass combustion returns to the atmosphere and is used by plants during photosynthesis. Thermo-chemical conversion of biomass into clean products such as gases and/or liquid fuel is more efficient and economical than its direction combustion. Among available methods to transform biomass thermochemically, pyrolysis and gasification of biomass are the most promising. Pyrolysis is the solid fuel feedstock's thermal decomposition into smaller fragments such as gases, liquid, and solid char under a medium temperature, and an inert environment. The major advantage of the pyrolysis process is the reduction in the transportation cost of the liquid fuel when compared to the gasification products [2]. Bio-char obtained from pyrolysis could be applied as an alternative modifier of soil due to its substantial absorbing characteristic. Also, the sequestration of biochar can significantly reduce the CO_2 concentration in the environment and can prove to be an important role-playing entity in the market of carbon capture setup worldwide. Even after years of research, the fundamental scientific understanding of biomass pyrolysis is not matured enough primarily due to the absence of detailed models that can describe its chemistry and transport phenomena. The major challenges in the development of these descriptions arise due to the multiphase, multi-scale conversion techniques, and the complexity of feedstock. A thorough understanding of the mechanism of elementary reactions can open new avenues for process optimization in an accurate sense, and adaptability of feedstock.

Coal pyrolysis has been used efficiently for the generation of bio-oil and syngas for more than 100 years although considerable efforts to optimize the yield of liquids, solids, and gas fractions are given in the last few decades. Biomass pyrolysis is a relatively new research topic where a significant amount of work has been done only in recent years. Coal has a small H/C and O/C ratio as compared to biomass [3]. During co-pyrolysis of biomass with coal, the higher hydrogen donor ability, and high oxygen content of biomass facilitate reactivity of coal pyrolysis reaction. Thus, the process of co-pyrolysis of biomass with coal may be a useful way for the improvement of product yield particularly in the case of a low-grade coal feedstock. Various researchers have reported on the enhancement and synergetic effects of different types of coal and biomass feedstock. It is found out that the interaction of biomass-coal may improve the product yield obtained from the pyrolysis of coal alone [4]. Wu et al. [5] studied the interaction of bituminous coal with various biomass varieties like pine sawdust and residue straw using the Thermo-gravimetric analyzer (TGA) and concluded that using biomass blending with the coal enhanced the product yield. The synergetic effects are also confirmed by Senobe et al. [6] in the co-pyrolysis investigation utilizing lignite coal and corncob biomass using a reactor of fixed bed type. The occurrence of synergy in TGA and the fixed-bed reactor was observed by Park et al. [7] utilizing different blends of sub-bituminous coal and sawdust. Apart from the pyrolysis characteristics of a feedstock, its kinetic parameter also plays an important role to estimate its efficient utilization. Thermogravimetric analysis (TGA) is precisely a reliable technique of studying the kinetics of coal, thermal events, and biomass which mainly provides measurements of loss in terms of sample weight as a function of temperature and time. Various methods are proposed by researchers for the estimation of kinetic variables from curves and datasets related to TGA. These methods can be broadly divided into model-free and model-fitting methods where model-fitting method are not particularly useful for non-isothermal TGA data as it gives one order higher values [8]. The characteristics of pyrolysis and kinetics of high-ash Indian coal and Lignocellulosic biomass are investigated in the present work using TGA under the inert atmosphere. Using single first order reaction model (Friedman method), kinetic parameters like frequency factor (A) and activation energy (E_a) are obtained. The study's findings may be used to optimize the thermochemical conversion processes, including pyrolysis and gasification, and for the development of suitable technologies.

2. Materials and methodology

Indian sub-bituminous coal is used as the feedstock for the experimental investigations in the thermogravimetric analyzer. The coal feedstock was received in the size of 20-100 mm, which is ground in a hammer mill and sieved using a vibratory sieve shaker up to the maximum particle size of 150 μm. Rice husk is used as the biomass, which is ground using a mixer to a suitable distribution of particle size (<150 μm). The coal and biomass blends are prepared in blending ratios of 0:100, 25:75, 50:50, 75:25, and 100:0 in terms of weight percentage. Table 1 listed the data of proximate and ultimate analysis of coal and biomass. The higher calorific value (HHV) of the coal is calculated using Dulong and Petit equation as; $\mathrm{HHV} = 33823\mathrm{C} + 144249\left(\mathrm{H} - {}^{\mathrm{O}}/_{8}\right) + 9418 \times \mathrm{S}\ \frac{\mathrm{KJ}}{\mathrm{kg}}$. Here, the mass fractions (%) of carbon, hydrogen, oxygen, and sulfur, are represented by C, H, O, and S respectively which are taken from ultimate analysis (Table 1).

IOP Conf. Series: Materials Science and Engineering 1146 (2021) 012014 doi:10.1088/1757-899X/1146/1/012014

Table 1. Proximate and ultimate analysis of the biomass and the coal (As received)

Proximate analysis	Biomass (% wt.)	Coal (% wt.)	Ultimate analysis	Biomass (% wt.)	Coal (% wt.)
Moisture	7.36	2.0	Carbon	39.061	48.37
Volatiles	73.33	7.1	Hydrogen	5.567	1.23
Fixed carbon	6.48	45.6	Oxygen	34.27	1.65
Ash content	12.83	45.3	Nitrogen	0.911	1.11
HHV	17868.96	15062.68	Sulphur	-	0.34

The pyrolysis experiments are performed in a Thermo-Gravimetric Analyzer (make: TGA/DSC METTLER TOLEDO). Figure 1 shows the TGA experimental setup. For carrying out a typical experimental run, 15±0.2 mg samples were placed and the rate of heating was maintained at 10°C /min from ambient temperature to 1000°C in N_2 environment at a constant purge rate of 4 mL/min. lab facilities at the CESE, IIT Kanpur, India are used to conduct experiments. Experiments are carried out at different temperatures (30–1000°C) and blending ratios (0, 25, 50, 75, and 100 %) of coal and biomass. The temperature range is selected such that it covers drying and devolatilization processes till the ash softening temperature suitable for fluidized beds.

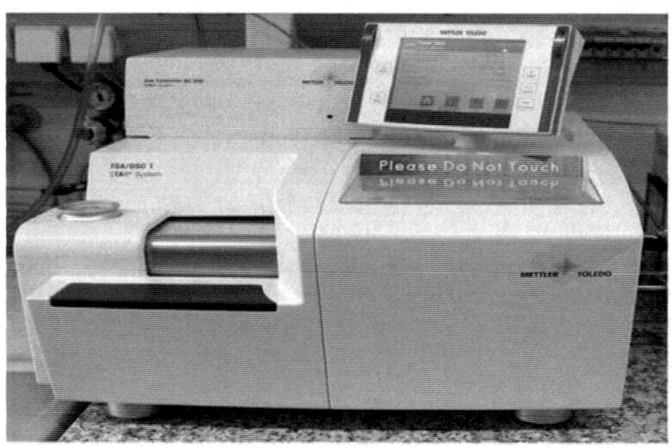

Figure 1. Photograph of the Thermo-Gravimetric Analyzer used in the study.

3. Kinetics parameters

The pyrolysis process of coal and biomass is assumed by a series of first-order reactions. Therefore, the following equation is used to calculate the reaction rate using the Arrhenius equation concerning a heterogeneous medium:

$$\frac{d\alpha}{dt} = Ae^{\frac{-E_a}{RT}} f(\alpha) \tag{1}$$

Here in Eq. (1), the conversion degree is represented by α, the reaction temperature is represented by T, and time is represented by t. Notations R, E_a, and A represents the ideal gas constant (8.314 J/mol K), activation energy, and the pre-exponential factor. The reaction model function is denoted by $f(\alpha)$, and the fuel's combustion rate at the temperature T is denoted by $\alpha(T)$, and it is evaluated by data obtained from TG analysis using the following equation;

$$\alpha = \frac{m_o - m_T}{m_o - m_f} \qquad (2)$$

Here, m_o and m_f denote the solid fuel sample's initial and final mass, and the sample's mass at temperature T is denoted by m_T. The function $f(\alpha)$ can be represented by Eq. (3).

$$f(\alpha) = (1 - \alpha)^n \qquad (3)$$

where, n denotes the reaction order. For an experiment of non-isothermal type with a constant heating rate ($\beta = dT/dt$), rate of reaction as a function of temperature can be written in the form of Eq. (4);

$$\frac{d\alpha}{dT} = \frac{A}{\beta} e^{\frac{-E_a}{RT}} f(\alpha) \qquad (4)$$

For the estimation of activation energy, the Friedman method is used. This Friedman analysis can be expressed with the help of the following equation;

$$ln(\beta \frac{d\alpha}{dT}) = ln(Af(\alpha)) - \frac{E_a}{RT} \qquad (5)$$

The activation energy values can be obtained from the slope of $\ln(\beta \frac{d\alpha}{dt})$ with – 1/T (K).

4. Results and discussion

4.1 Pyrolysis characteristics of coal and biomass

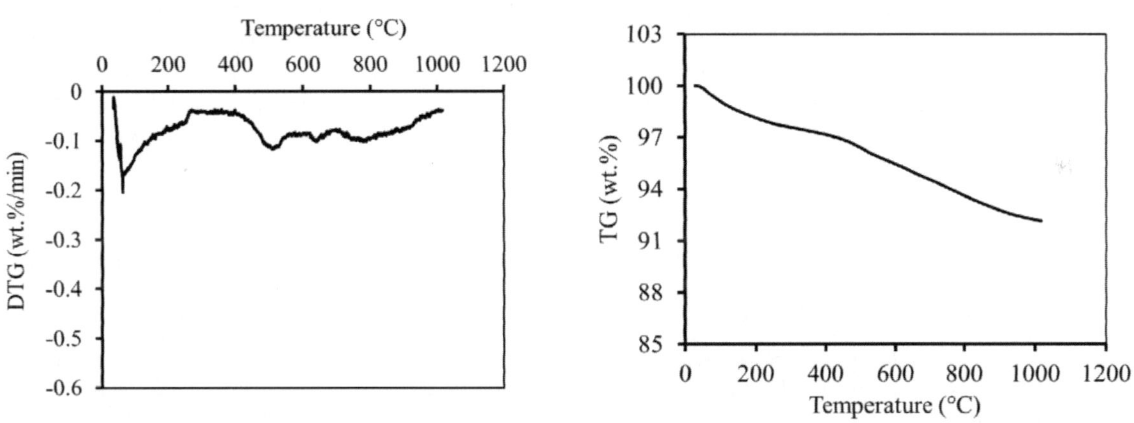

Figure 2. The DTG (left), and the TG (right) curves for coal pyrolysis at a heating rate of 10°C/min.

The DTG and the TG curves of coal are shown in figure 2. The used coal has small moisture and volatile content which is evident from the significantly lower weight loss observed from the TG plot. No drastic change is visible in the TG curve and a gradual, continuous decay is observed which corresponds to the moisture removal and devolatilization processes. A well-known three-stage analysis technique for low-rank coal pyrolysis is widely recognized. Surface moisture/water removal is considered as the first stage; which occurs mainly due to some depolymerization conversion reactions. In the second stage also called the reaction stage, 70%– 80% of weight losses occur followed by the third stage where the residue decomposition rate is even smaller. From the TG curve for coal, it may be said that the moisture discharge happens between the temperature limit of 100°C to 215°C. The value of residue ratio which remains constant after the moisture removal process can be observed from the TG curve. For a specific value of temperature, the DTG curve plot can be used to assess the value of the weight loss rate. DTG curve peak value denotes the temperature which accounts for the maximum loss in weight of the sample. It can be observed from figure 2 that, the maximum weight loss occurs around 510°C temperature for high-ash coal. The temperature of more than 250°C marks the beginning of thermal decomposition of the sample whereas a continuous decomposition takes place at higher temperatures till the maximum examined value of the temperature (~1000°C). The final weight of sample residue obtained during the pyrolysis is the result of both entities, that is, the partial oxidation of the sample material and the volatile released from the sample.

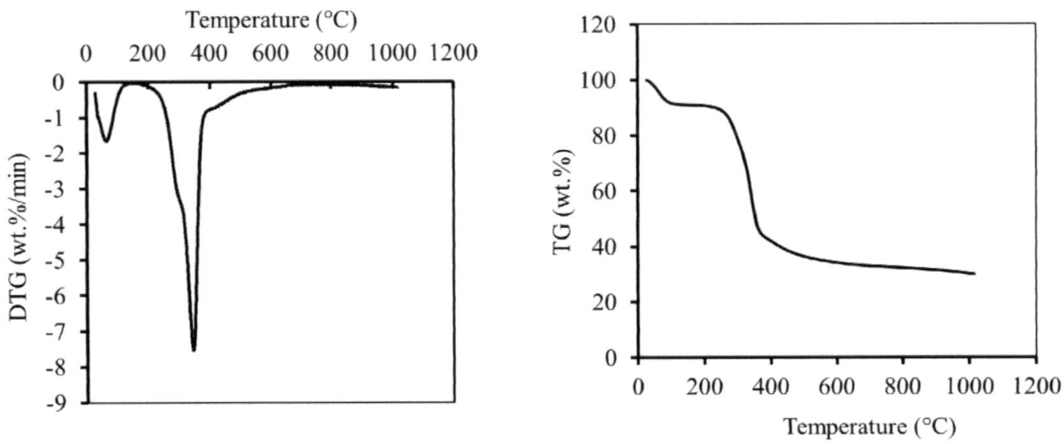

Figure 3. The DTG (left), and the TG (right) curves for biomass pyrolysis at the heating rate 10°C/min.

The effects of temperature on rice husk pyrolysis behavior are shown in figure 3. DTG and TG curves are plotted to determine the drying and devolatilization characteristics of lignocellulosic biomass material. Two distinct weight loss curves are visible for rice husk pyrolysis from the TG curve as shown in figure 3, which can be associated with the drying and the devolatilization. The start of the first major weight-loss event mainly due to the drying process can be distinguished around the temperature of 100°C, which becomes constant after 150°C. The second drastic weight loss event takes place around the temperature of 250°C. Hemicellulose decomposition begins to take place after 200°C and can go up to the temperature of 310°C, whereas cellulose decomposition occurs after a temperature of 310°C. The slope of the DTG curve changes at 390°C, which signifies the maximum rate of lignin decomposition [9], or we can say that the maximum weight loss for rice husk takes place at this temperature.

4.2 Pyrolysis characteristics of biomass-coal blends

Figure 4 shows the DTG & TG curves of various values of rice husk and coal blends. The obtained curves of DTG and TG for pure biomass and coal are also shown here. TG analysis shows that by increasing the biomass content in the coal-biomass blends, the fraction of residual weight decreases. The biomass portion present in the blend influences the liquid, solid, and gas distribution. Char yield decreases with the blending ratio of coal-biomass, whereas the yield of gas and liquid content increases. Whereas the yield of gas and liquid content increases. The immobile structural coal phase is generally composed of aromatics with cross-links joined together by C=C bonds of heavy strength and 1000 KJ/mol bond energy [10].

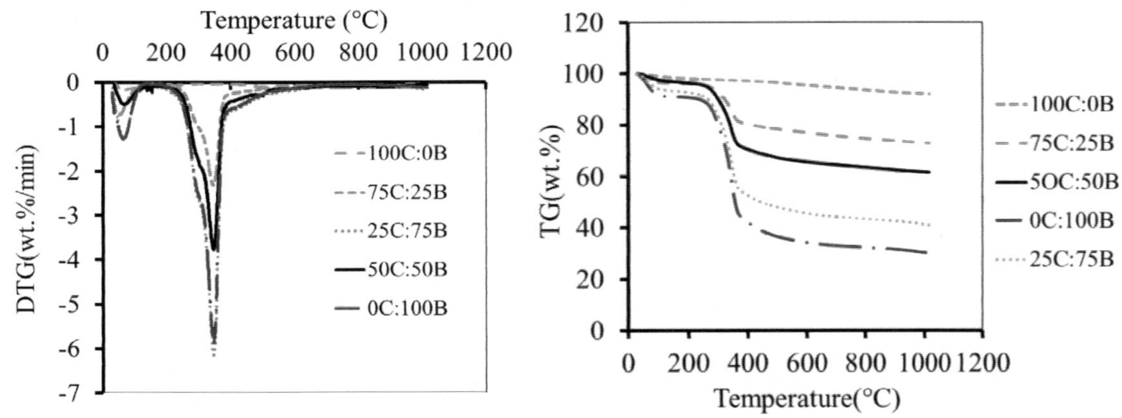

Figure 4. DTG (left), and TG (right) curves for different coal-biomass blends pyrolysis at the heating rate 10°C/min.

The breakage of these bonds is easier in the presence of the macromolecular structure of hemicelluloses, cellulose, and lignin present in the biomass, comparing heating it alone. The biomass components are bonded with each other by relatively weak ether bonds with an approximated bond energy of 380–420 KJ/mol, which results in a much quicker decomposition of biomass than coal [10]. TG curve explains these characteristics where a higher weight loss occurs in the case of higher values of blending ratios. The DTG curve displayed higher peaks for larger values of coal/biomass blending ratio.

4.3 Synergetic or Additive behavior

To examine whether any synergy occurred or not, the weight loss (TG) curve obtained from blending experiments is compared with that obtained from the individual biomass and coal feedstock experiments multiplied with their respective fractions. The weight loss (W_{cal}) due to the additive rule can be written as follows;

$$W_{cal} = m_c W_C + m_b W_b$$

where m_c and m_b are mass fractions of coal and biomass in the blends, whereas, W_C and W_b represents the weight losses of coal and biomass during their separate pyrolysis. The weight loss (W_{cal}) of the blend is calculated for 25C:75B, 50C:50B, and 75C:25B blends, and its comparison of TG curve with that obtained directly from blend pyrolysis experiments is shown in figure 5. The comparison between the two TG curves depicts slight differences, and there seems to be a little interaction between coal and biomass in the blend during heating. The low heating rate (<50°C) during the experiments could be the main reason for the insignificant interactions between the two blends. Lower heating rates inhibit interactions of the different blend components and separate their devolatilization process.

Table 2. Kinetic parameters for the pyrolysis of biomass and high-ash Indian coal.

Rate of Heating	E_a (KJ/mol)	A (min^{-1})	E_a (KJ/mol)	A (min^{-1})
	Biomass		Coal	
10°C per min	205.45	3.103 E4	68.23	187.37
20°C per min	203.38	3.213 E4	67.45	188.43
30°C per min	201.56	3.116 E4	65.23	187.29

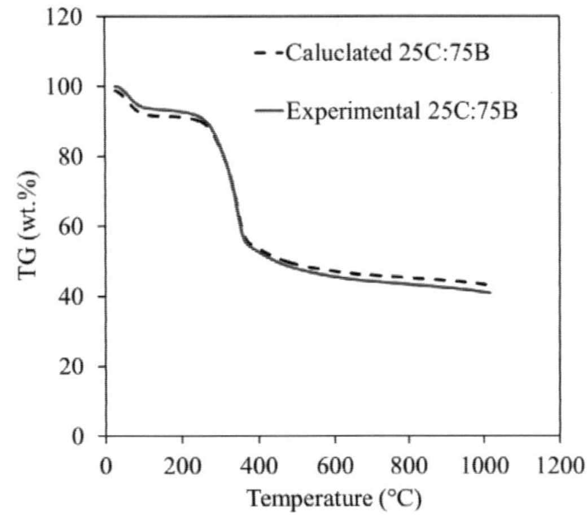

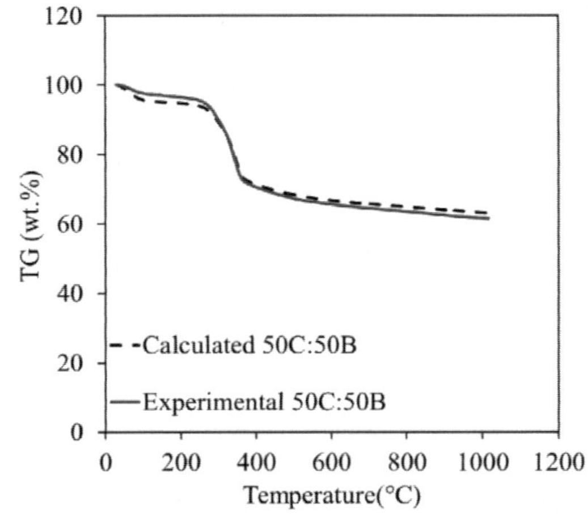

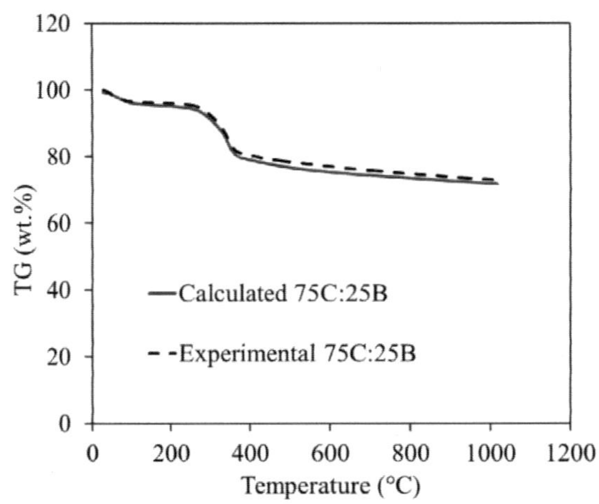

Comparison :oal and biomass blends.

Figure 5.

4.4 Estimation of kinetic parameters

An iso-conversational model-free technique known as the Friedman method is used to estimate the kinetic parameters for pyrolysis of biomass and high-ash Indian coal in an inert atmosphere of nitrogen at three distinct heating rates. In the present study, we apply this model-free technique in the form of a reaction with first-order kinetics. So, the graph shows a straight-line relationship in between $ln(\beta \frac{d\alpha}{dt})$ and $-$ *1/T (K)*. The mean estimate, when calculated for activation energy for biomass, was found to be in the range of 203.463 KJ/mol, whereas, for coal, the mean activation energy lies around 66.97 KJ/mol. Table 2 shows the values of frequency factor and activation energy that are calculated for biomass and coal samples at three different heating rates. It is found that the activation energy of both fuels decreases with the heating rate. The values of activation energy and correlation coefficient (R^2) with conversion degree (ranges from 0.1 to 0.9) for biomass and high-ash coal are listed in Table 3. The high value of correlation coefficient (R^2) suggests that our model fits the data well. The variation of the activation energy with the conversion degree (α) is shown in figure 6.

Table 3. Activation energy (E_a) and Coefficient of determination (R-Square) values with conversion degree (α).

	Biomass		Coal	
Conversion degree (α)	E_a	R^2	E_a	R^2
0.1	187.3	0.996	64.23	0.997
0.2	203.45	0.997	62.45	0.998
0.3	204.56	0.996	65.78	0.997
0.4	204.32	0.998	65.34	0.997
0.5	204.23	0.997	66.78	0.998
0.6	203.87	0.997	67.23	0.998
0.7	203.65	0.998	68.34	0.996
0.8	205.45	0.997	69.23	0.997
0.9	210.45	0.996	71.23	0.996

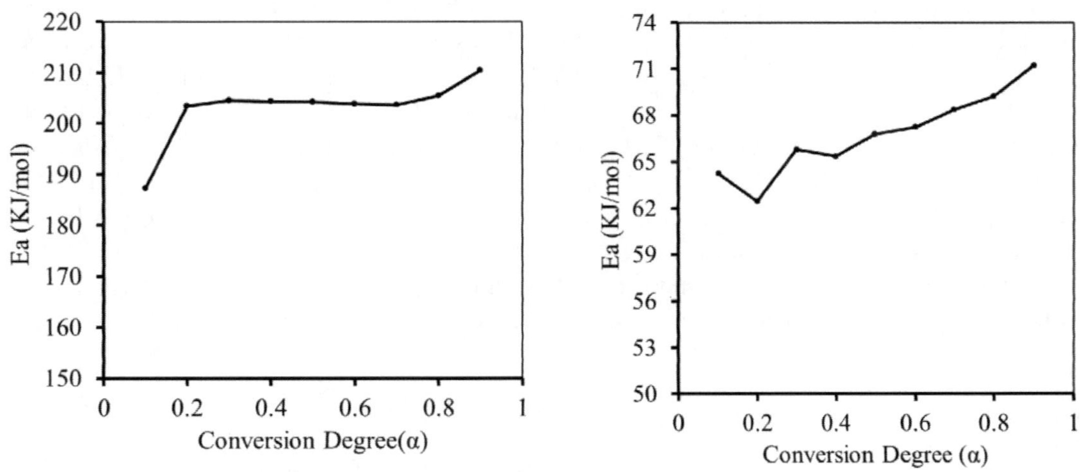

Figure 6. The activation energy of biomass (left), coal (right) as a function of conversion degree.

5. Conclusions

Thermogravimetric analysis is carried out to examine pyrolysis characteristics of blended biomass and coal in different proportions. Rice husk is used as the biomass with a low-grade, high-ash Indian coal. Pyrolysis characteristics are evaluated in terms of weight loss and differential weight loss curves. Synergetic interactions due to blending are examined by comparing them with the TGA data obtained from unary biomass and coal pyrolysis. During the individual pyrolysis of the coal and biomass, it is found out that the high-ash Indian coal incurs a very low weight reduction with no distinct peaks visible in the DTG curve. In contrast, biomass demonstrated a high reduction in the weight with an increase in temperature accompanied by the presence of well-defined devolatilization peaks. For both the fuels, the pyrolysis conversion increased as the temperature increased. For the biomass-coal blends, the TG curve showed that the weight reduction increased with increasing biomass content in blends. The calculated and experimental TG curves of the blends indicated that no significant synergetic interactions take place between both the fuels. The mean value of coal activation energy for coal were estimated as 66.97 KJ/mol which suggest High-ash Indian coal can be used for power generation and in industrial operation.

IOP Conf. Series: Materials Science and Engineering 1146 (2021) 012014 doi:10.1088/1757-899X/1146/1/012014

References:

[1] Perera F., Pollution from Fossil-Fuel Combustion is the Leading Environmental Threat to Global Pediatric Health and Equity: Solutions Exist. Int J Environ Res Public Health. 2017;15(1):16. Published 2017 Dec 23. doi:10.3390/ijerph15010016
[2] Kan T, Strezov V, and Evans T J, Lignocellulosic biomass pyrolysis: A review of product properties and effects of pyrolysis parameters. 2016 Renewable and Sustainable Energy Reviews 57 1126-1140.
[3] Sadhukhan A K, Gupta P, Goyal T, and Saha R K, Modelling of pyrolysis of coal–biomass blends using thermogravimetric analysis. 2008 Bioresource technology 99(17) 8022-8026.
[4] Wu H-x, Li H-b, and Zhao Z-l, Thermogravimetric analysis and pyrolytic kinetic study on coal/biomass blends. 2009 Journal of Fuel Chemistry and Technology 37(5) 538-545.
[5] Linlin S, Shiqing C, and Haiqing Z, Experimental study on pyrolysis behaviors of coal and biomass blending. 2006 Acta Energiae Solaris Sinica 27(8) 851.
[6] Sonobe T, Worasuwannarak N, and Pipatmanomai S, Synergies in co-pyrolysis of Thai lignite and corncob. 2008 Fuel processing technology 89(12) 1371-1378.
[7] Park D K, Kim S D, Lee S H, and Lee J G, Co-pyrolysis characteristics of sawdust and coal blend in TGA and a fixed bed reactor. 2010 Bioresource technology 101(15) 6151-6156.
[8] Du W, Wang G, Wang Y, and Liu X, Thermal degradation of bituminous coal with both model-free and model-fitting methods. 2019 Applied Thermal Engineering 152 169-174.
[9] Fábio Roberto Vieira, Carlos M. Romero Luna, Gretta L.A.F. Arce, Ivonete Ávila. "Optimization of slow pyrolysis process parameters using a fixed bed reactor for biochar yield from rice husk", Biomass and Bioenergy, 2020
[10] Quan C and Gao N, Copyrolysis of biomass and coal: a review of effects of copyrolysis parameters, product properties, and synergistic mechanisms. 2016 BioMed research international 2016.

Acknowledgments:
The authors acknowledge a financial grant from the MHRD-IMPRINT scheme (Gov. of India), Project No: 7986.

IOP Conf. Series: Materials Science and Engineering 1146 (2021) 012015 doi:10.1088/1757-899X/1146/1/012015

Investigations of heat source models in transient thermal simulation of pulsed MIG welding of AA6061-T6 thin sheet

Sachindra Doshi[*1,2]**, D B Jani**[3]**, A V Gohil**[4] **and C M Patel**[5]

[1]Gujarat Technological University, Ahmedabad, Gujarat, India
[2]Government Engineering College, Bhavnagar, Gujarat, India
[3]Government Engineering College, Dahod, Gujarat, India
[4]Shantilal Shah Engineering College, Bhavnagar, Gujarat, India
[5]School of Engineering, RK University, Rajkot, Gujarat, India
*Corresponding author e-mail address professordoshi@gmail.com

Abstract. The automobile and aviation industries are exhaustively working on weight reduction by replacing steel with aluminium alloy. However, joining the aluminium alloy body panel is remain a big challenge because of the inherent properties of aluminium. The dynamic behaviour of arc plasma in the pulsed MIG welding process is a multifaceted heat mass transfer phenomenon involving chemical, electro-magnetic, and fluid dynamics. Numerical simulation using FEA techniques are greatly helpful in understanding responses of weld bead and material behaviour. Simultaneously it reduces time and cost consuming experimental trial and error approach. The approximation of the welding arc heat source is paramount for precise prediction in the welding process simulation model. The article focuses on the comparison, suitability and effects of various moving heat source models by developing the transient thermal simulation of pulsed MIG welding of 1mm thick AA6061-T6 sheet in ANSYS workbench. Moving heat source models were developed as ANSYS APDL command subroutine. It is observed that for 1mm thin sheets Gaussian surface heat source model is also valid. The difference in temperature field is not prominent in comparison with Goldak's double ellipsoidal heat source model.

1. Introduction

Aluminium alloy was first time used during the 1980s in Mazda RX-7, Japan for its hood. Subsequently, Aluminium alloy sheets (AA sheets) swiftly used in sports cars, luxury sedans and the aviation sector. The application of aluminium alloy is growing at a moderate pace even after several advantages of the application of AA sheets in the automobile and aviation sector. The major reason for hindered growth is considerably different properties for joining compare to usual material steel. *"Exceptional properties of aluminium alloy like, high thermal conductivity; high solidification shrinkage; oxide formation on the surface; high coefficient of thermal expansion; high solubility of hydrogen when in the molten state; and relatively wide solidification-temperature ranges which influence the aluminium welds and make it difficult to weld."* [1] On other hand, the application of AA sheet in the automotive, marine and aerospace sector is demanding exponential growth because of its recyclability, less weight, excellent strength to weight ratio.

Heat input, hydrogen solubility, welding of different grades and thickness are challenges in joining thin AA sheet. [2] MIG welding provides controlled heat input. It is the most preferred method for welding thin sheet at a lower temperature. However, a challenge in controlled steady molten metal transport is an unfavourable factor for the MIG joining. Although higher ampere gives controlled spray metal transfer which provides a higher temperature that demands slow welding speed. The end of the 20th century witnessed phenomenal development in MIG welding techniques by researchers and implemented by industries. They utilized the pros of the MIG joining and reduced its process-induced cons by variation of the process. Pulsed MIG welding is successfully used for joining thin sheets by effective and synergic control. [3]

In the pulsed MIG welding process welding parameters have a significant role in developing quality weld bead. Identification of correct welding parameters demands exhaustive experiments which are time-consuming as well as costlier. Numerical simulation through finite element analysis (FEA) can be effectively used for evaluating the welding process. The phenomenal growth in computing technology has predicted the welding process and post-process through FEA. A good agreement between experimented and predicted FEA results is well established. Validation of the designed FEA problem is essential. Heat source modelling is prime input to the FEA of the welding process and selection of the correct heat flux distribution model is the key to the formation of the right FEA model. Comparative study of transient thermal analysis of various heat source models for fusion welding of the thin sheet is rare in the literature. The behaviour of a thin sheet of thickness 1mm is different than a thick sheet Therefore, an endeavour is made to develop a comprehensive transient thermal simulation model in FEA software ANSYS to investigate the effects of different heat source model in pulsed MIG welding of AA6061-T6 1mm thick sheet. The findings will add knowledge in the domain of numerical simulation of the joining process and selection of heat source model.

2. FEA Software

Since the mid of 20th-century, welding simulation is widely used. This simulation approach was programmed in FORTRAN, C, C Sharp and C++ programming tools. With the exceptional progress in computing technology, at present lots of FEA software exists. Simulation software are having integrated CAD Packages, advanced mesh techniques, and different packages of thermo-mechanical analysis. [4] FEA software which researchers have used for simulating the welding process can be summarized as follow.

- ANSYS (ANSYS Inc.),
- MATLAB (MathWorks),
- NASTRAN (MSC Software),
- COMSOL (COMSOL Inc.),
- ABAQUS (Dassault Systèmes),
- ArcWeld (ABB),
- WELDSYS (ESI Group)
- Virfac (ALTAIR) and
- MARC (Marc Analysis Research Corporation),

3. Heat source models

It is crucial to understand the way heat is supplied to the welded zone and heat flux is distributed in the welded zone. The dynamic behaviour of arc in the pulsed MIG joining is a complex heat mass transfer phenomenon. It involves chemical, electro-magnetic, and fluid dynamics. Moreover solid, liquid, gas and plasma condition of material found in tiny welding zone through simultaneous existence of thermal, mechanical, and electrical process as shown in figure 1. [4]

The optimum combination of pulse parameters in pulsed MIG welding gives high reliability of defect-free weld bead for the structural component. Trial and error method of setting up pulse parameters indeed not result in better combination and will end up to the poor quality of weld bead. Numerical simulation using FEA techniques reduces time and cost centric experimental trial and error approach. It helps in understanding responses of process variables over critical performance factors. The output of this simulation tools significantly depends on the provided transient thermal data of the process.

The simulation model builds by the provided data on surface heat flux model, right boundary conditions, accurate base material and filler material thermo-mechanical properties.

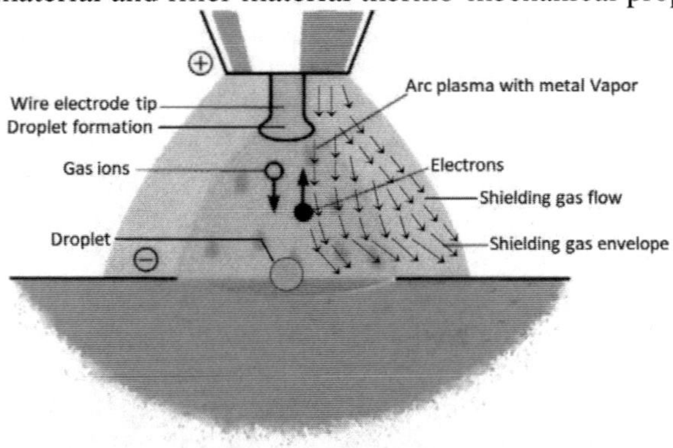

Figure 1. Welding zone in the MIG welding process

The heat source can be a point heat source, line heat source or plane heat source. Correct heat source model selection governs the efficiency of modelling of MIG welding of thin sheet. Many heat source distribution models were developed since the 1960s. such as *"2D Gaussian surface heat source model, 3D Circular disc heat source model, Single ellipsoidal heat flux distribution, Double ellipsoidal heat source model), 3D Gaussian conical heat source model and Uniform surface heat flux model."*

Table 1. Comparison of various heat flux models. [5-15]

Heat source model	Gaussian surface heat source model	Double ellipsoidal heat source model	3D Gaussian conical heat source mode	Uniform surface heat flux model
Principle of model	Heat is dispersed as per the Gaussian curve path.	The heat source was divide into the front and rear sections, each of which separates heat flux distribution as shown in the figure has.	The conical shape represents the arc and for that reason rationale for demonstrating the welding process. Conical shape with compact lower cone radius of and large upper cone radius.	A Uniform surface heat flux model can be formed with a circular, square or rectangular heat source. It is an uncomplicated faster method for shallow penetration
Characteristic curve	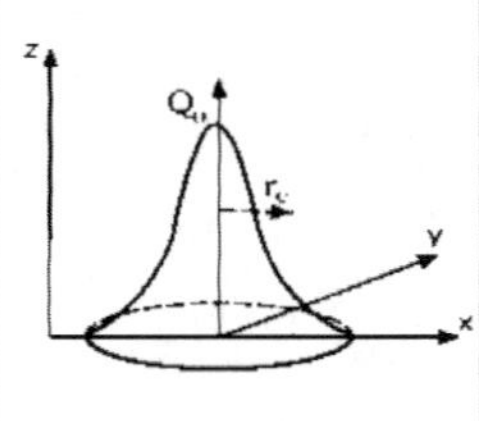	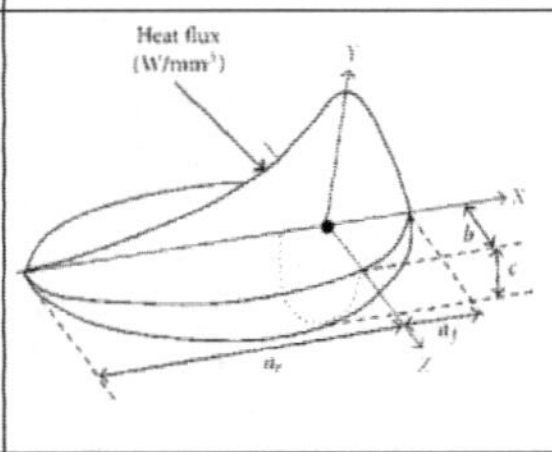	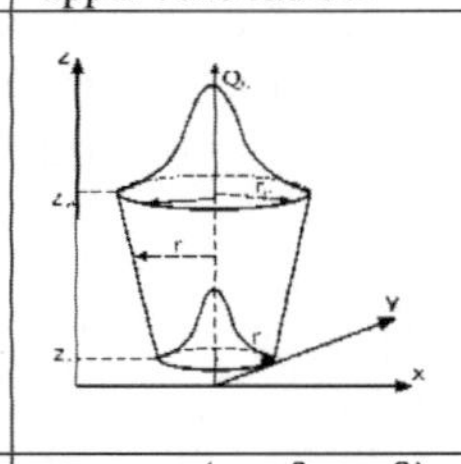	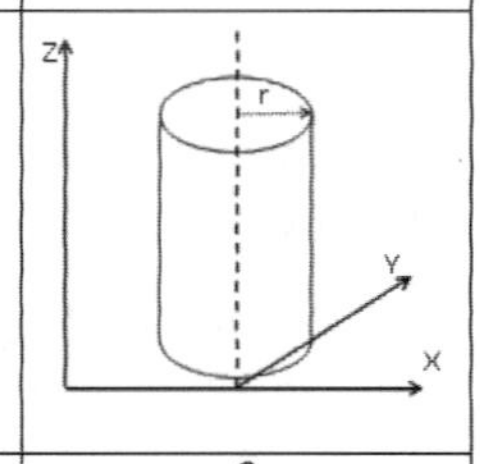
Heat source description $Q(x,y,z) =$	$Q_0 exp\left(-\frac{x^2+y^2}{r_e^2}\right)$ Q_0 = Amount of heat introduced by the source, *x, y,* and *re* = geometrical parameters of the heat source position.	$q_f = \frac{6\sqrt{3} f_f I \eta Q}{a_f b c \pi \sqrt{\pi}} exp\left(-\frac{3x^2}{a_f^2} - \frac{3y^2}{b^2} - \frac{3z^2}{c^2}\right)$ $q_r = \frac{6\sqrt{3} f_r I \eta Q}{a_r b c \pi \sqrt{\pi}} exp\left(-\frac{3x^2}{a_r^2} - \frac{3y^2}{b^2} - \frac{3z^2}{c^2}\right)$ q_f, q_r = Volumetric heat flux density in front and rear part of the model respectively. Q = Total introduced heat a_f, a_r, b, c = length of the front and the rear part, width and depth, of the estimated molten pool respectively	$Q_0 exp\left(-\frac{x^2+y^2}{r_0^2(z)}\right)$ $r_0(z) = r_e + \frac{r_i - r_e}{z_i - z_e}(z - z_e)$ Q_0 = Maximum value of volumetric heat flux density r_e, r_i = upper and lower cone radius respectively z_e, z_i = cone length parameters, x, y,z = point coordinate	$\frac{Q}{\pi r^2}$ r = radius of heat source Q = Total introduced heat
Suitability	This model is regularly applied in FEA of high power densities joining process i.e. laser, plasma, and microplasma welding	Widely used heat source model in commercial FEA software for arc welding.	It is suitable for simulations of deeper penetrating joining process like electron beam welding and laser beam welding.	It is similar to the Gaussian circular disc model, suitable for deep penetration.

Out of this many approaches following four heat flux model are more appropriate for thin sheet MIG welding. As the thin sheet is representing a plain surface it is necessary to investigate differences between the 2D surface model and the 3D heat flux model. These four primary heat flux models are elaborated in Table 1.

4. Methodology

In the present study, FEA analysis of pulsed MIG welding of AA6061-T6 thin sheet is carried out using different heat source models in ANSYS R19.3. ANSYS Parametric Design Language (Ansys APDL) is used to develop subroutine command for different heat flux distribution models. Following table 2 narrate the details of parameters selected for the present transient thermal analysis. The FEA strategy adopted is as per the three-stage FEA viz. Preprocessing, Solution and Post-processing.

Table 2. Numerical simulation details.

Material (Sheet metal)	AA6061-T6
Thickness	1.0 mm
Peak current	45A
Background current	15 A
Voltage	15
Welding speed	0.0032 m/sec

4.1. FEA assumptions

Following FEA assumptions are made while creating the FEA model.

1. The initial temperature of the FEA model is 28°C.
2. The thermo-mechanical properties of the AA6061-T6 are taken from the material library of ANSYS.
3. The pulsed MIG welding efficiency is considered as 75%
4. Sheet metal surface convection effect is considered.
5. The molten metal weld pool flow is neglected.
6. Weld bead material is considered the same as AA6061-T6
7. Weld bead is considered as part of the model.

4.2. Governing equations and heat source model

The source of energy is from arc plasma during the pulsed MIG Welding process. Heat conduction in the weld bead area is heat flux distribution from a hot area to a cold area. This can be presented as a linear relationship with temperature gradient and can be expressed as a partial differential equation as mentioned below.

$$\frac{\partial}{\partial x}\left(k(T)\frac{\partial T}{\partial x}\right)+\frac{\partial}{\partial y}\left(k(T)\frac{\partial T}{\partial y}\right)+\frac{\partial}{\partial z}\left(k(T)\frac{\partial T}{\partial z}\right)+q_i=\rho c\left(\frac{\partial T}{\partial t}\right)$$

Where k, T, q_i, ρ, c, v, w represent thermal conductivity, temperature, external heat supplied to the body, density, specific heat of the medium.

In the present work, all four models as mentioned in table 1 is adopted for calculating heat flux. Comparison of the distribution of volumetric heat flux in each of the modelled is derived using the heat source description as mentioned earlier.

4.3. Preprocessing

A butt joint design is prepared to join two thin sheets of AA6061-T6 as per American Welding Society standard code AWSD9.1 in ANSYS space claim. [16] Fine meshing was done in a transient thermal analysis system. SOLID90 mesh element type was selected, as it is a higher-order version thermal element SOLID70. The element has 20 nodes with a single degree of freedom, temperature, at each node as shown in figure 2. The solution o/p is in form of Nodal temperature included in the overall nodal solution.

Fine meshing was done throughout the sheet metal as shown in figure 3, in which total 72117 nodes were created from 21928 elements for the analysis.

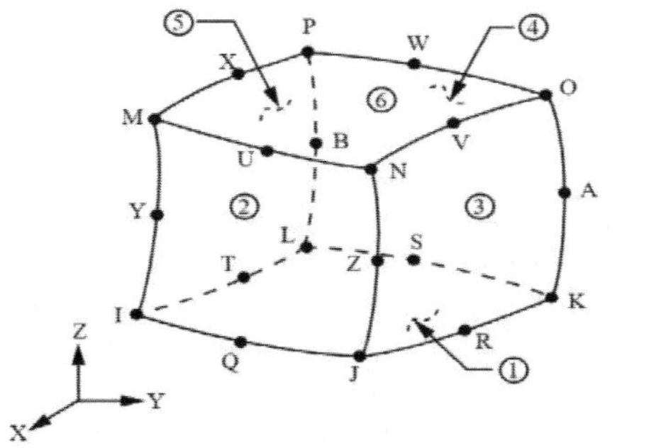

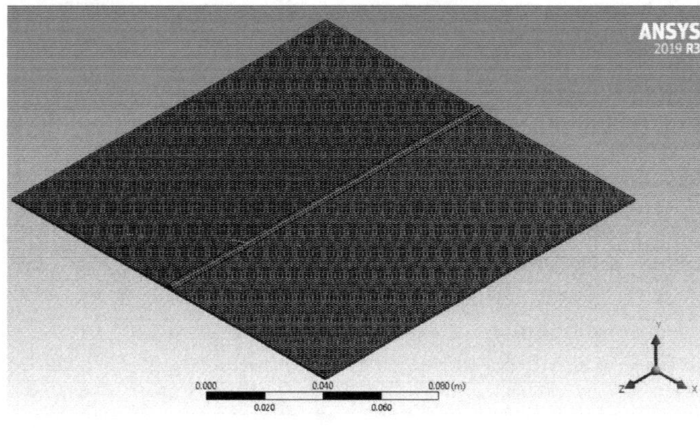

Figure 2. Solid90 element

Figure 3. Mesh model of the butt welding of AA6061-T6 sheet

4.4. Solution - defining initial conditions and boundary conditions

The heat dissipation through convection is considered between sheet metal and surrounding as mentioned in Figure 4. The ambient temperature is considered as 30° C. Four temperature probes are placed on sheet metal as shown in Figure 5 for measuring and comparing temperature profile. Coordinate system origin is placed between two sheet metal and at the centre of the weld bead.

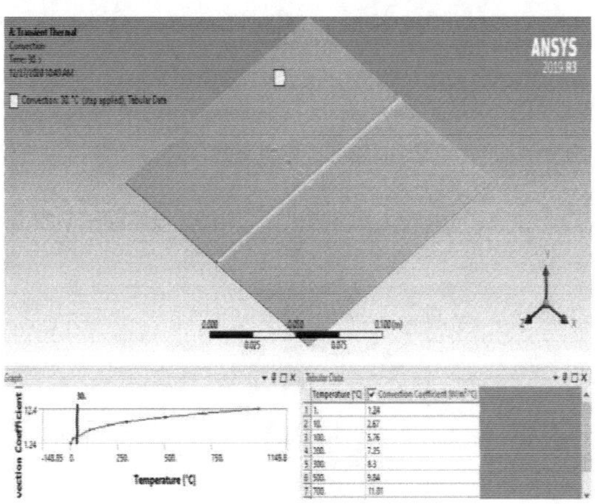

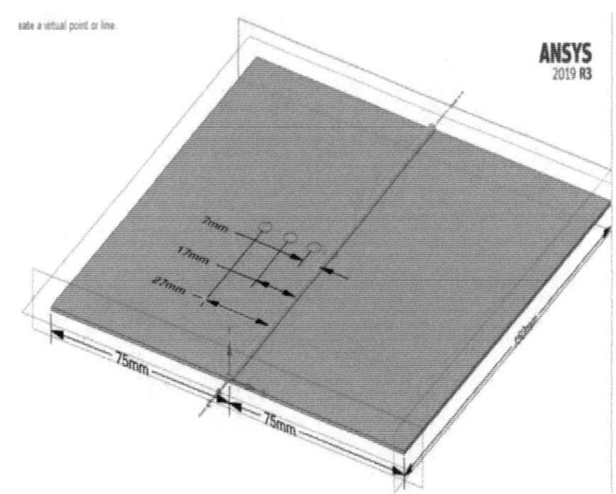

Figure 4. Convection in sheet metal

Figure 5. Temperature probe position in sheet metal

The moving heat source movement started from the origin of the coordinate system and ends at the sheet metal end on the z-axis.

Welding parameters as mentioned in table 3 is considered while calculating arc power.

Arc power = Q= ηVI

Where V and I are the input voltage and mean current and η is welding process efficiency considered as 75%. The calculated arc power is fed as input to various heat source models for solving temperature distribution in a sheet while welding. The Gaussian surface heat source model is available as Ansys ACT extension by Ansys Inc., whereas the other three models were coded in ANSYS APDL.

Table 3. Welding parameters

Peak current (I_p)	Background current (I_b)	Mean Current (I_m)	Voltage (V)	Speed (mm/sec)	Heat Flux (Joule/second)
45	15	30	15	3.2	360

5. Result and discussion

The transient thermal simulation was carried out to measure transient temperature distribution while pulsed MIG welding of AA6061-T6 thin sheet of 1.00mm. For measuring the temperature at various points on the sheet, the temperature probe was mounted at 7mm, 17mm, 27mm from the weld line and one probe is mounted on the weld line as shown in figure 5. Moving heat source moved with the input speed of 0.0032m/s in the welding direction over the weld line, simultaneously heat dissipation started over the sheet surface. Figure 6 depicts the temperature development on the entire four probes during the total movement of heat flux till the end of the sheet. The simulated temperature observed at various points on the sheet metal during the adoption of various heat source models is narrated in table 4.

Table 4. Temperature measurement

Sr. No.	Model	Max. Temperature in ℃			
		Probe 1	Probe 2	Probe 3	Probe 4
1	2D Gaussian surface heat source model	730	428	259	171
2	Double ellipsoidal heat source model	802	587	370	251
3	3D Gaussian conical heat source model	785	534	342	225
4	Uniform surface heat flux model	819	596	357	231

It was observed that the maximum temperature on the weld line was observed in the uniform surface heat flux model. Whereas minimum temperature was observed in the 2D Gaussian surface heat source model because of consideration of surface plane only. However, the difference between these heat sources is marginal.

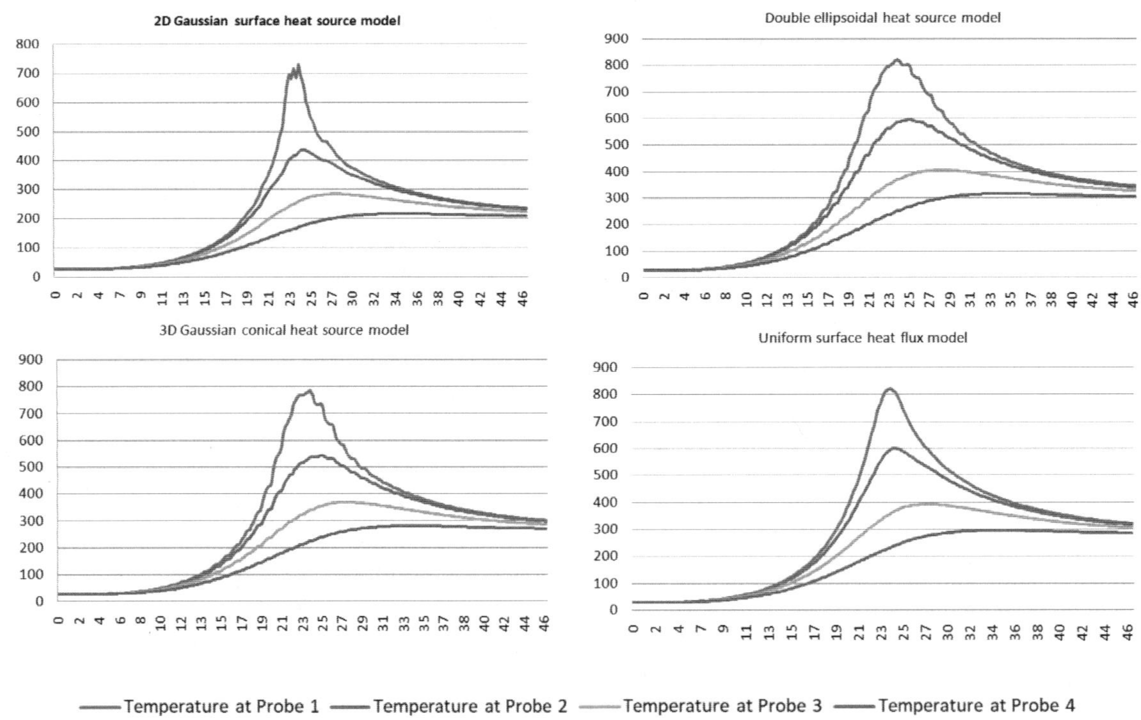

X-axis - time in seconds and Y-axis -temperature in ℃

Figure 6. Temperature distributions at various probes

The heat-affected zone is large in the Double ellipsoidal heat source model. It can be said that the rate of reduction of temperature at probe 4 is minimum in the Double ellipsoidal heat source model. The temperature at probe 4 is 31.87% higher in Double ellipsoidal heat source model compare to minimum temperature at 2D Gaussian surface heat source model. Whereas in contrast temperature at probe 1 is only 8% % higher in Double ellipsoidal heat source model compare to minimum temperature at 2D Gaussian surface heat source model. Therefore heat flux distribution is fast and the heat-affected zone is large in the Double ellipsoidal heat source model which is in agreement with experimental results of previous work. The development of heat affected zone can be easily predicted on the base of the temperature profile on the sheet metal at the end of the process. The temperature profile for the various heat source models is compared in figure 7. The heat dissipation pattern change as the heat source assumptions change. The elliptical temperature distribution pattern is visible in the double ellipsoidal heat distribution model whereas a small elliptical profile is visible in a 3D conical heat source model and a circular progressive temperature profile is visible in the other two models. The largest heat-affected zone with faster heat flux distribution is observed in the double ellipsoidal heat source model.

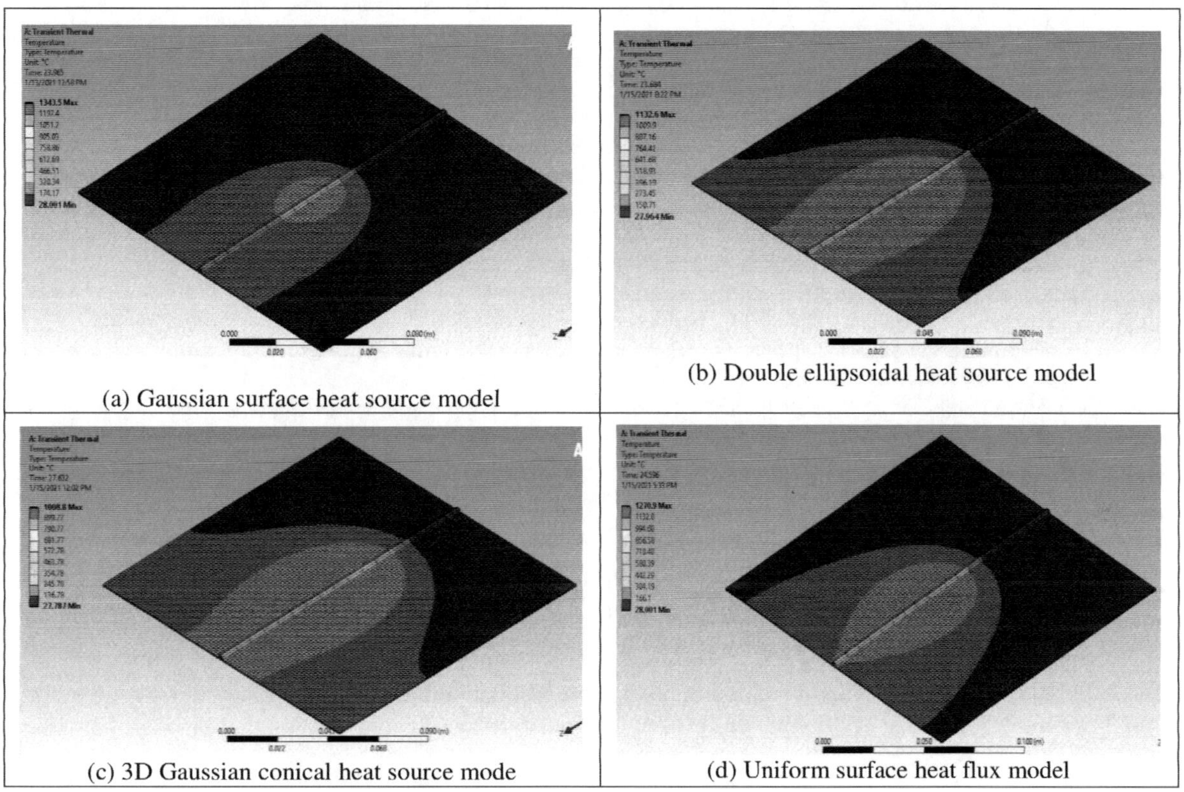

(a) Gaussian surface heat source model

(b) Double ellipsoidal heat source model

(c) 3D Gaussian conical heat source mode

(d) Uniform surface heat flux model

Figure 7. Comparison of transient heat distribution profile

6. Conclusion

In this investigation of the heat source models, ANSYS APDL code is used for transient thermal simulation of the welding process. The 1.0mm thin AA sheet 2D heat source model stands valid because of the smaller thickness and good thermal conductivity of sheet metal. It was observed that the heat distribution pattern matched the heat source models because of the vital role of conduction in heat transfer during welding. The difference between the temperature profile of heat affected zone between the 2D heat source model and the 3D heat source model is trivial because of the very less thickness of sheet metal. However, 3D Goldak's heat source result is having a good agreement with the past experimented data of the research article. The geometrical parameters of various heat source models also play a vital role in deriving temperature profile. Such geometrical parameters validation with the actual welding process is necessary as results may vary significantly with change in geometrical parameters of the heat source model. Therefore extensive studies and experiments required for the validation and consideration of the temperature profile generated by various heat source models in the future.

7. References

[1] Praveen, P. and Yarlagadda, P.K.D.V., 2005. Meeting challenges in welding of aluminum alloys through pulse gas metal arc welding. *Journal of Materials Processing Technology*, *164*, pp.1106-1112.

[2] Doshi, S.J., Gohil, A.V., Mehta, N.D. and Vaghasiya, S.R., 2018. Challenges in Fusion Welding of Al alloy for Body in White. *Materials Today: Proceedings*, *5*(2), pp.6370-6375.

[3] Goldak, J.A. and Akhlaghi, M., 2005. Computer simulation of welding processes. *Computational Welding Mechanics*, pp.16-69.

[4] Doshi, S., Jani, D.B., Gohil, A.V. and Patel, C.M., 2020. Progress in the arc plasma temperature measurement and its finite element analysis in pulsed MIG welding process. *Materials Today: Proceedings*.

[5] Pavelic, V., 1969. Experimental and computed temperature histories in gas tungsten arc welding of thin plates. *Welding Journal Research Supplement*, *48*, pp.296-305.

[6] Rosenthal, D., 1946. The theory of moving sources of heat and its application of metal treatments. *Transactions of ASME*, *68*, pp.849-866.

[7] Friedman, E., 1975. Thermomechanical analysis of the welding process using the finite element method.

[8] Andersson, B.A.B., 1978. Thermal stresses in a submerged-arc welded joint considering phase transformations10

[9] Goldak, J., Chakravarti, A. and Bibby, M., 1984. A new finite element model for welding heat sources. *Metallurgical transactions B*, *15*(2), pp.299-305.

[10] Wahab, M.A. and Painter, M.J., 1997. Numerical models of gas metal arc welds using experimentally determined weld pool shapes as the representation of the welding heat source. *International journal of pressure vessels and piping*, *73*(2), pp.153-159.

[11] Kamala, V. and Goldak, J.A., 1993. Error due to two-dimensional approximation in heat transfer analysis of welds. *WELDING JOURNAL-NEW YORK-*, *72*, pp.440-s.

[12] Mackerle, J., 2001. FEM and BEM analysis and modelling of residual stresses: A bibliography (1998–1999). *Finite elements in analysis and design*, *37*(3), pp.253-262.

[13] Marques, E.S., Silva, F.J. and Pereira, A.B., 2020. Comparison of Finite Element Methods in Fusion Welding Processes—A Review. *Metals*, *10*(1), p.75.

[14] Park, H.J., Kim, D.C., Kang, M.J3. and Rhee, S., 2013. The arc phenomenon by the characteristic of EN ratio in AC pulse GMAW. *The International Journal of Advanced Manufacturing Technology*, *66*(5-8), pp.867-875.

[15] Chakravarti AP, Goldak JA and Rao AS, 1985. Thermal analysis of welds. In: Proceedings of the international conference on numerical methods in thermal problems. Swansea.

[16] Sheet Metal Welding Code, AWS D9.1M/D9.1:2006, American Welding Society

Measurement of thermal diffusivity for food products under natural convection cooling

Md. Reyaz Arif[1], Zahid Farhan[1] and Md. Azhar[2]

[1]Department of Mechanical Engineering, Zakir Husain College of Engineering & Technology, Aligarh Muslim University, Aligarh-202002, Uttar Pradesh, India
[2]Department of Mechanical Engineering, Maulana Mukhtar Ahmad Nadvi Technical Campus, Malegaon-423203, Maharashtra,
Corresponding author's e-mail address: reyazarif@zhcet.ac.in

Abstract. In the present communication, experimental analysis has been done for evaluation of relationship between the thermal diffusivity with temperature of Potato and Brinjal (a spherical shaped food products) subjected to a natural convection. Free convection is a type of flow of motion of fluid which is not generated by an external source but by some parts of the fluid being heavier than other parts. The analysis simulates the one-dimensional Fourier equation experimentally, applicable to the regular shapes of the product (cylindrical and spherical shaped products). The experimental setup consists of a deep freezer maintained at 263 K and 1.013 bar pressure. Variation of product temperature inside the product has been measured along the radial direction of the shape.

Keywords: Natural convection; Pre-cooling; Thermal diffusivity; Skin temperature; Spherical food products.

1. Introduction

Being healthy is not an overnight phenomenon. Natural products are healthier than man made products [1]. A diet high in fresh fruits and vegetables in the form of salad can help to protect us from many diseases [2]. Natural products are biodegradable in nature so improper food storage can lead to several problems like growth of bacteria and moulds [3]. Fruits and vegetables are highly sensitive in nature and their spoilage costs wastage of billions of dollars annually worldwide [4]. Due to negative impact on the environment include air pollution, climate changes, soil and water pollution. The cost in transport sector growth dramatically which is neither by the consumer nor by the transport services but due to environment. A study say that road transport contributes up to 92% of all the cost when compared to other transport modes. Post-harvest losses of fruits and vegetables are more serious in developing countries than those in developed countries.

Food preservation is a technique used to prevent food from spoiling. It also increases shelf life of food products. It includes method such as drying [5], irradiation [6], pasteurization [7] and the addition of chemical additives [8]. The very first stage of food spoilage can be detected through its appearance, foul. Smell, colour etc. Major factor of food spoilage is due to change in pH values and changes in climatic conditions like temperature, air etc. [9]. Zhang et al. [10] have found the thermal properties of

pea fiber and potato pulp and seen the effect of these property of extruded starch thermoplastics. Kostaropoulos and Saravacos [11] have presented the thermal diffusivity of granular and porous foods at low moisture content. Arif et al. [12] presented the experimental methods for determination of thermal diffusivity of food products in forced convection environment.
The studied carried so far on the food preserving system, they have presented the effect of thermal properties for few fruits and vegetable. However deep investigation is missing in the literature. Also, the analysis on the Brinjal is lacking in the literature. Additionally, in the literature most of the works are reported on non-veg products. The aim of this work is to develop a simple method to determine the thermal diffusivity of selected fruits and vegetables (Potato and Brinjal) as a function of surface temperature. To the best of author knowledge, the present work will contribute the literature in the field of food and fruit preservation system.

2. Experimental Setup and Procedure

Figure 1(a) show the experimental setup of the deep freezer for natural convection. The inside view is shown in Fig. 1(b). Food preservation methods is an interesting area in engineering aspects. Potato and brinjal with precise spherical geometry have been chosen as a sample for testing and experiment. They are washed with tap water to remove its dirt before performing the experiment. Dimensions of the sample has been measured using Vernier calliper.

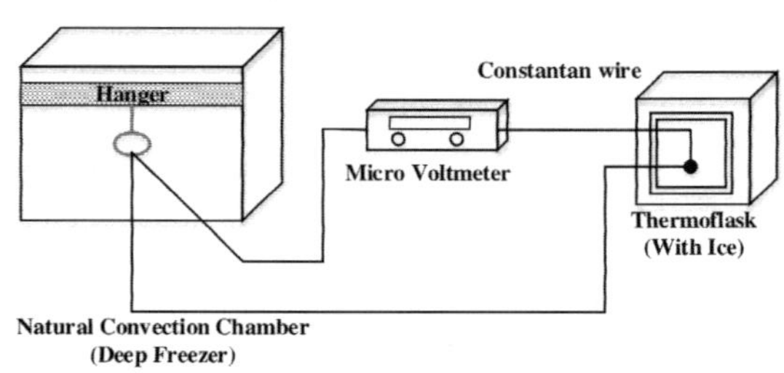

(a) (b)

Figure 1 (a) Photo of deep freezer for natural convection cooling **(b)** Schematic plot of the setup to measure the temperature of food products

This is an effective experimental setup consist of a thermo flask filled with ice. Temperature is measured with the help of calibrated copper constantan thermocouple which is pre connected with a digital DC microvolt diameter. Thermocouple is inserted into the potato in radial direction. The sample is hanged in a deep freezer which is initially maintained at -10°C. Observation is recorded uniformly at an interval of five minute. Temperature is measured at five different location from centre towards periphery. In order to simplify the analysis following assumption are made:

1. The rate of heat transfer is only along radial direction.
2. Losses due to transpiration from the surface of the product is assumed to be negligible
3. The food product is assumed to be Spherical and homogenous throughout.
4. There is no moisture in the cooling medium.
5. Specification of deep freezer are as follows: (Model-RQF-265(D), volume is 265 litre, recommended voltage stabilizer is -2 KVA, minimum temperature ranges from -35 to -40 ^{o}C, frequency is 50 Hz, compressor make: Tecumseh, Model: MCB 2410.

2.1. Experimental Procedure

The governing equation for cooling of solid food product (with characteristic dimension x), placed in air medium at constant temperature (T_∞) is essentially time-dependent heat conduction equation (or Fourier's equation) without internal heat generation and moisture loss.

$$\frac{\partial^2\theta}{\partial x^2}=\frac{1}{\alpha}\frac{\partial\theta}{\partial t} \qquad \text{where, } \theta\ (t)=\frac{T(t)-T_\infty}{T_I-T_\infty}, \qquad (1)$$

where, θ is the dimensionless temperature difference, T∞ is the temperature of cooling medium, TI is the initial temperature of the sample and, x = r/ro, where, ro (cm) is the radius of the sample under study, r (cm) is the position of thermocouple.

The left-hand side of equation (1), $\frac{\partial^2\theta}{\partial x^2}$ represents the double derivative of non-dimensionlized temperature difference with respect to x. This is calculated at the skin (x = 1) and for each value of time by using temperature variation across radius of the product. While, the differential on right hand equation (1) i.e., ∂θ/∂t represents single derivative of temperature variation at surface of the product with respect to time.

3. Result and Discussions

In the present work, the variation of thermal diffusivity with skin temperature and their empirical corelations are presented for spherical shaped potato and Brinjal in natural convection cooling. Temperature measurements along the radial direction during pre-cooling of food products in natural convection environments are carried out. These temperature distributions were plotted against radius ratios and were extrapolated up to the skin to predict skin temperature (refer Figs. 3 and 4). Figures 4 and 5 show temperature variation with time at different radius ratio for Potato and Brinjal respectively. It is observed that at particular time temperature decreases as radius ratio increases for both Potato and Brinjal. Figures 6 and 7 show the variation of thermal diffusivity with skin temperature for Potato and Brinjal respectively and their empirical correlations are developed and is given by Eq. (2) with the coefficients listed in Table 1.

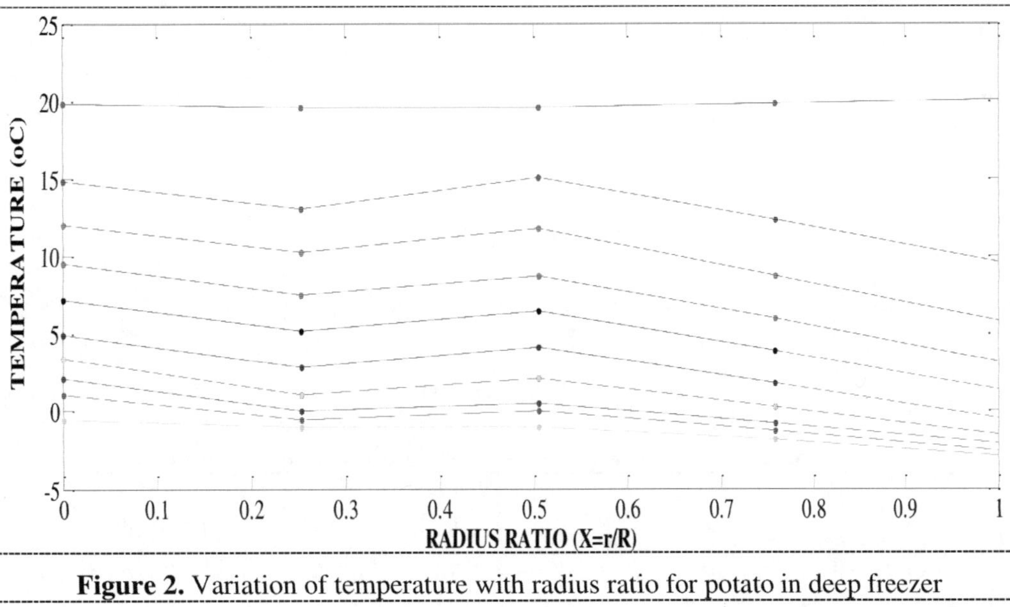

Figure 2. Variation of temperature with radius ratio for potato in deep freezer

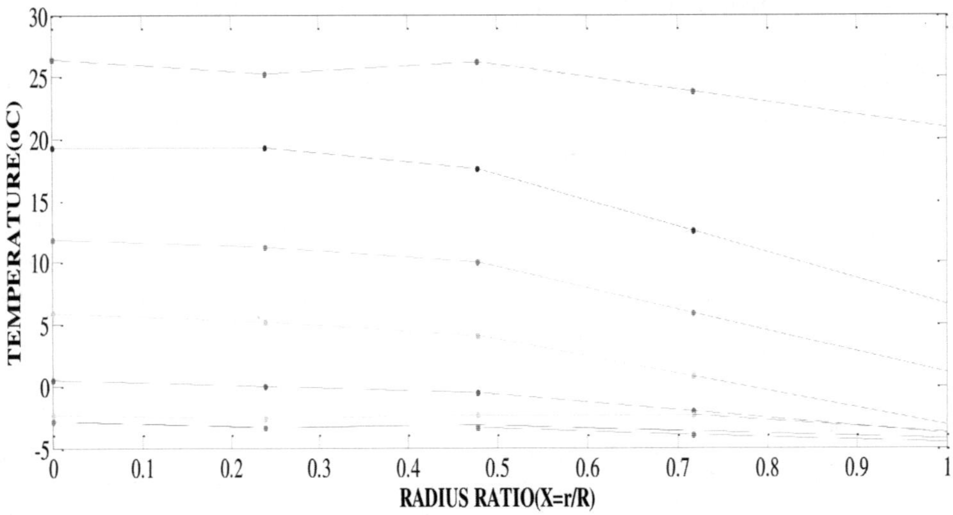

Figure 3. Variation of temperature with time for brinjal in deep freezer

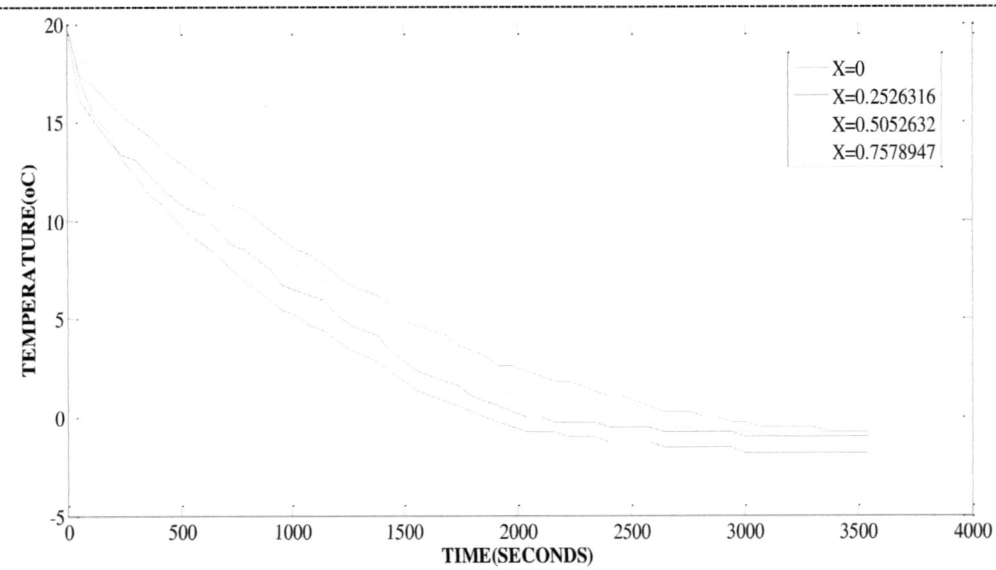

Figure 4. Variation of temperature with time for potato.

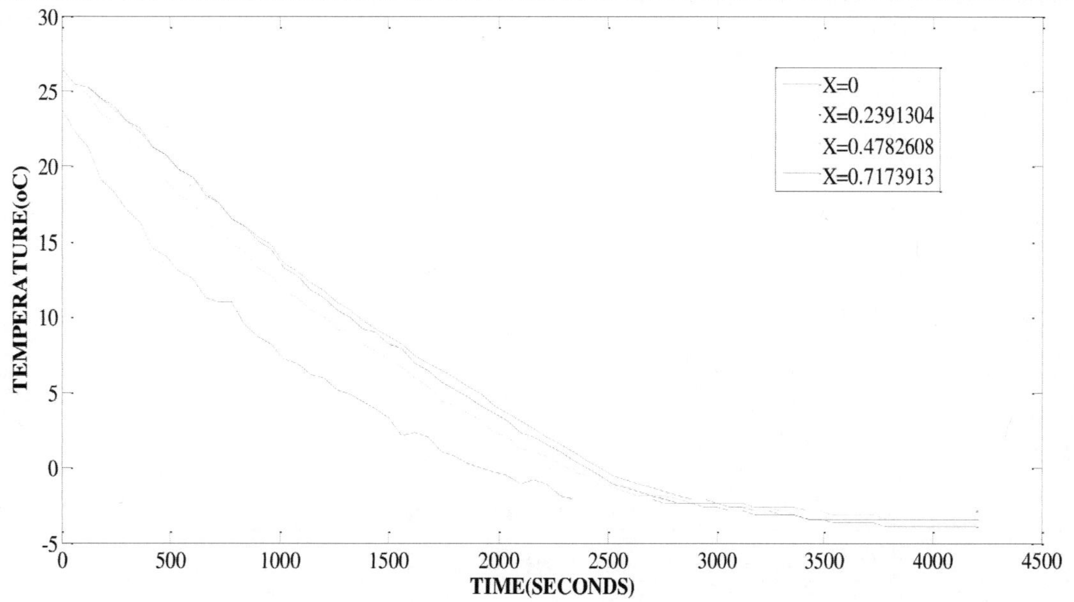

Figure 5 Variation of temperature with time for brinjal.

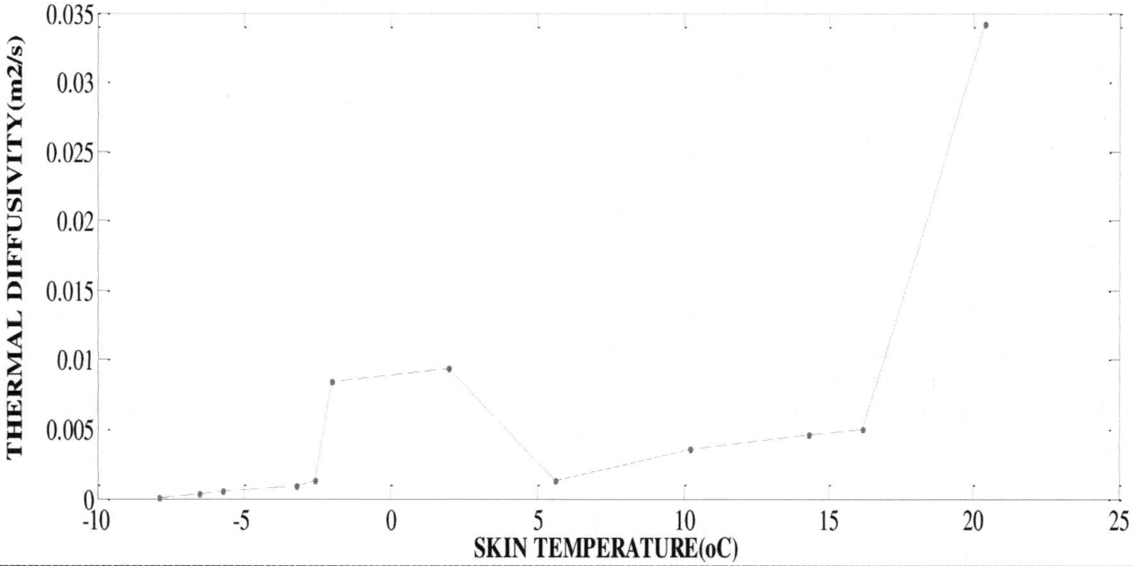

Figure 6 Variation of thermal diffusivity with skin temperature for potato.

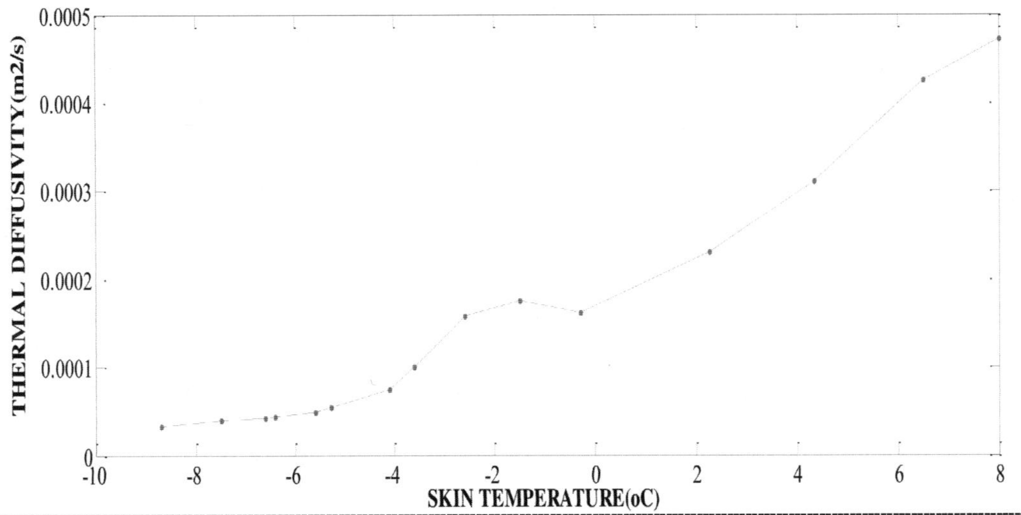

Figure 7 Variation of thermal diffusivity with skin temperature for brinjal.

The terms of the Fourier equation were evaluated and thermal diffusivity at the skin for each value of time was calculated. Finally, thermal diffusivity was plotted against temperature of the skin. It is a general observation for all the samples that at higher temperature thermal diffusivity shows strong dependence on temperature. But below -2°C, thermal diffusivity changes abruptly. The abrupt changes observed in thermal diffusivity of the products may be the result of one or more reason discussed below:

1. due to phase change of the various constituents of the products.
2. due to decrease in temperature of water.
3. thermal conductivity and specific heat variation of the product.
4. density variations below the freezing temperatures of the product.

However, the thermal diffusivity below -2°C temperatures illustrate abrupt variations and no regression curve closely represent the situation, due to this reason we have restricted the temperatures higher than -2°C. Cubical regression curves are found to be the best fit for thermal diffusivity (α) and are therefore fitted to data point. The thermal diffusivity (α) with temperature in general form are represented as;

$$\alpha = AT^3 + BT^2 + CT + D \quad (2)$$

where the regression coefficient A, B, C and D for different samples are given in Table 1.

If a table is divided into parts these should be labelled (a), (b), (c) etc but there should only be one caption for the whole table, not separate ones for each part.

Table 1. Coefficient of the products/sample

	Coefficients			
Nucleus	A	B	C	D
Brinjal	1.8E-07	6.8E-07	1.8E-06	8.6E-05
Potato	5.8E-06	-2.0E-05	-4.6E-04	2.5E-03

4. CONCLUSIONS

In this study variation of thermal diffusivity with temperature for potato and brinjal are measured in natural convection regime. Thermal diffusivity is obtained using one-dimensional (1D) Fourier's equation. The data obtained is shown graphically. Its shows a linear relationship of thermal diffusivity with temperature. Result obtained in the present study confirm that the variation of thermal diffusivity with temperature for proper geometry of food sample is simple and effective. Such temperature profile helps to understand various result over food product like deterioration rate shelf life, storage time for particular perishable food product. Also, the empirical relation has been developed for potato and brinjal under natural convection environment that will be help to those who are working in this area.

References

[1] Jiménez-Colmenero F, Carballo J and Cofrades S 2001 Healthier meat and meat products: Their role as functional foods *Meat Sci.* **59** 5–13
[2] Augustin M A, Sanguansri L, Fox E M, Cobiac L and Cole M B 2020 Recovery of wasted fruit and vegetables for improving sustainable diets *Trends Food Sci. Technol.* **95** 75–85
[3] Marsh K and Bugusu B 2007 Food packaging - Roles, materials, and environmental issues: Scientific status summary *J. Food Sci.* **72**
[4] Spang E S, Moreno L C, Pace S A, Achmon Y, Donis-Gonzalez I, Gosliner W A, Jablonski-Sheffield M P, Abdul Momin M, Quested T E, Winans K S and Tomich T P 2019 Food Loss and Waste: Measurement, Drivers, and Solutions *Annu. Rev. Environ. Resour.* **44** 117–56
[5] Masud M H, Ananno A A, Ahmed N, Dabnichki P and Salehin K N 2020 Experimental investigation of a novel waste heat based food drying system *J. Food Eng.* **281**
[6] Diehl J F 2002 Food irradiation - Past, present and future *Radiat. Phys. Chem.* **63** 211–5
[7] Meraj M, Mahmood S M, Khan M E, Azhar M and Tiwari G N 2021 Effect of N-Photovoltaic thermal integrated parabolic concentrator on milk temperature for pasteurization: A simulation study *Renew. Energy* **163** 2153–64
[8] Amit S K, Uddin M M, Rahman R, Islam S M R and Khan M S 2017 A review on mechanisms and commercial aspects of food preservation and processing *Agric. Food Secur.* **6**
[9] Steele R 2004 *Understanding and measuring the shelf-life of food*
[10] Zhang Y, Thompson M and Liu Q 2011 The effect of pea fiber and potato pulp on thermal property, surface tension, and hydrophilicity of extruded starch thermoplastics *Carbohydr. Polym.* **86** 700–7
[11] Kostaropoulos A E and Saravacos G D 1997 Thermal diffusivity of granular and porous foods at low moisture content *J. Food Eng.* **33** 101–9
[12] Arif M R, Arif S and Azhar M 2015 Thermal Diffusivity Measurement of Spherical Shaped Food Products under Forced Convection Cooling *Int. J. Innov. Res. Sci. Eng. Technol.* **4** 5521–7

Next generation air conditioner for sustainable cooling solutions

Saurabh Patel[1], Jatin Patel[1,*]

[1]Department of Mechanical Engineering, School of Technology, Pandit Deendayal Energy University, Gandhinagar (Gujarat), India.

Saurabh.pmtmt19@sot.pdpu.ac.in ,jatin.patel@spt.pdpu.ac.in*

Abstract. Generally in summer, air conditioning systems are essential for improving the human comfort and productivity. The mechanical vapor compression system (VCS) is probably the most used system. However, the refrigerant used in VCS is harmful to the environment and the initial cost is high. So, the aim of this study is to introduce a sustainable, energy-efficient and climate-friendly air conditioning system in which there is no refrigerant. The proposed system is the integration of two modules i.e., Direct evaporative cooling (DEC) module and vacuum-assisted isothermal membrane-based air dehumidification (VIMAD) module, where the DEC module will cool and humidify the entering air stream and then this cool and humidified air will enter into the VIMAD module and the air will get dehumidified. Finally, the output will be cool and dehumidified air at the exit of the system. In this paper, the proposed system is explained. Moreover, the basic principle of the DEC, the background of DEC and the basic principle of the VIMAD system, background on the types of membrane materials that can be used in VIMAD are explained. Also, the effects of various parameters on the performance of the DEC and VIMAD are presented.

Keywords: Air Conditioning, Direct evaporative cooling, vacuum membrane dehumidification, Isothermal dehumidification.

1. Introduction

Because of greenhouse gases and global warming, the ambient temperature of air is rising day after day. Staying in such a place where there is no proper air conditioning system is not easy. In order to convert the indoor situation to the human comfort zone, air conditioners play a major role worldwide. The inside relative humidity and air temperature play a crucial role in determining the human comfort conditions. Room air DBT of 25 °C and relative humidity of 55 % are declared to be in a human comfortzone [15]. The ASHRAE Standard 62.1-2016 also suggests that the HVAC (Heating, ventilation and air-conditioning) system must have the ability to keep the relative humidity between 30% and 60 % for the indoor climate [10].

In several tropical and humid regions, where the relative humidity of outside air remains continuously above 80-90 % for many days, high humidity will cause the high latent heat load which consumes the 20 to 40% of the total energy consumed by the HVAC system. Certainly, the dehumidification of highly humid air can reduce the energy consumption significantly [3]. It has been shown that the energy consumed by air conditioning system can be reduced by 20-64 % by incorporating appropriate dehumidification systems [5].

In addition, people are spending most of their lives in confined houses, so the indoor environment quality (IEQ) has a major influence on the health, efficiency and productivity of the people. So, it becomes necessary to control the relative humidity and temperature within the human comfort conditions of the environment. Otherwise, high temperatures and high level of humidity will restrict sweat evaporation, leading to irritation, discomfort and a decrease in productivity. The higher moisture content also provides an important breeding environment for the production of bacteria and fungi in residences that directly impactthe people's health [7,8]. ASHRAE guidelines suggest humidity ratios below 0.012 and relative humidity in inhabited areas between 30% and 60% to maintain the indoor human comfort conditions and lessens the risk of bacterial evolution [9,10].

HVAC sector research has reinforced the techniques for these air conditioning systems, but it is still being done in the similar way like years earlier, i.e., dehumidification and cooling of air with the air conditioners having VCS. In hot environments, dehumidification and cooling of air with the help of VCS require a considerable portionof the highest electricity usage. Moreover, the control of relative humidity is very hard with the VCS because the improvement in energy also minimizes the sensible heat load of the building (e.g., from improved insulation), but the latent heat load is not influenced [11]. Therefore, a green alternative air conditioning system is strictly needed due to the energy crisis and its bad environmental impact.

The DEC system is reinforced again at the present stage as, dueto its zero-emission features, energy efficiency, simplicity, energy conservation and environmental protection are two topics in all engineering areas. Compared to traditional A/C units, the reporting of electricity usage by the evaporative cooling systems (ECS) in a smaller residential building showed significant energy savings and improved comfort conditions for human beings [12].

The air is cooled but extremely moist from the outlet of the direct evaporative. Excessive humidity not only contributes to irritation, discomfort and affects the skin temperature of the body, but also promotes the growth of germs, which ultimately causes some health issues such as SARS [2]. Therefore, dehumidification is necessary. The conventional dehumidification technology, such as heat wheel powered desiccant dehumidification and condensing-based air dehumidification, consumes a lot of energy. Therefore, the .alternative introduced in the proposed system is vacuum assisted isothermal membrane-based air dehumidification. Moreover, hollow fiber module is preferred because it can sustain a larger pressure drop as compared to the flat plate type module and also this module does not require any kind of additional support because the tubes are made up of membranes having higher strength. Moreover, the hollow fiber membrane system exhibits the better packing density characteristic that contributes to smaller equipment size.

So, in the present work, the system which is the integration of DEC module and the VIMAD module have been presented. Moreover, the basic principle of DEC and its background, basic principle of, background of the membrane materials have been presented. Also, lastly the possible outcomes have been discussed.

2. Basic principle of the direct evaporative cooling (DEC) and the background of DEC

The basic concept behind the DEC is nothing but the transformation of the sensible heat in to the latent heat. In this system, with the help of air blower, the flow of air is forced into an enlarged liquid water surface area for the evaporation of water. So, the heat and mass transfer between the air and the water occurs and a large amount of heat is absorbed by the water from the air stream for its evaporation (i.e., enthalpy of vaporization) and the non-saturated air will become cooled. So, the dry bulb temperature (DBT) ofthe air will decrease. In short, the water will take some amount of the sensible heat from the air which will evaporate the water and becomes latent heat and the evaporated water is diffuses through the air [13] which will increase the humidity of the air. A schematic principle of DEC is indicated in Figure 1.

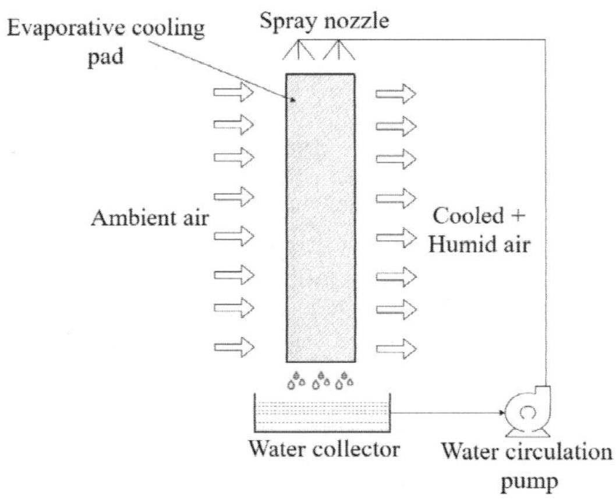

Figure 1. schematic principle of Direct evaporative cooling.

This process is ideally considered as isenthalpic. Theoretically, the wet bulb temperature (WBT) of the supplied air stream is the lowest temperature that can be achieved. In practice, wetted porous surfaces or evaporative cooling pads have a huge water surface area where contact with air moisture is achieved and the pad is moistened via spraying the water over the top portion of the evaporative cooling pads which has been kept vertically. The key types of active DEC systems used by various researchers are shown in table 1.

Table 1. Main types of direct evaporative cooling (DEC) systems.

System type	EvaporativeMedia	Cooling-effectiveness	Findings ofstudy	Investigators
Random media	Plastic fiber/foam or Excelsior which is supported by the frame made up of plastic.	>80%	- Shorter life. - Tough to clean	Banik and Ganguly [17] Tewari and Priyam [16] Tewari and Priyam [18] Reddy and Prasanna [19]
Rigid media	corrugated materials block: plastic, cellulose, fiberglass.	75-95%	- Higher initial cost. - Long life. Clean air.	
Remote pads	Rigid and random pads positioned on the structure's roof and walls	75-95%	- High energy consumption - Bacterial growth	

Deepak et al. [15] experimentally compared honeycomb & Aspen cooling pads and it has been found that the Honeycomb cooling pads have greater performance in terms of energy efficiency ratio and cooling power compared to Aspen cooling pads. Jain et al. [14] compared the cooling pad materials (coconut and palash fiber with the Aspen and khus fibers) in laboratory and concluded that the saturation effectiveness of pad with palash fibers is 26.31% and 13.2% higher than the saturation effectiveness of khus and aspen pads respectively. While saturation effectiveness of coconut fibers is 8.15% more than the saturation effectiveness of the khus pads and roughly equivalent to the saturation effectiveness of the aspen pads.

Based on available literature, it is found that DEC is an attractive option in India. But the cool air available from DEC system is very humid. So, it is not satisfying the air quality standard for human comforts. So, the alternative is to be explored for controlling the humidity of cool air. Therefore, as stated before, VIMAD is used for the air-dehumidification purpose in present work.

3. Basic principle of the VIMAD and background on the types of membranes materials

The schematic principle and the psychometric representation of the dehumidification process is shown in figure 2.

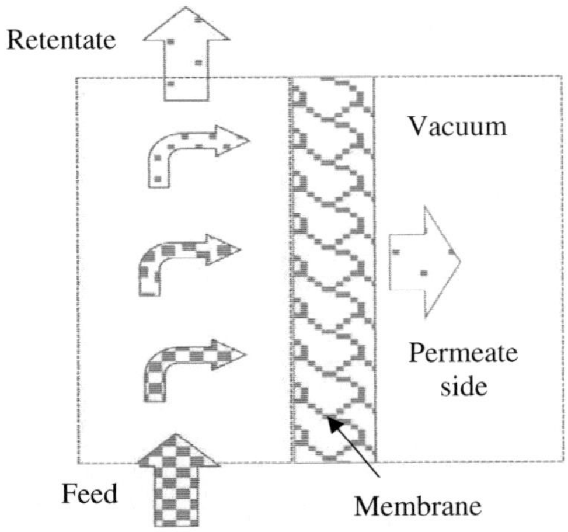

Figure 2. Schematic principle of VIMAD

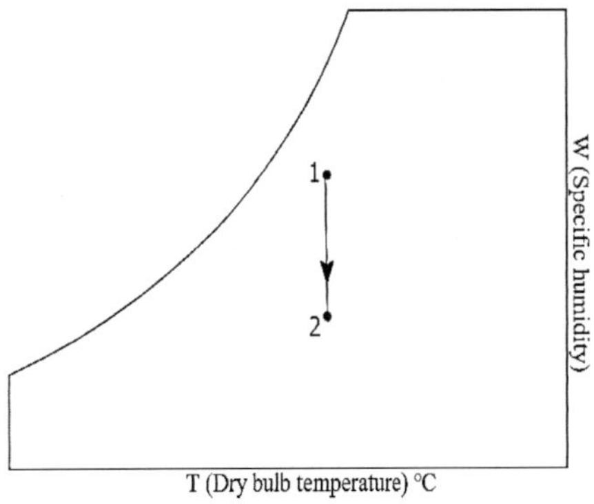

Figure 3. Psychometric representation of isothermal dehumidification process

The VIMAD membrane is a selective interface that enables only some molecules to penetrate through but restricts others because of its selectivity characteristic, which will result in isolation of the components from the gaseous mixture. Water vapor is extracted from the moist air by the membranes used in VIMAD. The gradient in the chemical potential between the two sides (i.e., feed and the permeate side) of the selective membrane is the driving force for vapor transfer. Because of chemical potential gradient, the permeate (here water particles) is passed through the membrane, which may be in the form of pressure gradient, concentration gradient, temperature gradient, or electric potential gradient. For the dehumidification processes, the first two gradients are most important. So, in present system the pressure gradient is used for the vapor transfer which can be created with the help of vacuum pump. As shown in figure 2, the water vapor particles from the feed side air is passed selectively on to the other side of the membrane and then the water vapor particles are pumped out by the vacuum pump. The remaining air is coming out for space cooling as the retentate stream. In short, the vapor is extracted from the air with the help of chemical potential gradient created in the form of pressure gradient with the help of vacuum pump. This process is isothermal dehumidification process which is shown in figure 3.

Control of the humidity in the cool air received from evaporative cooler is a key challenge. Various researchers over the worldwide have tried to solve this issue by various techniques. Isothermal membrane air dehumidification is studied critically. The review is tabulated in table 2.

Table 2. Main types of membrane materials.

Sr. no	Types of Membrane Materials	Investigators	Findings
1	Polymeric	Metz and Sybrand [27]	Investigated 19 types of polymer membranes for vapor permeation. Concluded that Particles of the water diffuse more rapidly than the molecules of the other gases.
		Bolto [20]	Concluded that the hydrophilic organic polymers improve the water solubility in the membrane.
2	Zeolitic	Xing and Rong [21] Zhang, Jian and Liu. [22]	Hydrothermal membrane growth technique was used to manufacture a thinner NaA zeolite membrane on the porous and flexible Ni substrate.
3	mixed matrix	Sanchez and Clément [23]	The mixed matric membrane has been shown to have improved reproducibility, cheaper and easier manufacturing process, efficient transportation characteristics and chemically and thermally stable.

4	Supported liquid membrane (SLM)	Kudasheva and Alina [24]	Designed SLM to slow down a liquidphase by capillary forces inside a poroussupported liquid membrane.
		Zhang and Li-Zhi [25]	LiCl salt was manufactured with composite membrane (PES as supportlayer and PVA as skin layer) andconcluded that the addition of LiCl in PVA cast solution would significantly increase vapor permeability.
		Thuan and Duc Bui [26]	The flat sheet type composite membrane composed of polymer and ceramic layers was developed and reported that the efficiency of the membrane-based air dehumidification doesn't rely on the humidity of the feed air, also its COP doesn't depend on the permeability of the membrane and the feed air velocity.

Isothermal membrane-based air dehumidification is found to extract the water particles from the moist air by using a selective barrier from the critical literature survey. With the help of this technique, only water vapor particles can be moved from a heavy concentration from the feed side of the membrane to the permeate side at a smaller concentration. This method is economic and has showed impressive energy efficient performance compared to other conventional dehumidification systems. This system is presently modern and in the testing process. In order to obtain lower power utilization as well as lower operating costs in Indian climate conditions, the system requires the understanding of the system efficiency, appropriate process analysis and design. So, the objective is to integrate the isothermal membrane-based dehumidifier to the traditional DEC system to control the humidity in a sustainable way.

4. Proposed system schematic diagram and explanation

Figure 4 displays the schematic of sustainable air conditioning system. The diagram has been prepared using 'draw.io' online software. The system consists of following components.

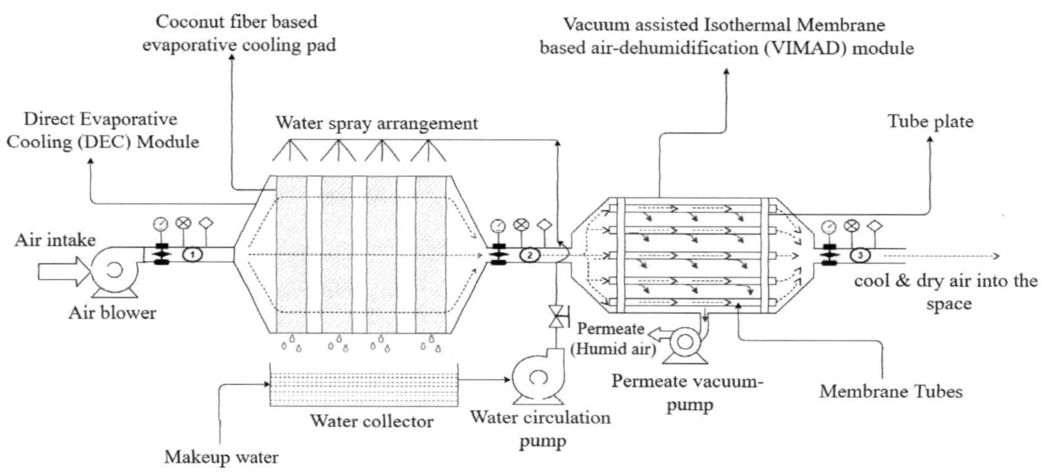

Figure 4. Schematic diagram of sustainable air conditioning system.

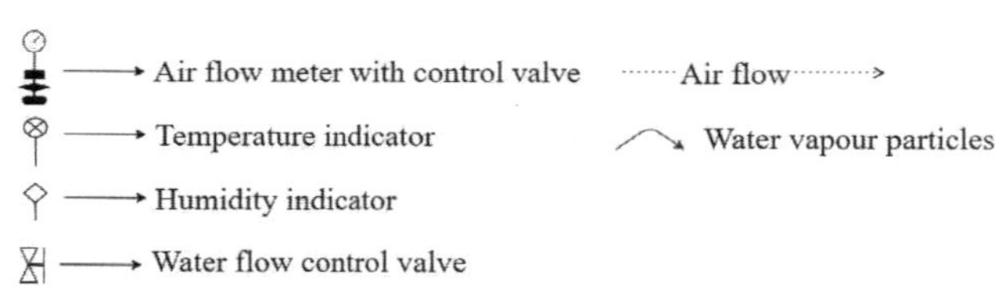

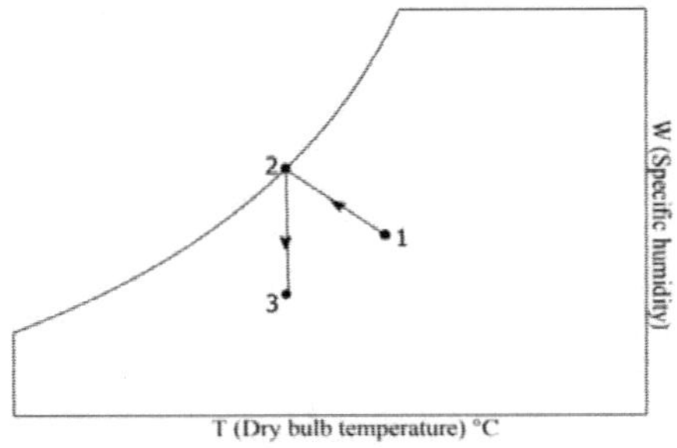

Figure 5. Psychometric representation of the two processes occurs in the proposed system

1) Direct evaporative cooling module
2) Temperature indicator
3) Vacuum assisted isothermal membrane-based air dehumidification module. (hollow fiber module type).
4) Humidity indicator
5) Air blower
6) Air flow meter with control valve
7) Pump for the water recirculation
8) Water flow control valve
9) vacuum pump
10) Water collector

As shown in figure 4, the atmospheric air is entered into the humidification module with the help of air blower. In the module, the air passes through the evaporating cooling pads through which the continuous flow of cooling water takes place with the help of water spraying arrangement and hence the humidity ratio and relative humidity of air will rise and the air will become cool as there is a direct contact between the air and water. This process is known as cooling and humidification of air. Which is indicated by process 1-2 in figure 5.

The cooled and highly humid air will enter into the vacuum-based membrane dehumidification module. The air will pass through the membrane tubes where the absolute pressure and the water vapor partial pressure will be higher than the shell side because the shell side is sealed properly to create a low pressure with the help of permeate vacuum pump. As a result, the particles of water in the air passing through the membrane tubes migrate from the tube side to the shell side (i.e., the shell having a low partial pressure of water vapor) of the interfaces (membrane tubes).

The vacuum pump will continuously remove the water vapor particles from the shell side to the atmosphere and maintain the low partial pressure of water vapor continuously. Hence the air flowing through the membrane tubes will get dehumidified at the constant dry bulb temperature. This process is known as isothermal dehumidification. Which is indicated by process 2-3 in figure 5. Air constituents apart from water (e.g., N_2, O_2, CO_2 etc.) will be significantly obstructed to some degree from traveling via the interfaces (membrane tubes).

5. System performance indicators

5.1 Performance indicators for DEC

The DEC effectiveness 'η' (also known as saturation effectiveness) is the ratio of the difference between temperature of the entering air (T_{in}) and temperature of outgoing air (T_{out}) and the maximum theoretical decrease in temperature which is nothing but the difference between the input air dry-bulb temperature (DBT) and input air wet bulb temperature (WBT). After all, the theoretical temperature at which air can be cooled inside the cooler is WBT. The efficiency of DEC is defined by the equation (1) [4],

$$\eta = \frac{T_{in} - T_{out}}{T_{in} - T_{wb}} \quad (1)$$

5.2 Performance indicators for VIMAD

Both efficiency of dehumidification and coefficient of performance (COP) of dehumidification are considered for evaluating the performance of a membrane device. The efficiency of dehumidification is considered in terms of moisture removed percentage can be determined by the equation (2) [1],

$$\text{Percentage of Moisture removed} = \frac{\omega_{in} - \omega_{out}}{\omega_{in}} \quad (2)$$

Where, 'ω_{in}' is the inlet humidity ratio and ω_{out} is the outlet humidity ratio of the dehumidification module.
The dehumidification COP (COP_{Latent}) is defined as the latent heat related to the vapor extracted (Q_{latent}) per vacuum pump work or the compressor work ($W_{V. Pump}$) and it is denoted by the equation (3) [1],

$$COPLatent = \frac{Q_{Latent}}{W_{V.Pump}} \tag{3}$$

Where, the air pressure is raised with the help of vacuum pump to an ambient pressure of 101 kPa from the pressure on the permeate side, so that the work needed by the vacuum pump '$W_{V. Pump}$'. It is possible to express the pump work by the equation (4) [26],

$$W_{V.Pump} = \frac{nRT}{\varepsilon} \ln\left(\frac{P_{atm}}{P_{Permeate}}\right) \tag{4}$$

Where, 'R' is the specific gas constant, 'n' is the cumulative moles in the mixture collected with the help of vacuum pump, 'T' represents the absolute temperature, 'P_{atm}' represents the ambient pressure, the permeate side pressure is represented by "$P_{Permeate}$' and 'ε' is the vacuum pump efficiency.

6. Possible outcomes

The effects of different parameters on the performance indicators of the DEC module and VIMAD module can be understood with the help of following literature presented in this work. For the DEC, the effect of thickness of pad 'δ' and feed air velocity 'v' on the cooling effectiveness have been shown. For the VIMAD, the effect of the transmembrane pressure, feed velocity of air, water vapor permeance, membrane selectivity (also known as water/air permeance ratio) and the relative humidity (RH) on the percentage of moisture removed and the dehumidification COP have been explained.

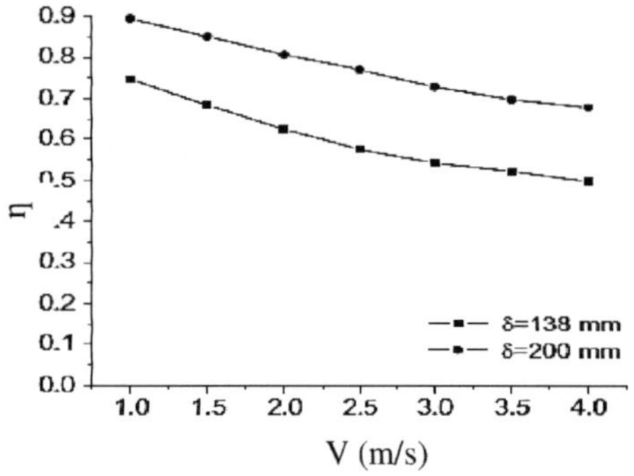

Figure 6. Effects of Pad thickness on the saturation effectiveness (η) of DEC [6]

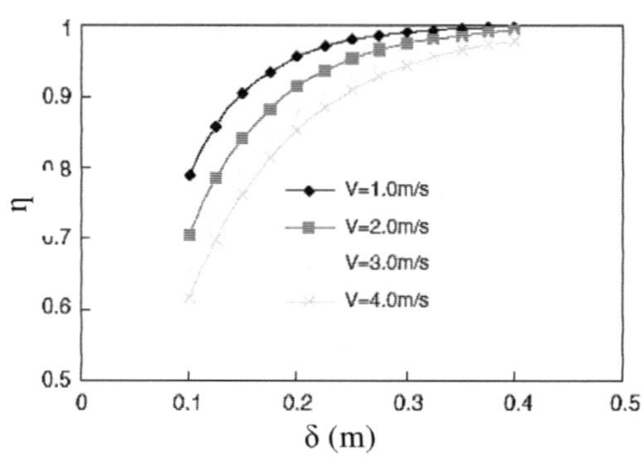

Figure 7. Effect of evaporative pad thickness and frontal air velocity on the saturation effectiveness [4]

6.1 Effects of frontal air velocity on the saturation effectiveness (η) of DEC

From Figure 6, It is clear that the higher frontal air speed 'v' reduces the interaction time between theair and the water layer, thus the transfer of heat from the air to the water decreases, ultimately reducingthe effectiveness of saturation 'η'.

6.2 Effects of Pad thickness on the saturation effectiveness (η) of DEC

Figure 7 shows that the saturation effectiveness 'η' is decreasing with the increase in frontal air speed 'v' at a given pad thickness 'δ'. This behavior is only because of reduction in the time of interaction among the water and the air stream. At a given frontal air speed 'v', the saturation effectiveness 'η' will rise with the rise in pad thickness 'δ', but it reaches to one, when pad thickness 'δ' attains some fixed value.

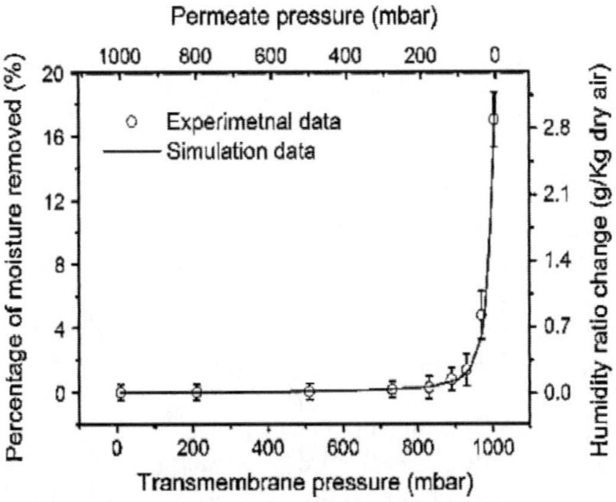

Figure 8. Effect of transmembrane pressure and permeate pressure on the % of moisture removed [26]

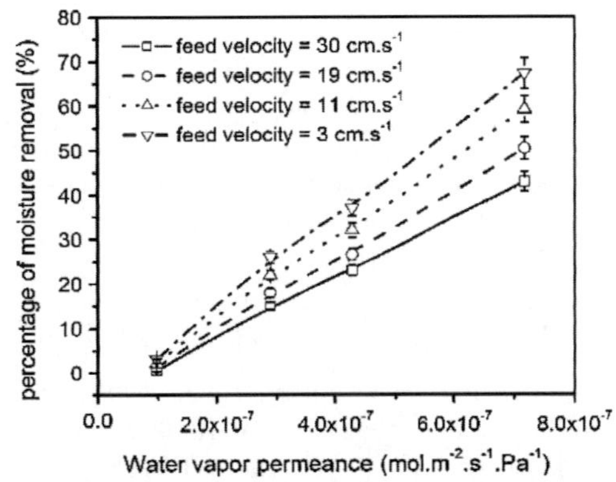

Figure 9. Effect of feed velocity and water vapor permeance on the % of moisture removed [26]

However, lower air speeds and thick evaporative cooling pads demand bigger fans, pumps and wetting systems. Moreover, supply of nearly saturated air can cause warping and corrosion of susceptible materials. The suitable evaporative cooling pad thickness should be selected depending upon the required cooling efficiency. From figure 7, the optimum air speed should be around 2.5 m/s for the required volume of air. The frontal area of a direct evaporative cooler can be determined using this suggested value.

6.3 Effect of transmembrane pressure and permeate pressure on the percentage of moisture removed

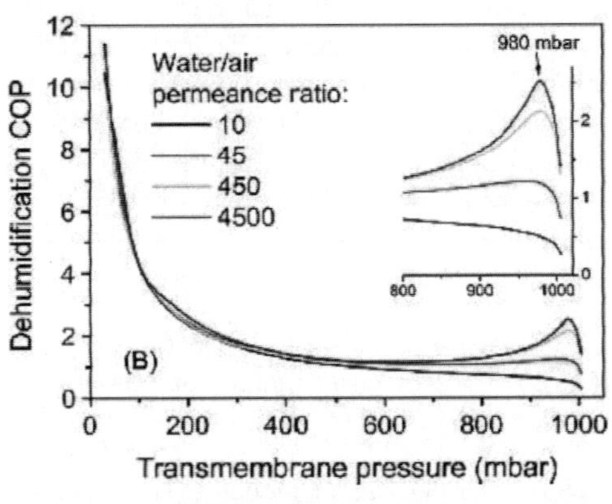

Figure 10. Effect of transmembrane pressure and water/air permeance ratio (membrane's selectivity) on the % of moisture removed [26]

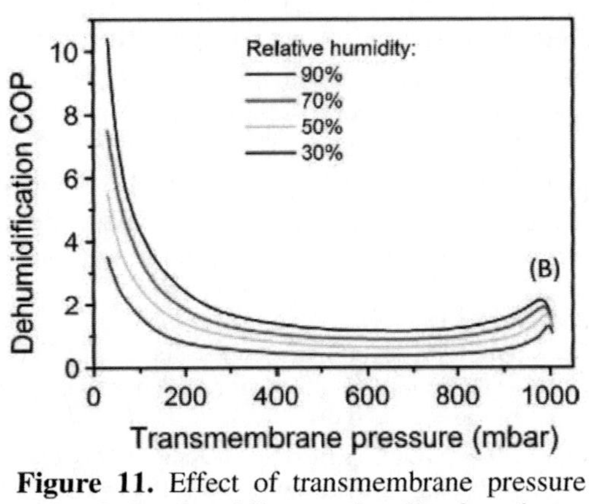

Figure 11. Effect of transmembrane pressure and relative humidity on the % of moisture removed [26]

From figure 8, with the rise in the transmembrane pressure, the amount of moisture removed increases. But the behavior is non-linear. It can be observed that the moisture removed percentage is low at lower transmembrane pressure and rises progressively as the pressure on the permeate side drop to 100mbar from the ambient pressure. As the pressure on the permeate side varies from 100 mbar to 5 mbar,the percentage of moisture removal increases sharply.

6.4 Effect of feed velocity and water vapor permeance on the moisture removed percentage
Figure 9 indicates that the moisture removed % raises with the rise in permeance of the water vapor for each feed velocity tested. Also, at certain water vapor permeability, the moisture removed % rises with lower feed velocity of air.

6.5 Effect of transmembrane pressure and water/air permeance ratio (membrane's selectivity) onthe percentage of moisture removed

It can be seen from the figure 10 that the COP of the dehumidification system does not change with the selectivity of the membrane for the transmembrane pressure below 500 mbar. At smaller transmembrane pressure, the COP rises. The COP rises sharply for the transmembrane pressure value below 200 mbar. It is evident from figure 10 that for a lower transmembrane pressure value, a maximum COP value of 11 is attainable. The COPs for transmembrane pressure levels above 500 mbar have substantial variations. Membranes with high selectivity achieve higher COPs for dehumidification.

6.6 Effect of transmembrane pressure and relative humidity (RH) on the percentage of moisture removed

As depicted in figure 11, at some certain transmembrane pressure, the COP of dehumidification increases with the rise in relative humidity. That means the system works more efficiently at the higher humidity condition. The primary parameters affecting the membrane efficiency and COP are shown in the table 3.

Table 3. Important aspects that affect the Efficiency of dehumidification and COP of dehumidification [26]

Factors	Efficiency of dehumidification	COP of dehumidification
Rise in the supply air speed	Decreasing	Change is Negligible
Rise in water vapor permeability	Increasing	Change is Negligible
Rise in the membrane selectivity tothe water vapor	Decreasing	Increasing
Rise in transmembrane pressure	Increasing	Decreasing
Rise in supply air RH	Change is Negligible	Increasing

7. Conclusion

The simple air conditioning system which integrates the direct evaporative cooling module and VIMAD module have been presented in this study in which natural and sustainable materials like coconut fiber-based cooling pad material has been suggested. Moreover, refrigerant have not been used unlike in case of conventional vapor compression, so it is totally climate friendly and so this technology holds a strong place in upcoming future. The study shows that the feed velocity of air and the pad thickness arethe two major affecting parameters to the DEC cooling effectiveness. The transmembrane pressure drops, feed velocity of air, water vapor permeance, the membrane selectivity and feed air humidity arefive major influencing parameters on the dehumidification performance and the dehumidification COP.

8. References

[1] Qu, M., Abdelaziz, O., Gao, Z., & Yin, H. (2018). Isothermal membrane-based air dehumidification: Acomprehensive review. *Renewable and Sustainable Energy Reviews*, *82*, 4060-4069.
[2] Zhang, L. Z., Zhu, D. S., Deng, X. H., & Hua, B. (2005). Thermodynamic modeling of a novelair dehumidification system. *Energy and Buildings*, *37*(3), 279-286.
[3] Yang, B., Yuan, W., Gao, F., & Guo, B. (2015). A review of membrane-based air dehumidification. *Indoor and Built Environment*, *24*(1), 11-26.
[4] Wu, J. M., Huang, X., & Zhang, H. (2009). Theoretical analysis on heat and mass transfer in a direct evaporative cooler. *Applied Thermal Engineering*, *29*(5-6), 980-984.
[5] Mahmud, K., Mahmood, G. I., Simonson, C. J., & Besant, R. W. (2010). Performance testing of a counter-cross-flow run-around membrane energy exchanger (RAMEE) system for HVAC applications. *Energy and Buildings*, *42*(7), 1139-1147.
[6] Wu, J. M., Huang, X., & Zhang, H. (2009). Numerical investigation on the heat and mass transfer in a direct evaporative cooler. *Applied Thermal Engineering*, *29*(1), 195-201.
[7] Dannemiller, K. C., Weschler, C. J., & Peccia, J. (2017). Fungal and bacterial growth in floor dust at elevated relative humidity levels. *Indoor air*, *27*(2), 354-363.
[8] Frankel, M., Bekö, G., Timm, M., Gustavsen, S., Hansen, E. W., & Madsen, A. M. (2012).

Applied and environmental microbiology, *78*(23), 8289-8297.

[9] ASHRAE S. Standard 55-2013. Thermal environmental conditions for human occupancy vol. 12. 2013 https://www.ashrae.org/technical-resources/bookstore/ standard-55-thermal-environmental-conditions-for-human-occupancy.

[10] ASHRAE S. Standard 62.1-2016. Ventilation for acceptable indoor air quality. 2016 https://www.ashrae.org/technical-resources/bookstore/standards-62-1-62-2.

[11] Rudd, A., & Henderson, H. I. (2007). Monitored indoor moisture and temperature conditions in humid climate us residences. *ASHRAE Transactions*, *113*(1), 435-449.

[12] Saman, W. Y., Percy, A., Sardelis, P., & McNab, J. (1995). A comparison between a conventional heat pump system and one incorporating heat recovery/evaporative cooling.

[13] Awafo, E. A., & Dzisi, K. A. (2012). Appropriate design of evaporative cooler for mango storage in the transitional and savannah zones of Ghana. *International Journal of Engineering Sciences Resea rch (IJESR)*, *3*(2), 687-695.

[14] Jain, J. K., & Hindoliya, D. A. (2011). Experimental performance of new evaporative cooling pad materials. *Sustainable Cities and Society*, *1*(4), 252-256.

[15] Bishoyi, D., & Sudhakar, K. (2017). Experimental performance of a direct evaporative cooler in composite climate of India. *Energy and Buildings*, *153*, 190-200.

[16] Tewari, P., Mathur, S., Mathur, J., Kumar, S., & Loftness, V. (2019). Field study on indoor thermal comfort of office buildings using evaporative cooling in the composite climate of India. *Energy and Buildings*, *199*, 145-163.

[17] Banik, P., & Ganguly, A. (2017). Performance and economic analysis of a floricultural greenhouse with distributed fan-pad evaporative cooling coupled with solar desiccation. *Solar Energy*, *147*, 439-447.

[18] Tewari, P., Mathur, S., Mathur, J., Loftness, V., & Abdul-Aziz, A. (2019). Advancing building bioclimatic design charts for the use of evaporative cooling in the composite climate of India. *Energy and Buildings*, *184*, 177-192.

[19] Thunga Venugopal Reddy and I.Prasanna. "Design and Working Model of Humidifier with Evaporative Cooling Pads." International Journal & Magazine of Engineering, Technology, Management and Research 2(8) (2015): 435-460

[20] Bolto, B., Hoang, M., Gray, S. R., & Xie, Z. (2015). New generation vapour permeation membranes. In *Pervaporation, Vapour Permeation and Membrane Distillation* (pp. 247-273). Woodhead Publishing.

[21] Xing, R., Rao, Y., TeGrotenhuis, W., Canfield, N., Zheng, F., Winiarski, D. W., & Liu, W. (2013). Advanced thin zeolite/metal flat sheet membrane for energy efficient air dehumidification and conditioning. *Chemical Engineering Science*, *104*, 596-609.

[22] Zhang, J., & Liu, W. (2011). Thin porous metal sheet-supported NaA zeolite membrane for water/ethanol separation. *Journal of Membrane Science*, *371*(1-2), 197-210.

[23] Sanchez, C., Belleville, P., Popall, M., & Nicole, L. (2011). Applications of advanced hybrid organic–inorganic nanomaterials: from laboratory to market. *Chemical Society Reviews*, *40*(2),696-753.

[24] Kudasheva, A., Kamiya, T., Hirota, Y., & Ito, A. (2016). Dehumidification of air using liquid membranes with ionic liquids. *Journal of Membrane Science*, *499*, 379-385.

[25] Zhang, L. Z., Wang, Y. Y., Wang, C. L., & Xiang, H. (2008). Synthesis and characterization ofa PVA/LiCl blend membrane for air dehumidification. *Journal of Membrane Science*, *308*(1-2), 198-206.

[26] Bui, T. D., Chen, F., Nida, A., Chua, K. J., & Ng, K. C. (2015). Experimental and modeling analysis of membrane-based air dehumidification. *Separation and Purification Technology*, *144*, 114-122.

[27] Metz, S. J., Van de Ven, W. J. C., Potreck, J., Mulder, M. H. V., & Wessling, M. (2005). Transport of water vapor and inert gas mixtures through highly selective and highly permeablepolymer membranes. *Journal of Membrane Science*, *251*(1-2), 29-41.

Numerical investigation on natural convection of hybrid nanofluid Al_2O_3 –MWCNT/water inside a vertical annulus

Raheel Usmani, F Hussain, SA Khan, NA khan, U Khan and S Husain
Dept. of Mech. Engg., Zakir Husain College of Engg. & Tech.,
AMU, Aligarh-202002, U.P, India.

Corresponding author's e-mail ID: raheelamu999@gmail.com;

Abstract. A numerical inspection on natural convection inside a vertical annulus interior of which water based (Al_2O_3 +MWCNT) (7:3) hybrid nanofluid has presently been carried out using CFD solver Ansys Fluent. Finite volume method (FVM) is used to perform the numerical simulations at constant heat flux of $3kW/m^2$and various volume fraction (0-0.04) of nanoparticle.The thermal conductivity and viscosity of hybrid nanofluid is computed by modified Maxwell model and Batchelor model respectively. The paper carried explicit analysis on effect of nanoparticle concentration, Nu number and heat transfer coefficient. The maximum enhancement in average heat transfer coefficient found to be 73.67% when gauged on volume fraction of 0.04.

Keywords: Natural Convection, Vertical Annulus, CFD, and Hybrid Nanofluid

1. Introduction

Natural convection occurs due to the density difference. Natural convection is generally preferred in power plants to avoid accidents by eliminating danger of pumps especially during shut off periods [1]. The more inputs are given through machines like pump and blower in case of the forced convection there for natural convection is a better choice when it comes to save the input energy. The design of the heat exchange devices has a significance part in increasing rate of heat transfer in natural convection. Annular geometries have wide application in various heat exchange devices. Various experimental and numerical studies are reported in literature concerning natural convection inside a vertical annulus. Husain et al. [2] performed mathematical inspection to analyse the effects of various geometrical parameters of annulus like radius ratio, aspect ratio and inclination on thermal and hydrodynamic performance. The numerical results extracted from study reveals that Reynolds number increases with inclination while Nusselt number decreases with power law. Husain and his team extended their work by conducting experiments on two phase flow of water inside vertical annulus and revealed that the length of the boiling zone increases with heat flux and observed that flow patterns are periodic in nature [3-5]. Mebarek –Oudina et al. [6] in his numerical study found that heat transfer increases as the length of the annulus decreases and further stated that shape and dimensions of gap inside annulus plays a significant role in increasing the heat transfer coefficient. Researchers have already made several attempt to enhance the coefficient of heat transfer in natural convection by altering the design of the heat exchangers. Kiwan and his co-workers [7] carried out numerical investigation on natural convection using porous wall in closed vertical annuli and observed that buoyancy effect is dominant at stages away from the entrance.

Many researchers were made attempts to enhance heat transfer rate by dispersing solid nanoparticles to conventional fluids and thus new fluid formed were coined by [8] as nanofluid. Researchers came up with the idea of regular and hybrid nanofluid to enhance coefficient of heat transfer. Hybrid nanofluid is formed by mixing two different materials inside the base fluid. The Buoyancy induced flow of regular nanofluid (Al_2O_3/water) and hybrid nanofluid (Cu-Al_2O_3/water) inside eccentric horizontal were examine by Tayebi and Chamkha [9]. The research concluded that average Nusselt number enhances more effectively in the case of hybrid nanofluid rather than regular nanofluid. Khan et al. [10-11] performed numerical study on heat and fluid flow of nanofluid (Al_2O_3/water) inside a vertical annulus by considering different models of viscosity and thermal conductivity. The research infers that contradiction in results concerning buoyancy induced flow of nanofluids in various geometries arises because of the uncertainty in computation of thermal conductivity and viscosity of nanofluids due to adoption of discrete models.

It can be summarized from literature that experimental and numerical studies reported in literature concerning buoyancy induced flow of hybrid nanofluid in annular geometries are limited. Therefore a numerical investigation of buoyancy induced flow inside a vertical annulus interior of which water based (Al_2O_3 +MWCNT) (7:3) hybrid nanofluid has presently been carried out using CFD solver Ansys Fluent.

2. Formulation of the problem

The annulus geometry (vertical) is selected as computational geometries shown in the figure 1. Geometry consist of two cylinder with different radius over one another on the same axis, the wall at outer is considered as non-heat transfer, while inner wall is fed with heat. The geometry has three parts lower, middle and upper. The outer diameter of the cylinder taken in geometry is 45mm and the inner diameter is38 mm which makes the annular gap of 3.5 mm. Overall height of the annulus is 1232mm which makes aspect ratio as 352. The water based hybrid nanofluid (Al_2O_3+MWCNT) flows from bottom to the top covering the whole length of the geometry against the gravitational force. The heat is supplied via heat flux of 3kw/m^{2}in the middle region of geometry from y = $y_1$to y_2.

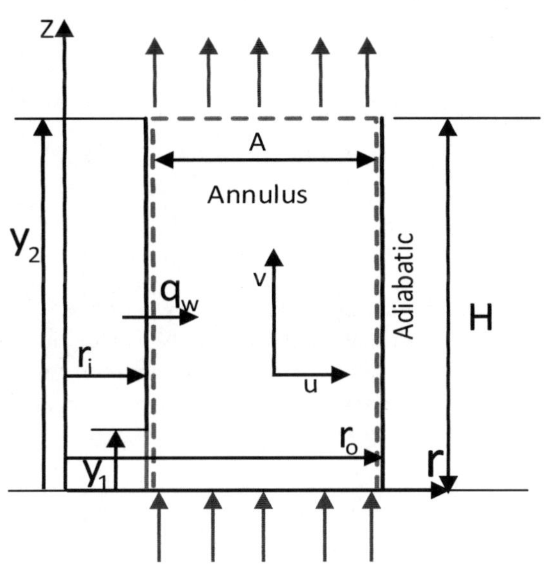

Figure 1. Computational geometry.

3. CFD methodology

3.1 Governing equations for single phase model

Things taken in consideration for making the analysis:

- The geometry is taken as two dimensional.
- Cylinders of the annulus lie on the same axis i.e Axisymmetric.
- The circulating fluid in the interior of the annulus is considered as Newtonian and dense.
- The property of the fluid at the interior of the annulus is independent of temperature *(Boussinesq Approximation).*
- Single phase approach is used to model the flow of nanofluids.

Continuity equation controls the energy transfer and circulation of nanofluid interior of the vertical annulus, Navier-Stokes equation and the energy equation which in their elliptic form are shown below:

$$\frac{1}{r}\frac{\partial(ru)}{\partial r}+\frac{\partial v}{\partial y}=0 \tag{3.1}$$

$$\left(u\frac{\partial u}{\partial r}+v\frac{\partial u}{\partial y}\right)=-\frac{1}{\rho_{hf}}\frac{\partial p}{\partial r}+v\left[\frac{1}{r}\frac{\partial}{\partial r}\left(r\frac{\partial u}{\partial r}\right)-\frac{u}{r^2}+\frac{\partial^2 u}{\partial y^2}\right] \tag{3.2}$$

$$\left(u\frac{\partial v}{\partial r}+v\frac{\partial v}{\partial r}\right)=-\frac{1}{\rho_{hf}}\frac{\partial p}{\partial y}+g\beta_{nf}(T-T_a)\ +v_{hf}\left[\frac{1}{r}\frac{\partial}{\partial r}\left(r\frac{\partial v}{\partial r}\right)+\frac{\partial^2 v}{\partial y^2}\right] \tag{3.3}$$

$$\left(u\frac{\partial T}{\partial r}+v\frac{\partial T}{\partial y}\right)=\lambda_{hf}\left(\frac{1}{r}\frac{\partial}{\partial}\left(r\frac{\partial T}{\partial r}\right)+\frac{\partial^2 T}{\partial y^2}\right) \tag{3.4}$$

Below mentioned framework are used for non dimensionalization:

$$R=\frac{r}{A},\ Y=\frac{y}{A},\ U=u\left(\frac{A}{\lambda_{hf}}\right),\ V=v\left(\frac{A}{\lambda_{hf}}\right),\ P=p\left(\frac{1}{\rho_{hf}}\right)\left(\frac{A}{\lambda_{hf}}\right)^2,\quad \theta=\frac{T-T_a}{\frac{|q_w|A}{k_{hf}}},$$

$$Aspect\ ratio, A_R=\frac{H}{A},\ Radiusratio, RR=\frac{r_o}{r_i},\quad Annular gap\ A=r_o-r_i$$

Hence, the commanding equation in non dimensionalized state will be:

$$\frac{1}{R}\frac{\partial(RU)}{\partial R}+\frac{\partial V}{\partial Y}=0 \tag{3.5}$$

$$U\frac{\partial U}{\partial R}+V\frac{\partial U}{\partial Y}=-\frac{\partial P}{\partial R}+Pr\left[\frac{1}{R}\frac{\partial}{\partial R}\left(R\frac{\partial U}{\partial R}\right)-\frac{U}{R^2}+\frac{\partial^2 U}{\partial Y^2}\right] \tag{3.6}$$

$$U\frac{\partial V}{\partial R}+V\frac{\partial V}{\partial R}=-\frac{\partial P}{\partial Y}+Ra\ Pr\ \theta+Pr\left[\frac{1}{R}\frac{\partial}{\partial R}\left(R\frac{\partial V}{\partial R}\right)+\frac{\partial^2 V}{\partial Y^2}\right] \tag{3.7}$$

$$U\frac{\partial\theta}{\partial R}+V\frac{\partial\theta}{\partial Y}=\frac{1}{R}\frac{\partial}{\partial R}\left(R\frac{\partial\theta}{\partial R}\right)+\frac{\partial}{\partial Y}\left(\frac{\partial\theta}{\partial Y}\right) \tag{3.8}$$

The set of commanding equations for buoyancy induced flow of water based hybrid nanofluid interior of vertical annulus must be solved synchronously. The non-dimensional state of length in axially is depicted as: lower adiabatic section is $Y = 0$ to $Y = Y_1 = 21$ and the middle heating zone is $Y = Y_1 = 21$ to $Y = Y_2 = 300$ and upper adiabatic zone $Y = Y_2 = 300$ to $Y = A_R = 352$. For radial direction $R_i = \frac{r_i}{A}$, and $R_o = \frac{r_o}{A}$.

Boundary Conditions:

(a) Inner wall ($R = R_i$):

i) Lower unheated zone:

$$U=V=0 \text{ and } \frac{\partial\theta}{\partial R}=0 \text{ for } 0\leq Y<Y_1$$

IOP Conf. Series: Materials Science and Engineering 1146 (2021) 012018 doi:10.1088/1757-899X/1146/1/012018

ii) Middle heated zone:

$$U=V=0 \text{ and } \frac{\partial\theta}{\partial R}=-1 \text{ for } Y_1<Y\leq Y_2$$

(b) Outer wall ($R = R_o$): $U=V=0$ and $\frac{\partial\theta}{\partial R}=0$ for$0<Y\leq A_R$

(c) Inflow (inlet) condition ($Y = 0$):

$$\frac{\partial U}{\partial Y}=\frac{\partial V}{\partial Y}=0 \text{ and } \theta=0 \text{ for } R_i<R<R_o$$

(d) Outflow (outlet) condition ($Z = A_R$):

$$\frac{\partial U}{\partial Y}=\frac{\partial V}{\partial Y}=0 \text{ and } \frac{\partial^2\theta}{\partial Y^2}=0 \text{ for } R_i<R<R_o$$

The heat transfer coefficient at any point can be numerated by:

$$h=q_w/(T_w-T_l) \tag{3.9}$$

The bulk fluid temperature is numerated by:

$$T_l=\frac{\int \rho C_{pl}\vec{V}TdA}{\int \rho C_{pl}\vec{V}dA} \tag{3.10}$$

where$\vec{V}$ is the resultant velocity.

Nu Number $Nu_y=\frac{q_w A}{k(T_w-T_b)_y}=\frac{hA}{k}$ (3.11)

3.2 Solution Methodology

The Ansys Fluent computational fluid dynamics (CFD) solver is used for simulation, and the Ansys ICEM CFD grid tool is used in mesh development. Figure 2 demonstrates the distribution of the grid in the full computational domain. In the axial direction and non-uniform in the radial direction, the grid created has a uniform dissemination .To determine the governing equations, the two-dimensional, segregated laminar solver of Ansys is used. For discretized convective terms of the momentum and energy equations respectively, the third order MUSCL and second order upwind scheme are used. The Least Square Gradient system is used to overcome terminology for diffusion and PRESTO! For the discretization of pressure terms, formulating is used. The decoupling of pressure velocity is carried out using the semi-implicit method for pressure-linked equations.

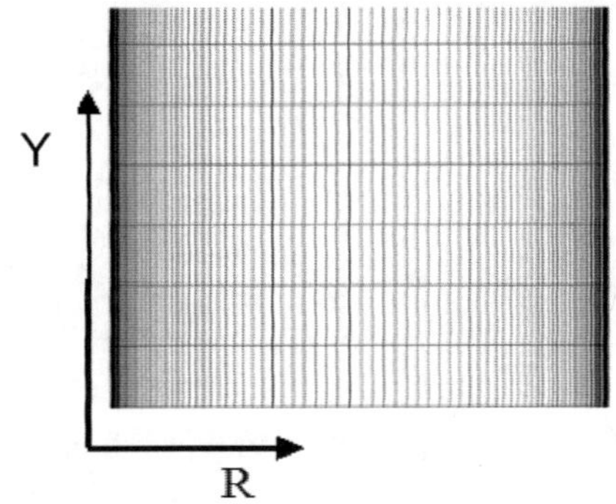

Figure 2. Representative Numerical grid.

3.3 Thermo physical Properties of Hybrid nanofluids

The hybrid nanofluid is formed by mixing nano-sized particles (MWCNT and Al_2O_3) and water in an appropriate volume fraction as given in Table 1. Table 2 shows the thermo-physical properties of nanoparticles (MWCNT and Al_2O_3) and water.

Table 1. Volume fraction of nanoparticles in hybrid nanofluid.

α_{MWCNT}	$\alpha_{Al_2O_3}$ in Al_2O_3/water	$\alpha = \alpha_{MWCNT} + \alpha_{Al_2O_3}$ in MWCNT+ Al_2O_3/water
0.003	0.007	0.01
0.006	0.014	0.02
0.009	0.021	0.03
0.012	0.028	0.04

Table 2. Thermophysical properties of nanoparticles and water.

Thermophysical properties	Base fluid (Water)	Nanoparticle (MWCNT)	Nanoparticle (Al_2O_3)
Thermal conductivity (W/mK)	0.613	3000	40
Density (kg/m^3)	997.1	2100	3970
Specific Heat (J/kgK)	4179	600	765
Volumetric expansion coefficient (1/K)	0.00021	-0.0000012	0.0000085
Viscosity (Kg/m-s)	0.001123	-	-

The effective density of hybrid nanofluid

$$(\rho)_{hf} = (1-\alpha)\rho_{bf} + \alpha\rho_{hp}$$

The heat capacitance of hybrid nanofluid as

$$(C_p)_{hf} = \frac{(1-\alpha)(\rho C_p)_b + \alpha(\rho C_p)_{hp}}{\rho_{hf}}$$

Volumetric expansion coefficient:

$$(\beta)_{hf} = \frac{(1-\alpha)(\rho\beta)_{bf} + \alpha(\rho\beta)_{hp}}{\rho_{hf}}$$

Following models are used to evaluate viscosity and thermal conductivity of hybrid nanofluid
Viscosity: Viscosity of hybrid nanofluid is evaluated using Batchelor [12] model:

$$\mu_{hf} = (1+2.5\alpha+6.2\alpha^2)\mu_{bf}$$

Thermal conductivity: To evaluate the thermal conductivity of hybrid nanofluid, equation of Maxwell model [13] is rewrite in the following form:

$$\frac{k_{hf}}{k_{bf}} = \frac{(1/\alpha)\sum_{np}\alpha_{np}k_{np} + 2k_{bf} + 2\sum_{np}\alpha_{np}k_{np} - 2\alpha k_{bf}}{(1/\alpha)\sum_{np}\alpha_{np}k_{np} + 2k_{bf} - \sum_{np}\alpha_{np}k_{np} + \alpha k_{bf}}$$

$$\rho_{hp} = \frac{\alpha_{mwcnt}\rho_{mwcnt} + \alpha_{Al_2O_3}\rho_{Al_2O_3}}{\alpha}$$

$$C_{p_{hp}} = \frac{\alpha_{mwcnt}C_{p_{mwcnt}} + \alpha_{Al_2O_3}C_{p_{Al_2O_3}}}{\alpha}$$

$$\beta_{hp} = \frac{\alpha_{mwcnt}\beta_{mwcnt} + \alpha\beta_{Al_2O_3}}{\alpha}$$

$$k_{hp} = \frac{\alpha_{mwcnt}k_{mwcnt} + \alpha_{Al_2O_3}k_{Al_2O_3}}{\alpha}$$

$$Concentation of hybrid\ nanoparticle \alpha = \alpha_{mwcnt} + \alpha_{Al_2O_3}$$

4. Results and Discussion

The numerical investigation of buoyancy influenced flow of water based hybrid nanofluid (Al_2O_3+MWCNT) interior of a vertical annulus has been carried at constant heat flux of 3.5 kW/m^2.The thermal performance of hybrid nanofluids in vertical annulus is judge by temperatures, heat transfer coefficient and Nusselt number as presented and discussed in subsequent section.

4.1 Wall and Liquid Temperatures

The numerical outcome is displayed in form of wall and liquid temperature. The bulk liquid and wall temperatures increases along axial length of the annulus for both water and hybrid nanofluid as shown in figure 3. The wall temperature seems to increase at faster rate in the lower heated zone and then axially, it increases linearly up to reaches a maximum value. It then decreases in the upper non heated region of the annulus. The bulk temperature of water and hybrid nanofluid increases linearly in the heated region and almost become constant in the upper adiabatic region as shown in figure. The lower heated region of annulus describes the developing boundary layer where temperature of wall rises rapidly. The remaining heated zone signifies the evolved flow zone where temperature profile of wall and fluid becomes corresponding. The wall and bulk liquid temperature diminishes with volume concentration of nanoparticle which can be seen in the fig. This is because adding nanoparticles increases the thermal conductivity of the base fluid. Thus, higher thermal conductivity of hybrid nanofluid provides opportunity to extract more heat from heated wall.

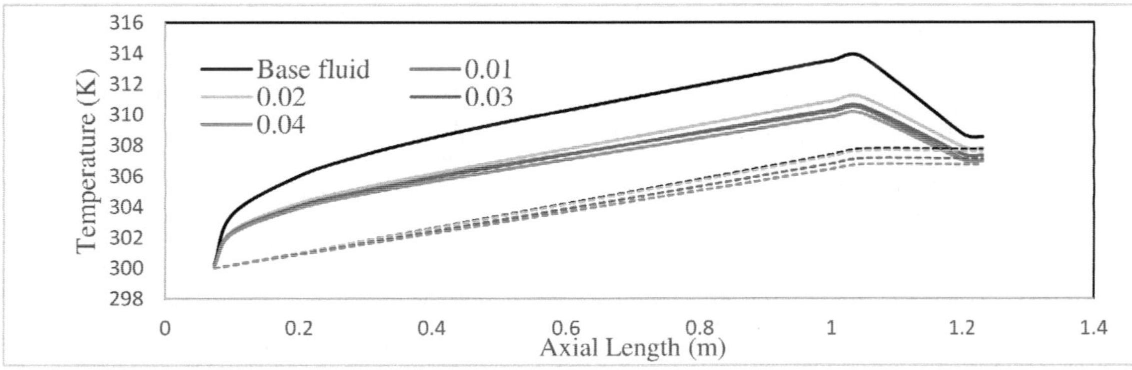

Figure 3. Contrast of wall and fluid temperature of water and hybrid nanofluid axially.

4.2 Heat Transfer coefficient

Local heat transfer coefficient is evaluated using equation 3.9. Since the wall and fluid temperature varies axially and so is the heat transfer coefficient for both water and hybrid nanofluid as shown in figure 4. It is seen from figure, heat transfer coefficient drips abruptly in the lower entry zone signifies evolved flow and afterwards it seems to be constant representing fully developed flow. Further, it can be concluded through graph that heat transfer coefficient increases with volume fraction of nanoparticles. The effect of volume fraction of nanoparticles on average heat transfer coefficient is shown in figure 5. Average heat transfer coefficient is enhanced by 73.67% in case of hybrid nanofluid as compared to pure water on dispersion nanoparticles with volume fraction of 0.04.

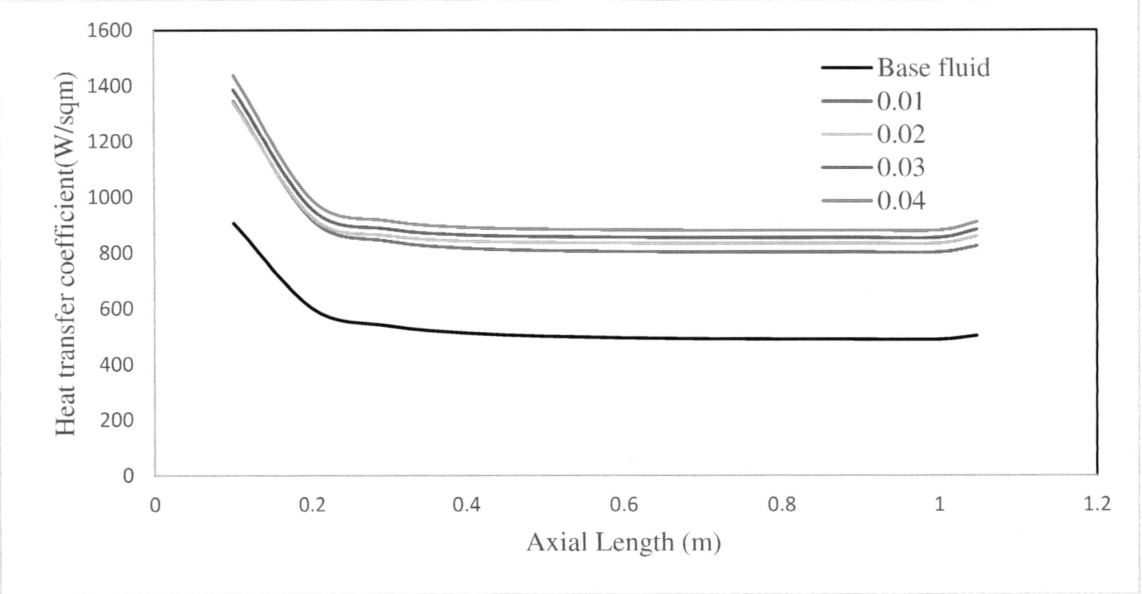

Figure 4. Contrast of local heat transfer coefficient axially.

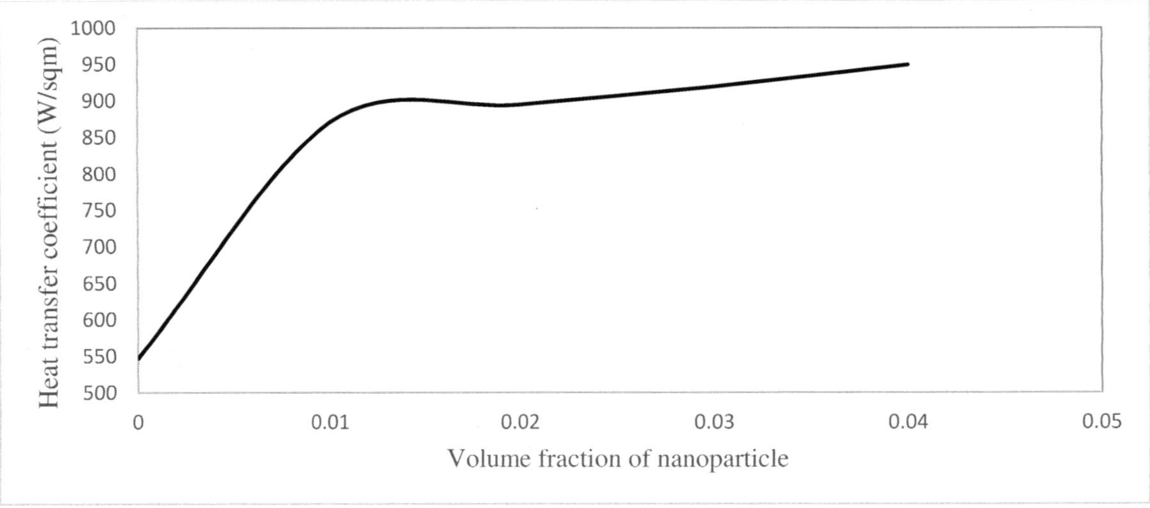

Figure 5. Variation of average heat transfer coefficient with nanoparticle concentration

4.3 Nusselt Number

The non-dimensional form of heat transfer coefficient is describes as Nu number that is calculated using equation 3.11. Contrast of Nu number axially is depicted in figure 6. The Nu number seems to drop abruptly in the lower entry zone signifies evolved flow region, afterwards it seems to be constant representing completely evolved flow. The Nu number estimated in case of hybrid nanofluid inside a vertical annulus is higher than pure water. However concentration of nanoparticles does not have significant effect on Nusselt number, because the commanding eqn. states that higher thermal conductivity of hybrid nanofluid suppresses the Nu number.

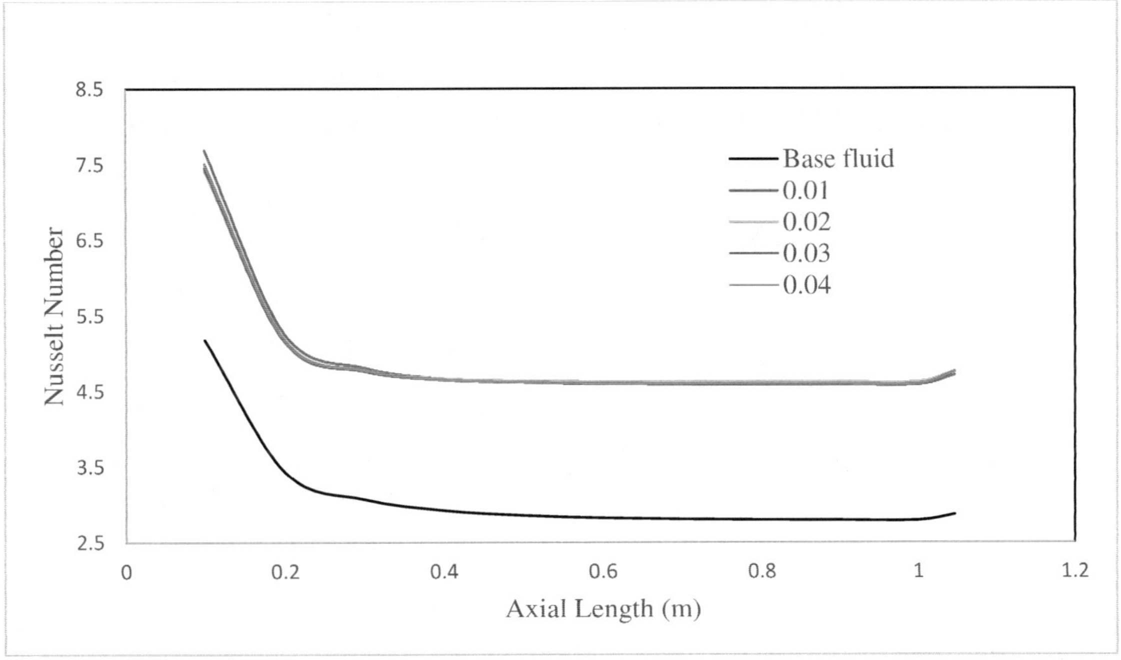

Figure 6. Contrast of Nu number axially.

5. Conclusion

The numerical inspection of buoyancy induced flow in vertical annulus interior of which water based hybrid nanofluid (Al_2O_3+MWCNT) has been carried out. The important points concluded from the present study:

- The lower heated region of annulus describes the developing boundary layer where temperature of wall rises rapidly. Whereas the remaining heated region signify the completely evolved region where temperature profile of wall and fluid becomes aligned.
- The local heat transfer coefficient increases with volume fraction of nanoparticles.
- Average heat transfer coefficient is enhanced by 73.67% in case of hybrid nanofluid as compared to pure water on dispersion nanoparticles with volume fraction of 0.04.
- The Nusselt number estimated in case of hybrid nanofluid inside a vertical annulus is higher than pure water.
- The concentration of nanoparticles does not have significant effect on Nusselt number of hybrid nanofluids.

IOP Conf. Series: Materials Science and Engineering 1146 (2021) 012018 doi:10.1088/1757-899X/1146/1/012018

NOMENCLATURE

Greek

A_f	Flow area	A	Annular gap (m)
c	Specific heat	λ	Thermal diffusivity (m^2/s)
d	Diameter of tube (m)	β	Expansion coefficient (K^{-1})
h	Heat transfer coefficient ($W\ m^{-2}\ {}^oC^{-1}$)	ρ	Density ($kg\ m^{-3}$)
		μ	Dynamic viscosity ($N\ s/m^2$)
k	Thermal conductivity ($W\ m^{-1}\ {}^oC^{-1}$)	θ	Non-dimensional Temperature
L	Length (m)	ν	Kinematic Viscosity ($m^2\ s^{-1}$)
g	Acceleration due to gravity ($m\ s^{-2}$)	***Dimensionless parameters***	
p	Pressure (Pascal)	A_R	Aspect ratio (L/δ)
q_w	Heat flux (W/m^2)	Nu	Nusselt Number $\left(\frac{hA}{k}\right)$
r	Radial distance (m)	Pr	Prandtl Number $\left(\frac{\nu}{\delta}\right)$
R	Non-dimensional Radial distance	Ra	Rayleigh Number $\left(\frac{g\beta\lvert q_w\rvert A^4}{k\nu\lambda}\right)$
T	Temperature (${}^oC$)	RR	Radius ratio $\left(\frac{r_o}{r_i}\right)$
Ri	Richardson Number		
u	Radial velocity ($m\ s^{-1}$)	***Subscripts and Superscripts***	
U	Non-dimensional radial velocity	*a*	Ambient
		h	Heated
v	Axial velocity ($m\ s^{-1}$)	*i*	Inner
V	Non-dimensional axial velocity	*o*	Outer
y	Axial distance (m)	b*f*	Basefluid
Y	Non-dimensional axial distance	*hf*	Hybrid Nanofluid

References

1. Mustafa, J., Husain, S., & Siddiqui, M. A. (2017). Experimental studies on natural convection of water in a closed-loop vertical annulus. *Experimental Heat Transfer*, ***30*(1), 25-45.**
2. Husain, S., Siddiqui, M. A., & Khan, S. A. (2019). Effect of geometrical parameters on natural convection of water in a narrow annulus. *Progress in Nuclear Energy*, ***112*, 146-161**.
3. Husain, S., Siddiqui, M. A., & Ahmad Khan, S. (2019). Wall boiling in a vertical annulus: Effect of inlet sub cooling and mass flow rate. *Numerical Heat Transfer, Part A: Applications*, ***75*(11), 776-793.**
4. Husain, S., Siddiqui, M. A., & Khan, S. A. (2019). Visualization of flow patterns of water in open-ended vertical annulus during natural convection flow. *Journal of Flow Visualization and Image Processing*, ***26*(3), 209-222.**
5. Husain, S., Khan, S. A., & Siddiqui, M. A. (2021). Wall boiling of Al_2O_3-water nanofluid: Effect of nanoparticle concentration. *Progress in Nuclear Energy*, ***133*, 103614.**
6. Mebarek-Oudina, F. (2017). Numerical modeling of the hydrodynamic stability in vertical annulus with heat source of different lengths. *Engineering science and technology, an international journal*, ***20*(4), 1324-1333.**
7. Kiwan, S., & Alzahrany, M. S. (2007). Effect of using porous inserts on natural convection heat transfer between two concentric vertical cylinders. *Numerical Heat Transfer, Part A: Applications*, ***53*(8), 870-889.**
8. Choi, S. U., & Eastman, J. A. (1995). *Enhancing thermal conductivity of fluids with nanoparticles* (No. ANL/MSD/CP-84938; CONF-951135-29). Argonne National Lab., IL (United States).
9. Tayebi, T., & Chamkha, A. J. (2016). Free convection enhancement in an annulus between horizontal confocal elliptical cylinders using hybrid nanofluids. *Numerical Heat Transfer, Part A: Applications*, ***70*(10), 1141-1156.**
10. Ahmad Khan, S., & Altamush Siddiqui, M. (2020). Numerical studies on heat and fluid flow of nanofluid in a partially heated vertical annulus. *Heat Transfer*, ***49*(3), 1458-1490**.
11. Khan, S. A., Siddiqui, M. A., & Husain, S. (2020). Numerical Studies on Thermally Induced Flow of Nanofluid in a Vertical Annulus. In *Proceedings of International Conference in Mechanical and Energy Technology* (pp. 87-100). Springer, Singapore.
12. Batchelor, G. K. (1972). Sedimentation in a dilute dispersion of spheres. *Journal of fluid mechanics*, ***52*(2), 245-268.**
13. Maxwell, J. C. (1873). A treatise on electricity and magnetism (Vol. 1). Oxford: Clarendon Press.

Numerical study on turbulent characteristics of wall jet in a quiescent environment over a plate in motion

V M Behera[1,] and S K Rathore[2]

[1]National Institute of Technology Rourkela, Odisha 769008, India

[2]National Institute of Technology Rourkela, Odisha 769008, India

Corresponding author E-mail: isushilrathore@gmail.com;

Abstract. The present article reports the numerical analysis of a two dimensional turbulent wall jet flow over a horizontal moving plate in a quiescent environment. The simulation is done using the low-Reynolds number turbulence model proposed by Yang and Shih [21]. The turbulent characteristics are observed for various velocity ratio of plate with respect to jet (U_p) from $0-2$ keeping Reynolds number at jet inlet to be constant at 15000. The Reynolds averaged Navier Stokes $(RANS)$ equations are discretized using the finite volume method and solved in steady state condition. The pressure correction technique applied to couple the velocity and pressure terms is SIMPLEC (semi-implicit pressure linked equation - consistent). The variation of jet half-width and the decaying local maximum velocity (U_{max}) for the bottom plate moving with high velocity $(U_p > 1)$ shows dominancy of plate motion over the jet stream in the flow field. The velocity contour diagram shows the velocity distribution in the flow field. The Reynolds stresses induced in the wall proximate regions are directly proportional to the relative motion of the flow field but the magnitude of Reynolds stress away from wall decreases by increasing the plate velocity. The plate motion significantly affects the turbulence characteristics and the flow field in the viscous laminar zone of the wall region.

Keywords. turbulence, low-Reynolds model, wall jet, near wall region, plate velocity

1. Introduction

Turbulent jets have enormous area of applications that includes cooling of electronic devices, cooling in a continuous casting process and hot rolled sheets in steel industries, prevention of icing problem in aircraft at higher altitude [1-3] and so on. One such type of turbulent jet which has numerous applications is the wall jet. If the jet stream coming out of the nozzle flows tangentially over the flat surface, such that boundary layer effect is observed in the fluid layer at the wall, it is called as wall jet flow.

Many experimental work have been performed in past to study turbulent wall jet flow behaviour. The flow parameters like jet half-width, local maximum axial velocity and Reynolds shear stresses variations are studied by Rostamy et al. [4] in an experiment and later by Tang et al. [5]. A large number of researches have been reported on turbulent impinging jet flow over moving surface. The plate motion strongly influences the mean velocity of the fluid stream along with the turbulent parameters like the Reynolds stresses and turbulence kinetic energy as reported by Chattopadhyaya and Saha [6]. Sharif and Banerjee [7] investigated the impinging jet on a moving plate by varying Reynolds number and found that skin friction coefficient does not respond to variation in Reynolds number however it's magnitude appeared to be higher and more stable at lower plate velocity and is lower for higher plate velocity for a given Reynolds number. Kadiyala and Chattopadhyay [8] using transient SST model studied the slot jet impinging over moving plate case and found that plate motion has significantly influenced the flow field as well as the heat transfer rate rather than the Reynolds number.

Numerous computational work with various types of turbulence models such as - Anderson and Spall [9] conducted a numerical investigation using standard k-ε model along with Reynolds stress model, SST model by Shivankar et al. [10], standard k-ε turbulence model by Vishnuvardhanrao and Das [11], Singh et al. [12] worked with wall jet on wavy surface, and so on has been reported. The high Reynolds number turbulent models use wall functions in wall proximity regions and perform poorly in case of flow separation regions. To overcome these drawbacks some researchers developed low-Reynolds number (LRN) turbulence model by modifying the $k-\varepsilon$ two-equation. Later, Kechiche et al [13] investigated turbulent wall jet flow by applying LRN turbulence models of Herrero et al. [14], Nagano and Hishida [15], and Chien [16] and concluded that LRN models can predict the turbulence parameters in near wall regions accurately. Rathore and Das [17-19] worked on turbulent offset and wall jet flow using different turbulence models like standard $k-\varepsilon$, two LRN models (namely LS model [20] and the YS model [21]) and later on included $k-\omega$ SST model and concluded that YS model could readily capture the flow field parameters in the viscous sub-layer region very accurately.

The above literature reveals that many investigations have been performed in past years on turbulent wall jets using various turbulent models. But so far no investigation on turbulent wall jet flow over a moving plate was reported. This paper aims at characterizing the turbulent wall jet flow on a moving plate.

2. Mathematical formulation

2.1. Problem definition

A two-dimensional, steady incompressible wall jet stream in a quiescent atmosphere is considered as shown in Fig. 1. The fluid properties are taken constant and effect of gravity is neglected. The jet inlet velocity is non-dimensionally taken as $U_0 = 1$, and the inlet Reynolds number is 15000. The plate is moving with a constant velocity which can be expressed in the non-dimensional form (U_p) as the ratio of plate to jet velocity. The turbulent flow behavior in the near wall regions are observed for the moving plate and the stationary plate.

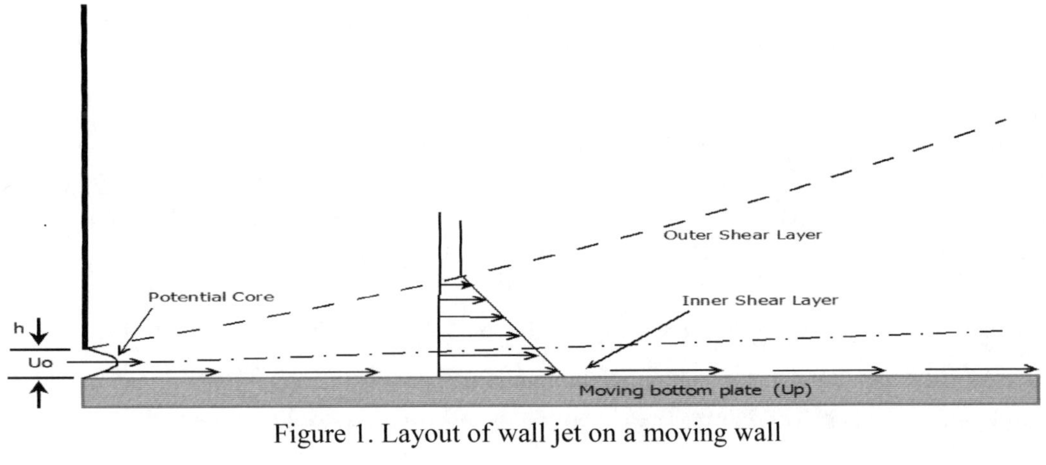

Figure 1. Layout of wall jet on a moving wall

2.2. Numerical methodology

The *RANS* equations are discretized using FVM. The diffusion terms in RANS equations are discretized by central differential scheme whereas for convection terms, power law scheme is applied. The mean velocity gradients and turbulent Reynolds stresses are related by applying Boussinesq model. The eddy viscosity equation as given by Kolmogorov-Prandtl used expresses the turbulent viscosity (ν_t) in terms of turbulent kinetic energy (k_n) and rate of dissipation (ε_n). The pressure and velocity terms are related by adopting SIMPLEC (semi-implicit pressure linked equation - consistent) by Doormaal and Raithby [22]. Algebraic equations are solved by applying Tri-diagonal matrix algorithm by Malalasekhara [23].

2.3. Governing equations

The equations used are given in a dimensionless form:

$$U_i = \frac{\overline{u_i}}{U_0}, X_i = \frac{x_i}{h}, \; P = \frac{\overline{p} - P_0}{\rho U_0^2}, \; \nu_{t,n} = \frac{\nu_t}{\nu}, \; \varepsilon_n = \frac{\varepsilon}{U_0^3/h}$$

Continuity equation:

$$\frac{\partial U_i}{\partial X_i} = 0 \qquad (1)$$

Momentum equation:

$$U_j \frac{\partial U_j}{\partial X_j} = -\frac{\partial\left(P + \frac{2}{3}k_n\right)}{\partial X_j} + \frac{1}{2}\frac{\partial}{\partial X_j}\left[\left(1 + \nu_{t,n}\right)\left(\frac{\partial U_i}{\partial X_j} + \frac{\partial U_j}{\partial X_i}\right)\right] \qquad (2)$$

Turbulent Kinetic energy equation:

$$U_j \frac{\partial k_n}{\partial X_j} = \frac{1}{Re}\frac{\partial}{\partial X_j}\left[\left(1 + \frac{\nu_{t,n}}{\sigma_k}\right)\left(\frac{\partial k_n}{\partial X_j}\right)\right] + G_n - \varepsilon_n \qquad (3)$$

Rate of Dissipation equation:

$$U_j \frac{\partial \varepsilon_n}{\partial X_j} = \frac{1}{Re}\frac{\partial}{\partial X_j}\left[\left(1 + \frac{\nu_{t,n}}{\sigma_k}\right)\left(\frac{\partial \varepsilon_n}{\partial X_j}\right)\right] + G_n C_{1\varepsilon} f_1 \frac{\varepsilon_n}{k_n} - C_{2\varepsilon} f_2 \frac{\varepsilon_n^2}{k_n} + E_n \qquad (4)$$

Where, $E_n = (2\nu_{t,n}/Re^2)(\partial^2 U_i/\partial X_j{}^2)$

Production of shear:

$$G_n = \frac{\nu_{t,n}}{Re}\left(\frac{\partial U_i}{\partial X_j} + \frac{\partial U_j}{\partial X_i}\right)\frac{\partial U_i}{\partial X_j} \qquad (5)$$

Eddy viscosity:

$$\nu_{t,n} = C_\mu f_\mu Re \frac{k_n^2}{\varepsilon_n} \qquad (6)$$

YS model functions:

$$f\mu = (1 + 1/Ret)\,[1 - exp\,(-aRe_y - bRe_y^3 - cRe_y^5)]$$

$$f = f_2 = \sqrt{(Re_t\,(1 + Re_t))}$$

Where, $a = 1.5 \times 10^{-4}, b = 5.0 \times 10^{-7}, c = 1 \times 10^{-10}$

2.4. Boundary conditions

The jet inlet velocity in the non-dimensional form is taken as $U = 1$, $V = 0$ and the turbulent kinetic energy at jet inlet, $k_n = 1.5I^2$ where I is the turbulent intensity taken to be 0.05 and the dissipation rate, $\varepsilon_n = (k_n{}^{3/2} C_\mu{}^{3/4})/0.07$. No-slip condition of $U = U_p$ where $U_p = u_p/U_0$ and $V = 0$ is imposed on horizontal moving plate surface at the bottom of fluid domain and. At the vertical wall, $U = 0$ and $V = 0$ is imposed as the boundary condition. At walls, $k_n = 0$ whereas rate of dissipation is given as $\varepsilon_n = \frac{2}{Re}(\frac{\partial\sqrt{k_n}}{\partial n})^2$, where *n* represents dimensionless coordinate in normal direction. At the entrainment side (top side) of the domain, Neumann boundary condition ($\frac{\partial \Phi}{\partial Y} = 0$) is imposed where, $\Phi = U, V, k_n, \varepsilon_n$ and at the right side exit boundary out flow condition is set which is expressed as $\frac{\partial \Phi}{\partial X} = 0$.

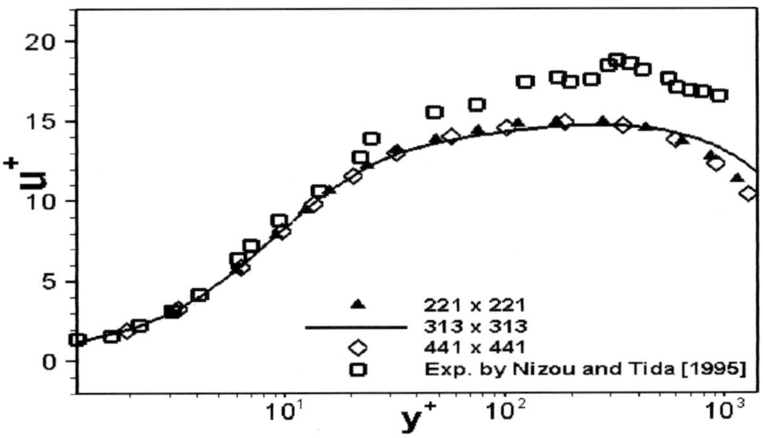

Figure 2. Validation with the experimental result (Nizou and Tida [24])

3. Grid independent test and validation

A non-uniform grid cluster is formed such that highly dense smaller size grid is generated within the jet inlet region as well as in the near wall area and as moved away the grid size increases. The first grid size is chosen such that, the $y^+ \leq 1.5$ in all cases. The fluid domain size is 90×50. The grid sizes considered are $221 \times 221, 313 \times 313$ and 441×441. From Fig 2, it is found that the grid size 313×313 would be the most suitable grid size from the computational cost view. The variation of $u^+ - y^+$ in the stationary plate case is validated with experimental result of Nizou and Tida [24]. The $u^+ - y^+$ profile is in good agreement in near wall region for $y^+ \leq 11$ but it deviates from the experimental result as moved away in logarithmic region.

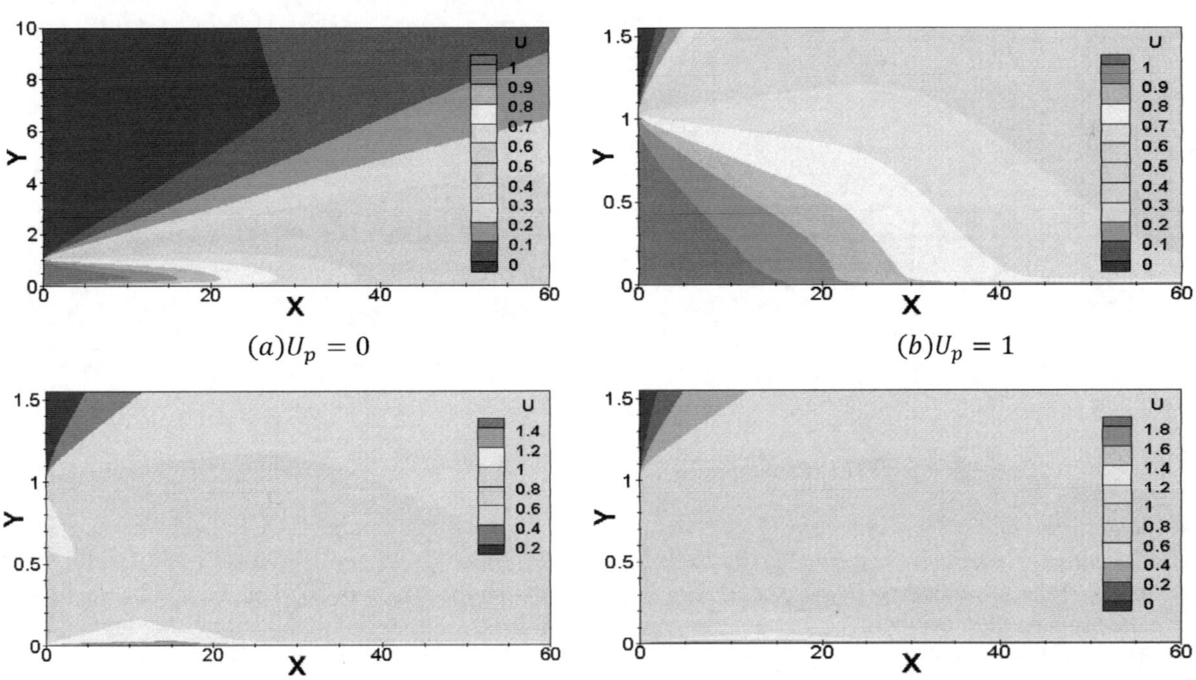

$(a) U_p = 0$ $(b) U_p = 1$

$(c) U_p = 1.5$ $(d) U_p = 2$

Figure 3. Zoomed view of U-component velocity contour in flow field

4. Results and discussion

4.1. Velocity contour diagram

The velocity contours are shown for $U-$ component of velocity, in $X-$ direction for stationary and moving plate with various plate to jet velocity ratios ($U_p = 1, 1.5$, and 2) in Fig. 3. The enlarged view of velocity contour illustrates the effect of moving plate in $X-$ direction. It can be observed from the contour diagram that $U-$component of velocity is maximum at the jet inlet in the stationary plate case whereas for the moving plate the maximum velocity is identified at the surface for the moving plate with $U_p = 1.5$, and 2.

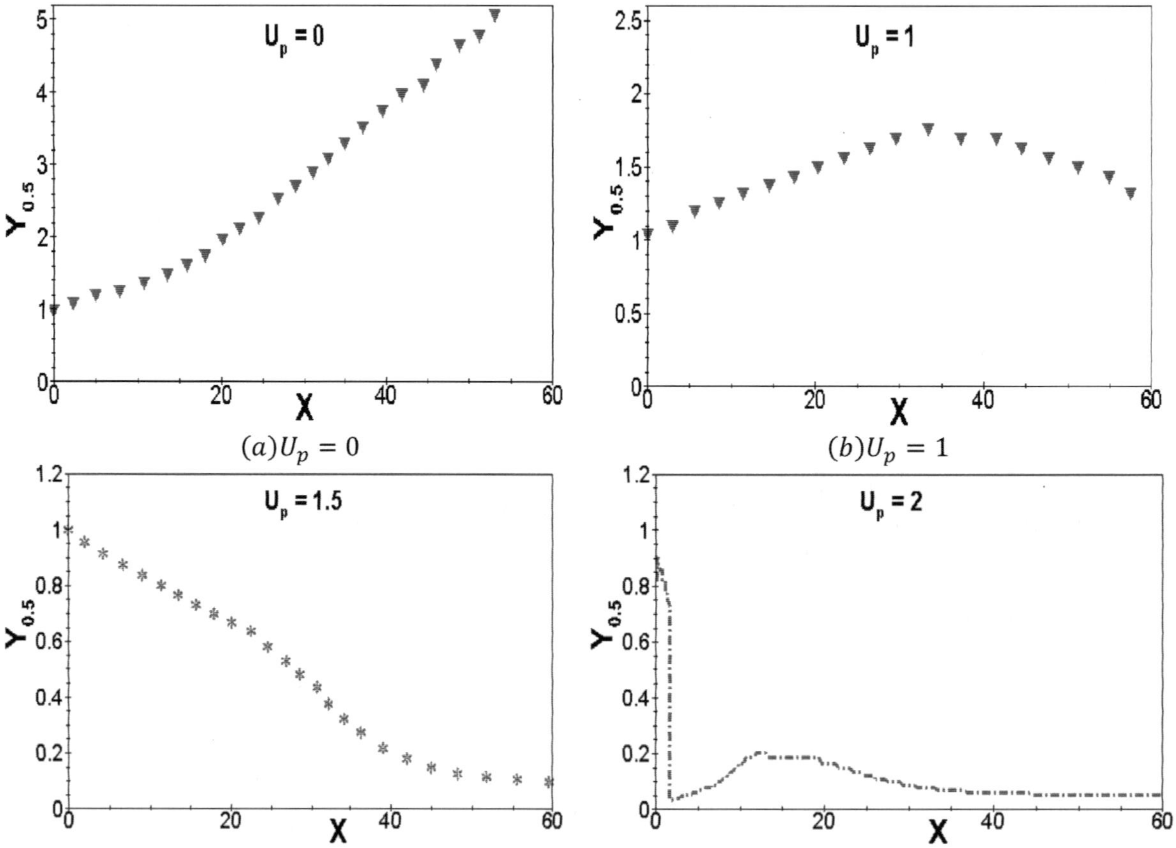

$(a) U_p = 0$ $(b) U_p = 1$

$(c) U_p = 1.5$ $(d) U_p = 2$

Figure 4. Axial variation of jet half-width for various cases of plate velocity

4.2. Spreading of jet half-width ($Y_{0.5}$)

Spreading of jet can be traced by jet half-width ($Y_{0.5}$) which gives the location of local velocity in vertical direction where, $U = U_{max}/2$. Figure 4, represents the variation $Y_{0.5}$ in the axial direction. It is observed that in a stationary plate, the jet half-width increase monotonically along the downstream. But for the cases of higher plate velocity where, $U_p \geq 1$, different patterns are observed. In case of moving plate ($U_p = 1$), the jet half-width increases to a peak point and then it decreases slowly at a distance far away in X-direction. For $U_p = 1$, the jet half-width decreases gradually in axial direction to a minimum point after which the rate of decrement becomes very slow. Whereas, for the moving plate ($U_p = 2$), the $Y_{0.5}$ shows a sudden fall near jet inlet. It then increases to a peak point and decreases in axial direction. The combined effect of plate velocity and jet velocity decides the profile of axial spreading of jet.

4.3. Distribution of local maximum velocity (U_{max}) in axial direction

The variation of (U_{max}) velocity in axial direction is shown in Fig. 5. For a stationary plate the velocity possessed at the jet inlet is highest but on moving away from the inlet in axial direction, the velocity decreases gradually due to interaction with the surrounding fluid. But for higher plate velocity cases where $U_p > 0.5$ the velocity traced in the profile is same as that of plate velocity. It is because velocity of fluid is highest at the plate surface due to no-slip condition imposed on the fluid layer in direct contact with the plate surface. Hence, for the plate to jet velocity ratio (U_p)≥ 1, the magnitude of velocity possessed by the fluid layer at the moving plate surface is maximum. It is observed further that the profile remains invariant while moving away in axial direction.

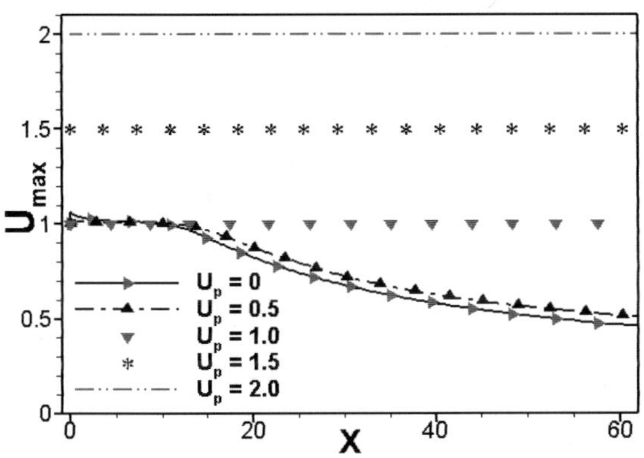

Figure 5. Axial variation of local maximum velocity

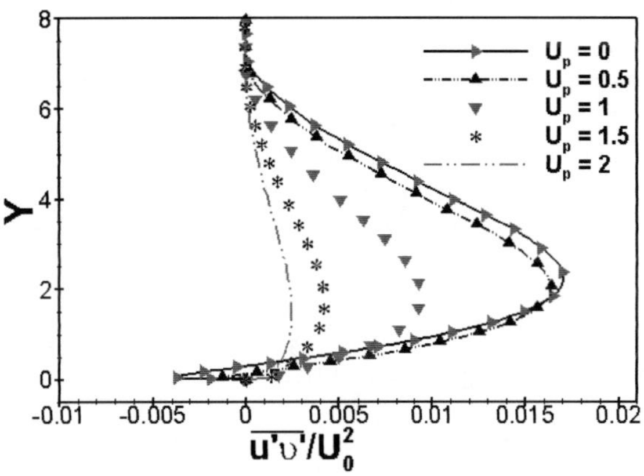

Figure 6. Reynolds shear stress variation

4.4. Distribution of turbulent Reynolds shear stress

The variation of Reynolds shear stress ($\overline{u'v'}/U_0^2$) along Y axis at axial location of $X = 30$ is shown in Fig. 6. At walls, the fluctuating velocity components ceases and hence the Reynolds shear stress ($\overline{u'v'}/U_0^2$) gets zero. A negative stress is developed near the plate surface in case of stationary plate. A similar negative profile can be seen for plate moving with $U_p = 0.5$ for the axial positions. It then increases and changes sign as moved away from the wall in $Y-$ direction, and becomes maximum in the region of local maximum mean velocity. Whereas, for the other cases of moving plate U_p= 1, 1.5 and 2, the Reynolds shear stress developed has a positive profile closer to wall which follows similar trend as that of stationary plate when moved away in vertical direction. When moved away towards the upper free shear layer, the stresses decreases and becomes minimum above the upper shear layer.

5. Conclusions

The turbulent flow characteristics for a wall jet stream flowing over a flat plate moving in same direction as that of jet stream have been investigated numerically using YS turbulence model proposed by Yang and Shih. The consequences of moving plate on the flow field are observed from the variation of velocity contours, jet half-width, local maximum velocity and distribution of Reynolds shear stress.

From the velocity contour, jet half-width ($Y_{0.5}$) and the maximum local velocity diagram (U_{max}) it was found that the maximum velocity is developed in the flow field in the fluid layer next to the

moving wall in case of higher plate to jet velocity ratio cases. It is because of the no slip condition of $U = U_p$ imposed on the fluid layer next to the solid moving plate. The jet half width diverges away from the wall in a stationary plate and for plate moving with lower velocity ratio. Whereas for higher plate velocity cases ($U_p > 1$), the jet half-width appears closer to wall which indicates that fluid speed is maximum at the wall. This indicates that flow field is greatly influenced by the velocity of the horizontal moving plate.
The Reynolds shear stress profile suggests that it is dominant in the flow field away from wall and in the flow region closer to wall surface it gets weaker and is zero at the wall for all cases. However, the Reynolds shear stress profile appears to possess a positive value closer to wall region in a moving plate. The magnitude of Reynolds shear stress decreases with increase in the velocity ratio(U_p). It may be because in stationary plate case and for $U_p = 0.5$, the mean velocity is maximum in the flow field away from the wall as verified by the jet half-width curve but for the cases of $U_p \geq 1$ the maximum velocity lies at the plate surface due to the no-slip condition hence the magnitude of $(\overline{u'v'}/U_0^2)$ decreases with the increase in velocity of plate. As the Reynolds shear stress is weaker within the thin viscous sub-layer region, the opposing shear force developed in high magnitude velocity field in this region is less in magnitude.
It is concluded from the observation of the jet half-width, local maximum velocity distribution and the turbulent Reynolds shear stress distribution that the flow field and turbulent characteristics are dominated by motion of the plate.

ACKNOWLODGEMENT

We gratefully acknowledge Science and Engineering Research Board (SERB), Department of Science and Technology, Government of India, for providing financial support to our research.

References

[1] Sengupta J, Thomas B G and Wells M A 2004 The use of water cooling during the continuous casting of steel and aluminum alloys *Metallurgical and Materials Trans. A* ***36*** pp. 187-204
[2] Al Ali A R and Janajreh I 2015 Numerical simulation of turbine blade cooling via jet impingement *Int. Conf. Applied Energy*, *Energy Procedia* **75** pp. 3220 – 3229
[3] Lund T S 1986 Augmented thrust and mass flow associated with two–dimensional jet reattachment *JIAA* **24** pp. 1964 –1970
[4] Rostamy N, Bergstrom D J Sumner D and Bugg J D 2011 The effect of surface roughness on the turbulence structure of a plane wall jet *Phy. Fluid* **23** pp. 31-36
[5] Tang Z, Rostamy N, Bergstrom D J, Bugg J D and Sumner D 2015 Incomplete similarity of a plane turbulent wall jet on smooth and transitionally rough surfaces *J. Turbulence* **16**, pp. 1076-1090
[6] Chattopadhyay H and Saha S K 2002 Simulation of laminar slot jets impinging on a moving surface *J. Heat Transfer* **124** pp. 1049-1055
[7] Sharif M A R and Banerjee A 2009 Numerical analysis of heat transfer due to confined slot-jet impingement on a moving plate *App. Thermal Engg.* **29** pp. 532–540
[8] Kadiyalaa P K and Chattopadhyay H 2018 Numerical analysis of heat transfer from a moving surface due to impingement of slot jets *Heat Transfer Engg.* **39** pp. 98-106
[9] Anderson E A and Spall R E 2001 Experimental and numerical investigation of two-dimensional parallel jets *J. Fluids Engg.* **123** pp. 401-406
[10] Shivankar S, Randive P R and Pati S 2020 Effects of undulated wall on the hydrodynamic and thermal transport characteristics of turbulent jet *Int. J. Thermal Science* **152** pp. 1-19
[11] Vishnuvardhanarao E and Das M.K 2008 Computational study of heat transfer in a conjugate turbulent wall jet flow at high Reynolds number *J. Heat Transfer* **130** pp. 1-7

[12] Singh T P, Kumar A and Satapathy A K 2020 Numerical study to enhance the heat transfer using sinusoidal wavy surface for turbulent wall jet *Numerical Heat Transfer A: Applications*, **77**, pp. 179-198

[13] Kechiche J, Mhiri H, Palec G L and Bournot P 2004 Application of low Reynolds number k–ε turbulence models to the study of turbulent wall jets *Int. J. Thermal Science* **43** pp. 201-211

[14] Herrero J, Grau F X, Grifoll J and Giralt F A 1991 Near wall k–ε formulation for high Prandtl number heat transfer *Int. J. Heat and Mass Transfer* **34** pp. 711-721

[15] Nagano Y and Hishida M 1987 Improved form of the k–ε model for wall turbulent shear flows *J. Fluids Engg.* **109** pp. 156-160

[16] Chien K Y 1982 Predictions of channel and boundary layer flows with a low-Reynolds number turbulent model *AIAAJ.* **20** pp. 33-38

[17] Rathore S K and Das M K 2013 Comparison of two low-Reynolds number turbulence models for fluid flow study of wall bounded jets *Int. J. Heat and Mass Transfer* **6** pp. 365-380

[18] Rathore S K and Das M K 2015 A comparative study of heat transfer characteristics of wall-bounded jets using different turbulence models *Int. J. Thermal Sci.* **89**, pp. 337-356

[19] Rathore, S K and Das M K 2016 Numerical investigation on the performance of low-Reynolds number k-ε model for a buoyancy-opposed wall jet flow *Int. J. Heat and Mass Transfer* **95** pp. 636 – 649

[20] Launder B.E and Sharma B I 1992 Application of the energy dissipation model of turbulence to the calculation of flow near a spinning disc *Lett. Heat Mass Transfer* **1** pp. 131–138

[21] Yang Z and Shih T H 1992 New time scale based k – ε model for near wall turbulence *NASA Lewis Research Center* **7** pp.1191–1198

[22] Doormaal J P V and Raithby G D 1984 Enhancements of the SIMPLE method for predicting incompressible fluid flows *Numerical Heat Transfer* **7** pp. 147-163

[23] Versteeg, H.K., Malalasekhara 2007 W., "*An Introduction to Computational Fluid Dynamics - The Finite Volume Method* (London: Pearson Education Limited)

[24] Nizou P Y and Tida T 1995 Transferts de chaleur et de quantit´e de mouvement dans les jets pari´etaux plans turbulents *Int. J. Heat and Mass Transfer* **38** pp. 1187-1200

IOP Conf. Series: Materials Science and Engineering 1146 (2021) 012020 doi:10.1088/1757-899X/1146/1/012020

On reducing lift and drag forces of a stationary circular cylinder using side-by-side pair of inward rotating cylinders

Uddhav S Pagar[1], Jijo Derick Abraham[1] and Trushar B Gohil[1*]

[1]Department of Mechanical Engineering, Visvesvaraya National Institute of Technology, Nagpur-440010, India

*E-mail address for correspondence: trushar.gohil@gmail.com

Abstract. Effect of side-by-side pair of inwardly rotating circular cylinders on the nature of drag and lift forces of the stationary circular cylinder is numerically investigated for Reynolds number Re=100. Non-dimensional rotation rates of cylinders in consideration are α= 0, 1, 3, and 5. Center to center gap spacing between rotating cylinders is taken as 6D (D is cylinder diameter), and the stationary cylinder is centrally placed in the gap. For the case of two side-by-side inward rotating cylinders, above a certain critical rotation rate, Chan *et al.* [1] reported the formation of a virtually elliptical body accompanied by vortex suppression. Drag on both the cylinders became almost zero after the formation of a virtually elliptical body. The present study aims to use the same concept of a virtually elliptical body to reduce the drag and lift forces on a stationary circular cylinder. At α =3, complete suppression of vorticity is observed, followed by the formation of a virtual elliptical body at α =5. With an increase in α from 0 onwards, the drag coefficient for the stationary cylinder reduces and becomes zero at α =5.

1. Introduction

Investigation of flow past rotating cylinders is of great significance from the point of view of wake dynamics and applications in flow control. Rotation of the cylinder substantially modifies the wake's behaviour compared to the stationary case due to the alteration in the boundary layers (before separation) and the detached shear layers. Numerous numerical and experimental studies have been carried out to document the flow behaviour for single and multiple rotating cylinders. Mittal and Kumar [2] numerically investigated the flow past rotating cylinder for low Reynolds number equal to 200. In their finite element based simulation, they considered the range of rotation rate α from 0 to 5. α is the non-dimensional rotational rate and is defined as the ratio of the surface velocity of the cylinder to the free-stream velocity, U_∞. For $\alpha < 1.9$, von Karman type vortex shedding was observed behind the cylinder. As the α was increased from 0, the wake became narrower, and the frequency of vortex shedding reduced. Finally, at $\alpha = 1.91$, vortex shedding was completely suppressed. Vortex shedding suppression prevailed for the range of $1.91 < \alpha < 4.34$, beyond which vortex shedding again resumed for $4.34 < \alpha < 4.7$. In this second range of instability, interestingly, only counter-clockwise vortices were observed compared to von Karman vortex street in the first range of $\alpha < 1.91$. Pralits *et al.* [3] analyzed the mechanism for first and second shedding modes for the single rotating cylinder at Re =100. They attributed suppression of vortex shedding around $\alpha = 1.9$ to the shear layer weakening for flow in the wake. The core of instability for the second shedding mode (for the range of $4.34 < \alpha < 4.7$) was observed in the advection of vorticity of flow from the lower portion to the stagnation point,

IOP Conf. Series: Materials Science and Engineering 1146 (2021) 012020 doi:10.1088/1757-899X/1146/1/012020

where vorticity will gather and then shedding happens. Kumar *et al.* [4] studied experimentally the flow past a single rotating cylinder at Re = 200, 300, and 400 in the range of $0 < \alpha < 5$. Their results agreed well with the numerical analysis of Mittal and Kumar [2], giving the first experimental evidence of second shedding mode with single signed vortices. The Strouhal number measurements in the second shedding mode showed a high dependence of vortex shedding frequency on rotation rates α compared to the Reynolds number. Munir *et al.* [5] carried out a three-dimensional numerical analysis below Re = 500 and defined different flow regimes for flow around a rotating cylinder. They observed that flow above Re =250 is inherently three-dimensional, and hence vortex suppression cannot be seen above Re =250. In three-dimensional flow, finger-shaped (elongated) and ring-shaped vortices were reported.

Many researchers have studied the side-by-side arrangement of rotating cylinders recently to check whether vortex suppression exists for such configuration or not. Yoon *et al.* [6] numerically studied flow past side-by-side rotating cylinders at Re =100 for $0 < \alpha < 2$ and gap spacing of 0.2D to 3D. For inward rotation case (i.e., top cylinder rotating clockwise and bottom one rotating counter-clockwise), they found that vortex suppression occurs for all gap spacing. They also mentioned a decrease in the mean drag coefficient and an increase in the mean lift coefficient as the rotation rate α is increased. Chan *et al.* [1] extended the work of Yoon *et al.* [6] by considering the effect of outward rotation (i.e., top cylinder rotating counter-clockwise and bottom one rotating clockwise) and increasing the range of rotation to $0 < \alpha < 5$ for Re = 100, 150 and 200. For inward rotation, α_{crit} (critical value of α) where vortex shedding gets completely suppressed, monotonically increased with an increase in gap spacing. The formation of a virtually elliptical body was reported beyond a certain critical rotation rate, wherein both cylinders get enveloped into elliptical shape fluid layers, and drag on both cylinders tend to zero. In-phase (i.e., asymmetric) vortex shedding mode was observed for inward rotation case before vortex suppression, whereas for outward rotation, anti-phase (i.e., symmetric) vortex shedding was observed. Kumar *et al.* [7] carried out experimental investigation for Re =100, 200, 300, 400 and 500 in the range $0 < \alpha < 5$ for three different gap spacing. They found good agreement with the results of Yoon *et al.* [6] and Chan *et al.* [1]. Vortex shedding suppression was quite clearly visible for the inward rotation case for all the cases considered, but for outward rotation of cylinders, vortex shedding was clearly evident only for the case of Re =100. Supradeepan and Roy [8] prepared a flow regime map for the side-by-side rotating cylinders for Re =100, α =0.5, 1.0, 1.5, and non-dimensional spacing of 1.1D to 3.5D. They defined eight different flow regimes based on vorticity contours, $\lambda 2$ criterion, and force coefficients for the considered range of parameters. Dou *et al.* [9] investigated the effect of the rate of cylinder rotation and Reynolds number (Re) on the stability regions of the flow. For stability analysis, they used the energy gradient theory by which energy gradient function K is defined. Through their work, they concluded that K is dominant in the separated shear layer region past two cylinders and hence is prone to instability. Schulmeister *et al.* [10] showed that drag on a stationary cylinder could be controlled using small rotating cylinders placed downstream to the cylinder surface.

Thus, we conclude that a side-by-side pair of rotating cylinders leads to the formation of a virtually elliptic body beyond a certain critical rotational speed. Whether such configuration can reduce the forces over a stationary body is the motivation behind the present study. The present study aims at finding the effect of the rotational rate of pair of cylinders on the force coefficients and nature of the wake of the single stationary cylinder.

2. Numerical details and methodology

The computational domain for the considered problem is as shown in figure 1. A stationary cylinder (C2) is centrally placed between the inward rotating pair of cylinders (C1 rotating clockwise and C3 rotating anticlockwise). All the cylinders are of identical diameter equal to D. Other dimensions of the computational domain are defined in terms of D and are shown in figure 1.

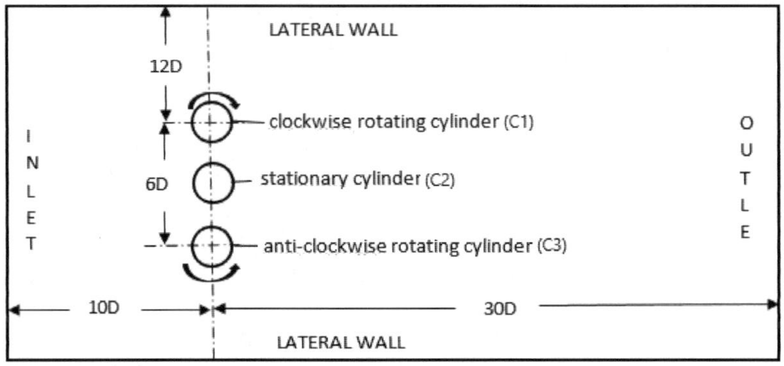

Figure 1. Computational domain (Dimensions not to scale).

Reynolds number is set to *Re* =100 for all considered cases. No-slip boundary condition is used on the surface of stationary cylinder C2, and rotational velocity of rotating cylinders C1 and C3 is defined in terms of α (α is the ratio of cylinder surface velocity to the free stream velocity, U_∞). α is given values equal to 0, 1, 3, and 5. A uniform velocity U_∞ is prescribed at the inlet. Free slip boundary conditions have been applied on the lateral walls of the computational domain. Zero velocity gradient and zero pressure boundary conditions are used at the outlet. A time step of 0.001 seconds is considered for all cases. The total number of mesh elements in the domain are 185K, with 200 nodes along the circumference of each cylinder and near-wall spacing of 0.001. Domain dimensions and mesh elements are selected based on the available literature.

Numerical simulation is performed by solving the unsteady incompressible Navier-Stokes equations. Finite volume-based open-source CFD tool kit OpenFOAM is used to solve the above governing equations. The PIMPLE algorithm, which is a combination of the SIMPLE (semi-implicit method for pressure-linked equations) algorithm and the PISO (pressure-implicit split-operator) algorithm, is incorporated to solve for velocity and pressure. Convection and diffusion terms are discretized using the second-order central differencing scheme.

3. Validation

For validation purpose, simulations were performed for two cases; the first case was a single rotating cylinder and the second case was a pair of inward rotating cylinders.

Table 1. Comparison of C_{Lmean} of a single rotating cylinder for different α.

α	C_{Lmean} (Present)	C_{Lmean} (Results from Mittal *et al.* [2])
0	0	0
1	-2.55	-2.5
3	-10.19	-10.08
5	-27.02	-26.90

For a single rotating cylinder case, the value of the mean lift force coefficient C_{Lmean} was compared for *Re* =200 with the results given by Mittal *et al.* [2] for varying non-dimensional rotation rate, α (see table 1). Excellent agreement was found quantitatively as well as qualitatively.

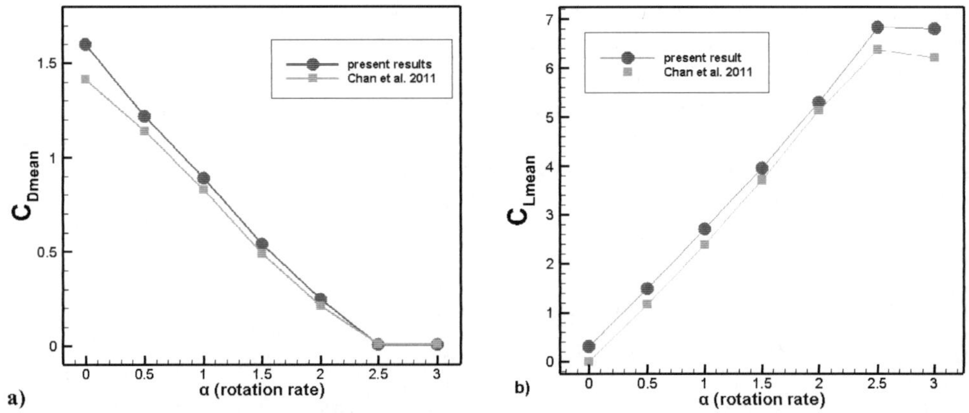

Figure 2. Comparison of C_{Dmean} and C_{Lmean} of the top cylinder with the results by Chan *et al.* [1] for the case of inward rotating cylinders.

For the case of two inward rotating cylinders, mean lift and drag force coefficients (C_{Dmean} and C_{Lmean}) of the top cylinder were compared at Re =100 and non-dimensional gap spacing of 1 for different values of α (see figure 2). Good agreement was found with the results by Chan *et al.* [1].

4. Results and discussion

Figure 3 shows the instantaneous streamlines and vorticity contours for α = 0, 1, 3, and 5 at Re = 100 and gap spacing of 3D between two consecutive cylinders. For α = 0, all 3 cylinders are stationary.

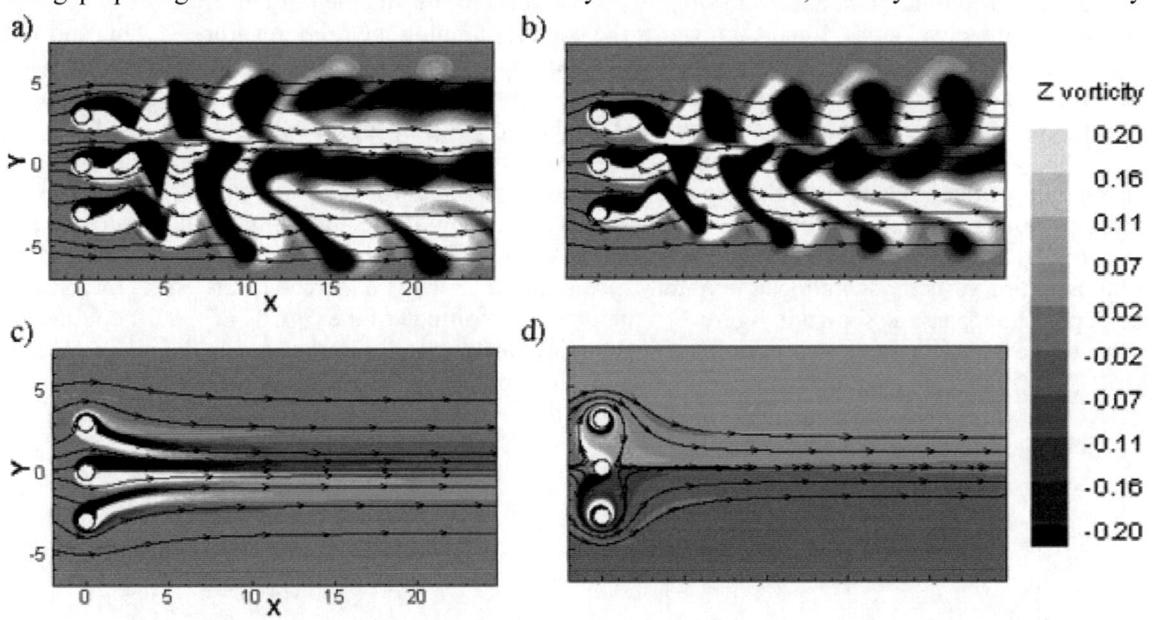

Figure 3. Instantaneous streamlines and vorticity contours for a) α=0 b) α=1 c) α=3 d) α=5.

The interaction of vortices is clearly evident in figure 3a. Due to asymmetry in the upstream velocity profile for the top (C1) and bottom cylinder (C3), vortex shedding starts prior to the middle cylinder (C2). Once the vortex shedding starts behind the C2, the merging of vortices of the same sign is observed. Merging follows periodic behaviour. First, the identical signed vortices of three cylinders merge, later merging happens for C2 and C3, followed by combining of vortices from three cylinders again. Lastly, the same signed vortices of C1 and C2 interact. While the vortices from any two cylinders interact, the vortex shedding behind the third cylinder follows the von Karman type of pattern. This periodic behaviour of merging of the same signed vortices modulates the sinusoidal time series of force coefficients of C2, showing the presence of secondary frequency as shown in figure 4a. Vortex shedding for α=1 follows the same pattern (figure 3b) with a slight increase in the secondary frequency accompanied by a decrease in the amplitude of lift forces. Reduction in the lift forces is due to the suppression of vorticity resulted because of cylinder rotation.

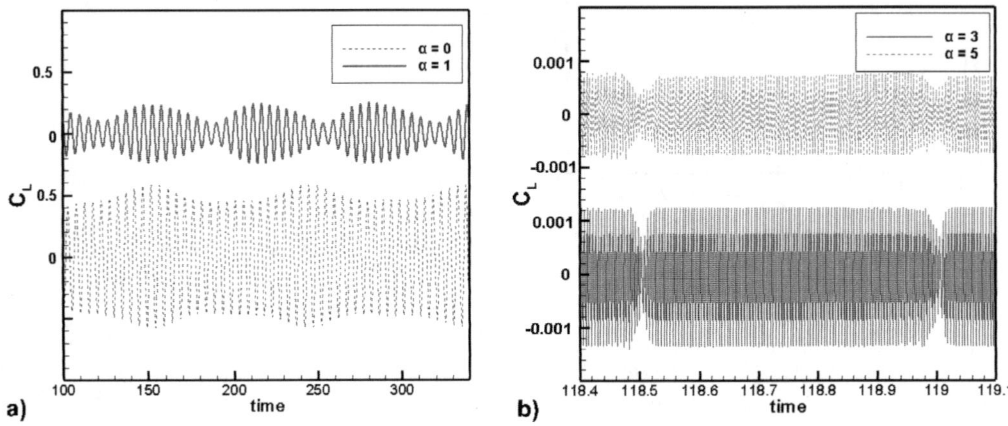

Figure 4. Time series of lift coefficients of middle cylinder C2. a) $\alpha = 0$ and 1 b) $\alpha = 3$ and 5

The nature of vorticity and streamlines changes significantly at $\alpha = 3$ and 5, as shown in figures 3c and 3d. Increased rotation rate leads to suppression of vorticity behind the C1 and C3. Rotation causes the shifting of the wake of C1 and C3 towards the center, pumping the fluid towards C2. This shifting of wake compresses the shear layer originating from the middle cylinder. The extent of compression is sufficient to completely suppress the vortex shedding behind the C2 at $\alpha = 3$. Also, the amount of fluid passing over the C2 reduces as the streamlines start shifting towards the outer edges of C1 and C3. At $\alpha = 5$, the system of three cylinders is entirely enveloped by elliptical-shaped fluid layers, not allowing any amount of fluid to enter into the gap and hence confirms the formation of a virtually elliptic body. Also, the vortex shedding is completely suppressed. Bypassing of incoming fluid and suppression of vorticity dramatically reduces the forces acting on the C2. It was expected that, for $\alpha = 3$ and 5, suppression of vorticity should lead to a non-fluctuating time series of force coefficients. But exactly the opposite happens, as shown in figure 4b. Though the amplitude of the signals of the lift coefficient is small, the frequency seems to have increased hugely. Intentionally, we have kept the range of time small (i.e., from 118.4 seconds to 119.1 seconds) to improve the visibility of the nature of signals. Otherwise, with increasing time range, signals look like a straight line. Power spectra of signals of lift coefficient for $\alpha = 3$ and 5 are shown in figure 5.

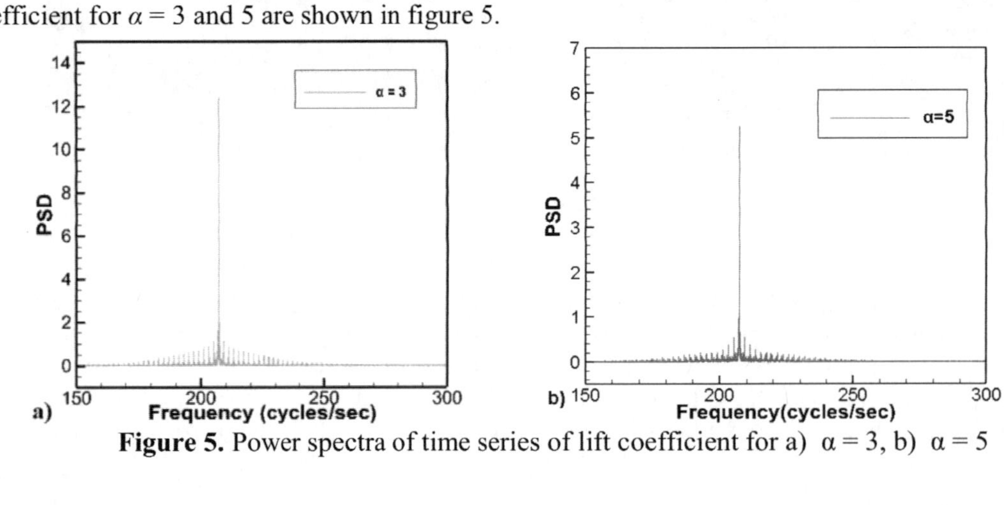

Figure 5. Power spectra of time series of lift coefficient for a) $\alpha = 3$, b) $\alpha = 5$

The peak at around 205 cycles per second confirms that C2 is subjected to this high frequency. These vibrations are attributed to localized opposite signed vorticity being imposed on C2 by the rotation of C1 and C3.

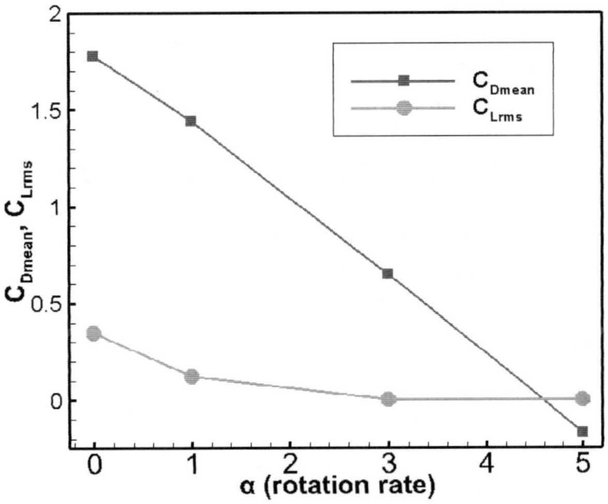

Figure 6. Variation of C_{Dmean} and C_{Lrms} of C2 with increasing α

Figure 6 shows the variation of force coefficients with increasing cylinder rotation rate. The mean coefficient of drag (C_{Dmean}) almost shows a linear reduction with α. It is important to note that C_{Dmean} becomes negative for $\alpha = 5$, which signifies pressure building up downstream of C2. The root-mean-square coefficient of lift (C_{Lrms}) also reduces with increasing α and becomes almost zero at $\alpha = 3$, remaining constant after that.

5. Conclusion

The numerical simulation was performed to understand the effect of side-by-side pair of inwardly rotating circular cylinders on the nature of drag and lift forces of the stationary circular cylinder placed at the center of pair of rotating cylinders. Reynolds number in consideration was 100, and non-dimensional rotation rates of cylinders were varied for $\alpha = 0$, 1, 3, and 5 with gap spacing of 6D between rotating cylinders. It is observed that both the mean drag coefficient (C_{Dmean}) and root-mean-square coefficient of lift (C_{Lrms}) for the middle cylinder (C2) reduce with increasing rotation rate. The virtually elliptic body is evident at $\alpha = 5$. C_{Dmean} reduces linearly with α and tends to zero for $\alpha = 5$. C_{Lrms} becomes almost zero for $\alpha = 3$ and remains constant after that. Time series of lift coefficients of C2 at $\alpha = 0$ and 1 shows the existence of two frequencies; primary frequency is associated with vortex shedding behind C2, whereas secondary frequency reflects the periodic merging of the same signed vortices of C2 and the rotating cylinders. Careful observation of time series of lift coefficients for $\alpha = 3$ and 5 reveals that C2 is subjected to periodic fluctuations of frequency as high as 210 cycles per second, which are highly undesirable.

References

[1] Chan A S, Dewey P A, Jameson A, Liang C and Smits A J 2011 Vortex suppression and drag reduction in the wake of counter-rotating cylinders *J. Fluid Mech.* **679** 343–82
[2] Mittal S and Kumar B 2003 Flow past a rotating cylinder *J. Fluid Mech.* **476** 303–34
[3] Pralits J O, Brandt L and Giannetti F 2010 Instability and sensitivity of the flow around a rotating circular cylinder *J. Fluid Mech.* **650** 513–36
[4] Kumar S, Cantu C and Gonzalez B 2011 Flow past a rotating cylinder at low and high rotation rates *J. Fluids Eng. Trans. ASME* **133** 41201-1-41201–9

[5] Munir A, Zhao M, Wu H and Lu L 2019 Numerical investigation of wake flow regimes behind a high-speed rotating circular cylinder in steady flow *J. Fluid Mech.* **878** 875–906
[6] Yoon H S, Chun H H, Kim J H and Ryong Park I L 2009 Flow characteristics of two rotating side-by-side circular cylinder *Comput. Fluids* **38** 466–74
[7] Kumar S, Gonzalez B and Probst O 2011 Flow past two rotating cylinders *Phys. Fluids* **23**
[8] Supradeepan K and Roy A 2015 Low reynolds number flow characteristics for two side by side rotating cylinders *J. Fluids Eng. Trans. ASME* **137** 1–10
[9] Dou H S, Zhang S, Yang H, Setoguchi T and Kinoue Y 2019 Effect of rotational speed on the stability of two rotating side-by-side circular cylinders at low reynolds number *arXiv* **27** 125–34
[10] Schulmeister J C, Dahl J M, Weymouth G D and Triantafyllou M S 2017 Flow control with rotating cylinders *J. Fluid Mech.* **825** 743–63

Performance enhancement of modified double slope passive solar still using different water-based nanofluids

Suresh Kr. Patel[1*], Vinay Kr. Singh[2], Mahindra Singh[1], Deepak Singh[1], Dhananjay singh[1]

[1]Department of Chemical Engineering, IET, Lucknow, U.P.-226021, India
[2]Department of Applied Chemistry, DSMNRU, Lucknow, U.P.-226017, India
E-Mail: suresh.ieju@bbdnitm.ac.in;

Abstract. The aim this study is to analyze the performance of a modified double slope passive solar still (MDSPSS) with and without using water-based nanofluids along with harvesting solar energy. For this, three types of water-based nanofluids used as Aluminum Oxide (Al_2O_3), Zinc Oxide (ZnO) and Copper Oxide (CuO) at different concentrations. This might be an alternative energy source for the existing conventional type of desalination. Consequently, analyzed physical and finalized the suitable water nanofluid for further experimental work. This paper describes a unit which is designed on simple principles of evaporation and condensation with no electricity being used. Thus we harness the heat radiated by the sun to accomplish the evaporation process. After that the condensation process will be start with natural temperature difference between heated and non-heated segment. This modified system with nano-fluid-based as a heat source has made to observed that the effect of different operating modes at climate conditions (month of March 2019) of Lucknow city, India. After experimental result comparison, found that the Aluminum Oxide (Al_2O_3) nanofluid has 36.55% higher productivity and through Copper oxide (CuO) and Zinc Oxide (ZnO), water nanofluid has 15.26% and 21.75% higher production rate respectively concerning still water. The payback period is also calculated for this modified system with nanofluids as 121 days, which is less than that of the unit with normal water.

Keywords: *Solar desalination, Nanofluids, Solar energy, Potable water, Productivity.*

1. Introduction

Today, water scarcity (or water crises) has become a worldwide issue and almost every country faces the problem of water crises. Water resources around the world have been continuously exploited and polluted. Energy crises, unsustainable cultivation, human anthropogenic activities, changes of dynamics of climate change; clouds formation, and hydrological cycle (or linkages among these) are the main cause of widespread water scarcity [1,2]. Because of the current circumstances, solar energy-based solar desalination system drags everyone's attention to address the global water shortage problems and providing a clean and sustainable option to the users without a carbon footprint on the environment [3,4] Solar distillation unit is a very simple and ecumenical for the production of potable/drinking water by solar energy. The basic working principle of this technique is evaporation process from basin surface to water and condensation process from water vapor to covered glass surface (lower part), so its production rate depends on both phenomena. It can be very useful especially in desert areas or where electricity not available. This technique is comparatively shameful to construct, maintenance is very less and reduces the shortage of drinking water supply. Due to its production rate, it using very limited, so it is the drawback of this process. [5, 6].
In the time-span of the last four-five decades, keeping all these aspects of solar stills, scientists and researchers working on various experimental and numerical techniques to augment the working temperatures and thus productivity rates. Therefore, different techniques and augmentation methods are employed on solar stills to enhance its productivity through deployment of inside/outside auxiliary systems to increase the heat (or temperature) input inside the solar stills such as PV/T collector [7,8], phase change materials (PCMs) [9,10], external reflector [11], PCMs with solar collector [12], additional condenser [13], and application of nanofluids [14,15,16,17]. Sahota and Tiwari [18] carried

out exergoeconomic and enviroeconomic analysis for the integrated solar still (double slope) system with and without helically coiled as heat exchanger using aluminum oxide (Al_2O_3), titanium oxide (TiO_2) and copper oxide (CuO) water-based nanofluids. Authors are observed that the exergoeconomic parameter as 0.151 kWh/Rs. (system without heat exchanger) and 0.103 kWh/Rs. (system with heat exchanger) laden with water (base fluid), respectively, at i = 4% and n = 50 years of the systems. Hassan et al. [19] revealed that the exergoeconomic and exergoenvironmental parameters for conventional solar distiller (accommodate sand in the basin) integrated with parabolic type collector (trough) are found to be 6.17 kWh/$ (at 12% interest rate and 10 years lifetime of the system) and 5.9 ton CO2/year, respectively, among different studied solar systems with their specifications. Sharshir et al. [20, 21] examined (experimentally) the exergoeconomic and exergoenviroeconomic explication of modified solar distiller loaded with Fe_3O_4 and CuO nanoparticles and with cotton hung pad in place of wick. Authors examined that the exergoeconomic, exergoenvironmental, and exergoenviroeconomic parameters are found as 6.939 W/$, 3.028 ton CO_2/year, and 43.906$/year, respectively, for modified solar still laden with CuO nanoparticle.

The unique aspect of this study is the utilization of different nanofluids to increase heat the water initially from the basin such that the water being heated by the sheet evaporates at an increased rate due to incoming water having an elevated temperature. The project depends upon the amount of sunlight that hits the black bottom surface. Considering this dependency the setup is set to pivot about in its given radial area, to obtain maximum sunlight throughout the day

Therefore, in this research, a detailed and complete conducted/performed on the modified double slope passive solar still (MDSPSS) with adding different metal oxides to enhance the yield and performance of the system. This study is carried out based on experimental data observed for the climatic condition of Lucknow (Latitude/Longitude: 26°84′ N/80°94′ E), Uttar Pradesh, India along with comparative analysis (with and without nanofluids). Furthermore, performance analysis and other aspects such as the cost of produced water (distilled) per litre along with the payback time is calculated and discussed in detail for the aforesaid systems.

2. Material and Methods

2.1. Fabrication and working of modified solar still

In this present work, the experimental setup of the modified solar still (single basin-type) is fabricated to detect the effect of incorporating modifications on the performance indicator such as enhanced productivity. The material used for the fabrication was Fiber Reinforced Plastic (fiber-reinforced polymer (NEMA Grade G-10/EP3) reinforced with glass fibers) and Acrylic (polymethyl methacrylate-PMMA). The principle behind using two material for construction is stem from the concept of ‘Equivalent Thickness’ for the equal rate of heat transfer/loss from the solar still system. The basin and north wall of solar still painted with black color are fabricated from a 5 mm thick FRP sheet. The schematic representation and photograph of MDSPSS are shown in figure 1 with thermocouples location on the system.

The basin has taken an area with 2 m^2 in fabricated double slope (MDSPSS). The basin surface and north wall are painted with black dye to absorbed the highest amount of sun rays. FRP sheet also acts as an insulation layer between ambient and solar still unit to retard the heat losses up to the limit. The three side walls (east, west, and south) of the solar still are made by 3 mm thick of transparent acrylic sheet and north wall by FRP sheet. Two glass covers with dimensions 1.03 x 1.03 x 0.004 m^3 are used and fixed in the frame along an inclination angle of 15° with respect to the horizontal surface. Window putty and silicon rubber gasket were used as a sealant fixing to reduces vapor leakage, air-void between the outside environment and edges of the system to reduce the upper and side heat losses from the modified solar still [22,23]. Condensed or distilled yield (water) trickles down toward small channels (troughs) attached at the inner side of walls and come out to the outlet point of each trough, and then received by measuring jar through PVC pipes. Two insertion holes are also given at the north wall (rear side) of the solar unit, one inlet for thermocouple wires and another for pouring contaminated water into the solar still by a pipe. This experimental setup is fixed on an iron supported stand at the terrace of Chemical Engineering Department, IET Lucknow (Latitude/Longitude: 26°84′ N/80°94′ E), U.P., India.

Figure 1. Pictorial view of experimental setup.

Figure 2 shows the points where thermocouples are placed in MDSPSS to measure thermal radiation on hourly basis at various locations like inside and outside walls & glass temperatures, water and basin temperatures of the still.

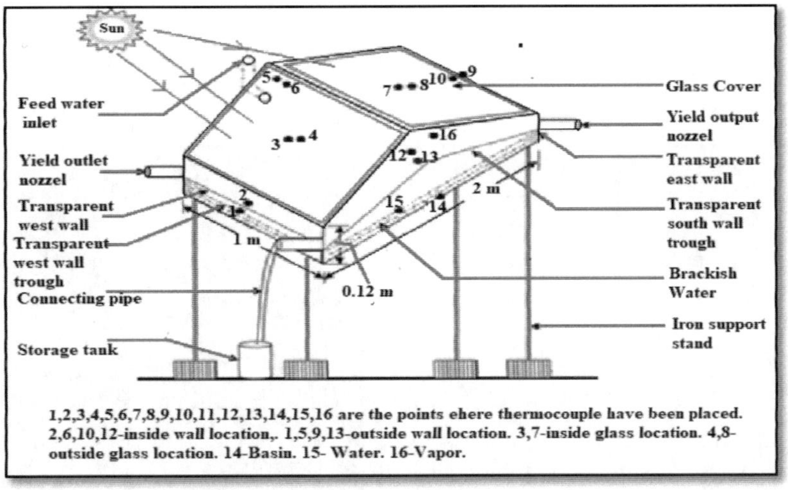

Figure 2. Schematic diagram of MDSPSS and location of thermocouples.

2.2. Equivalent Thickness Calculation

The concept of equivalent thickness is used to analyze the rate of heat transfer or reduction for fabrication of solar still's wall (except north wall) with 5 mm of FRP. The vital effective thickness can be calculated by using Fourier's law (Eqs. 1,2) with surface temperatures between T_1 and T_0 [4].

$$Q = \frac{K_{FRP}.A_{FRP}(T_1 - T_0)}{L_{FRP}} = \frac{K_{ACRY}.A_{ACRY}(T_1 - T_0)}{L_{ACRY}} \quad (1)$$

$$L_{FRP} = \frac{L_{FRP}.K_{ACRY}}{K_{FRP}} = \frac{0.2 \times 5}{0.351} = 3\text{ mm} \quad (2)$$

$K_{FRP} = 0.351$ W/m.K, $K_{ACRY} = 0.2$ W/m.K, Q = heat transfer rate by used FRP material and T_1 & T_0 = both side wall temperature. That's why 3.0 mm thickness of acrylic sheet (transparent) used in place 5.0 mm of FRP sheet for fabrication of South, East and West walls. Due to deviation of ambient temperature and solar radiation, the MDSPSS system is not considered as steady-state condition because wind velocity is assumed in steady-state. Therefore, for calculating equivalent thickness, considered the different inner side temperature of still such as (i) inner glass cover, (ii) filled water, (iii) basin surface, (iv) formed vapor, and (v) inner wall temperatures which are constant throughout for every hourly basis calculation.

2.3. Procedure for nanofluids preparation

In this study, different water-based nanofluids such as Aluminum Oxide (Al_2O_3), Copper Oxide (CuO) and Zinc Oxide (ZnO) have been taken depends on their thermal conductivity as well as economically suitable [24]. Proper preparation of nanofluid is an important factor for enhancing the thermal conductivity because these are behaved like hydrophobicity, prone to agglomerate jointly and very fast settle down. The sodium dodecylbenzene sulphonate (SDBS) is used as a surfactant for maintaining the stability along with suspension form [25]. After mixing, prepared a sample was stirred with a magnetic stirrer for up to 30 min and after that put on ultrasonicator for 2 hours. For good stability and thermal conductivity, the weight concentration should be less than 0.5% [14]. In this work, firstly, 50 ml of prepared water-based nanofluid samples were made with three concentrations as 0.2 %, 0.25%, and 0.3% and checked for those stability to find out the 0.25 % is a good concentration which has more stability to others and it tested with MDSSS. Figure 3 shows that the procedure for the preparation of nanofluid with flowing 2-way methods.

$$\text{Amount of nanoparticles (gm)} = \frac{\text{Concetration (\%)}}{100} \times \text{Vol. of water (ml)} \times \text{Density of nanoparticles} \left(\frac{\text{gm}}{\text{ml}}\right) \quad (3)$$

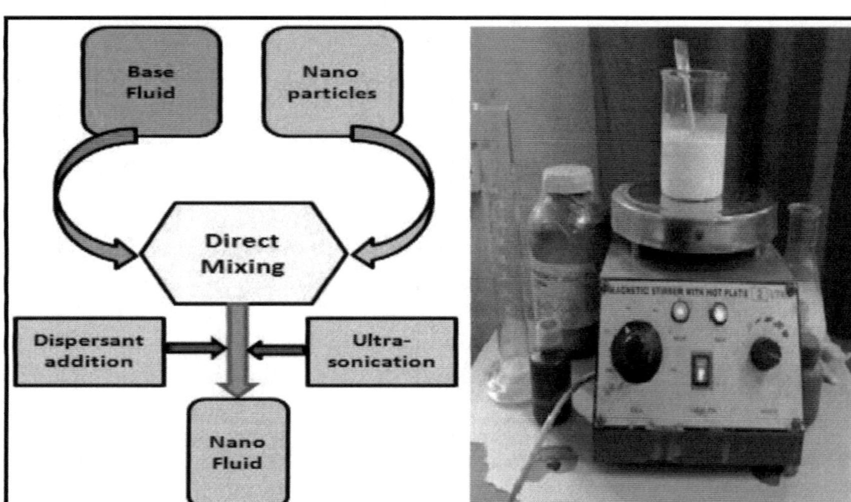

Figure 3. Stepwise procedure for the preparation of nanofluids.

Table 1. Thermal conductivity with different concentration.

S. No.	Nanofluids	Concentration	Thermal conductivity (W/m^2K)	Increment (%)
1	Aluminum Oxide (Al_2O_3)	0.25%	0.6435	11.35
2	Zinc Oxide (ZnO)	0.25%	0.6410	8.150
3	Copper Oxide (CuO)	0.25%	0.6318	6.15

2.4. Experimental procedure

Before carried out experimentation on the modified solar still, the basin of setup is properly cleaned and removed organic/inorganic pathogens, salts, etc. from the system. The experiential setup was orientated in an East-West direction to obtain the highest solar intensity. Then, contaminated (feed) water is poured into the basin and filled upto depth of 1.0 cm. The basin is filled with the required quantity of water at the desired water depth, one day before the start of the experiment to bring the system in a steady-state condition before starting the experimental measurements. On the day of experimental work, an hour before starting the experiments, the quantity and level of water depth was checked carefully along with proper inspection of any leakage, insulation between system parts, etc. The condensing glass covers were cleaned properly before the commencement of the experiments. Nanofluids also enter by inlet pipe and through drainage, a valve to collect spent nanofluids for further use. Experiments at 1.0 cm water depth were started at 07:00 a.m. (local time of Lucknow) in the morning and continued for 12 hours until 07:00 p.m. Experimental data (average value of experimental observations) for any typical day were taken at each one-hour time interval (hourly basis) and recorded for 12 hours (from 07:00 am to 07:00 pm). The experiments were performed in the month of March, 2019. During experimentation, various measuring instruments with their operating ranges are used with those accuracies and error limits which were mentioned in Table 2. In this study, Type –T Copper Constantan calibrated thermocouples are used. These thermocouples have been calibrated with the help of glass thermometer.

Table 2. Accuracy and error limits of measuring instruments.

S. No.	Instruments	Accuracy	Range	% error
1	Thermocouple	+ 0.1 °C	0-150 °C	0.5%
2	Digital temperature indicator	+ 0.1 °C	0-450°C	0.25%
3	Thermometer	+ 1 °C	0-100 °C	0.25%
4	Solarimeter	+ 1 W/m^2	0-25000 W/m^2	2.5%
5	Measuring Jar	+ 10ml	0-1000 ml	10%

3. Results and Discussion

Table 1 represented that the thermal conductivity of used nanofluids with 0.25% concentration along with base water and their % increases. It has been observed that Aluminum Oxide (Al_2O_3) gives the maximum thermal conductivity as compared with Zinc Oxide (ZnO), Copper Oxide (CuO), and water. Figure 4 depicts that the effect of basin fluid temperature with solar time at 0.25% concentration of different nanofluids and maximum temperature was found as 78.2°C through Al_2O_3 at 15:00 hr, similarly through ZnO and CuO were found as 71.5°C & 66.7°C at 16:00 hr, respectively. Similarly, figure 5 depicts that the effect of productivity with solar time at 0.25% concentration of different nanofluids along with base water and maximum productivity was found as 1780 ml through Al_2O_3 at 15:00 hr, similarly through ZnO and CuO were found as 1650 ml & 1510 ml at 15:00 hr, respectively. It is seen that the temperature variation and production rate depend on the change of thermal conductivity and convective heat transfer coefficient of nanofluids. It means, the rate of potable water production is a function of their thermal conductivity.

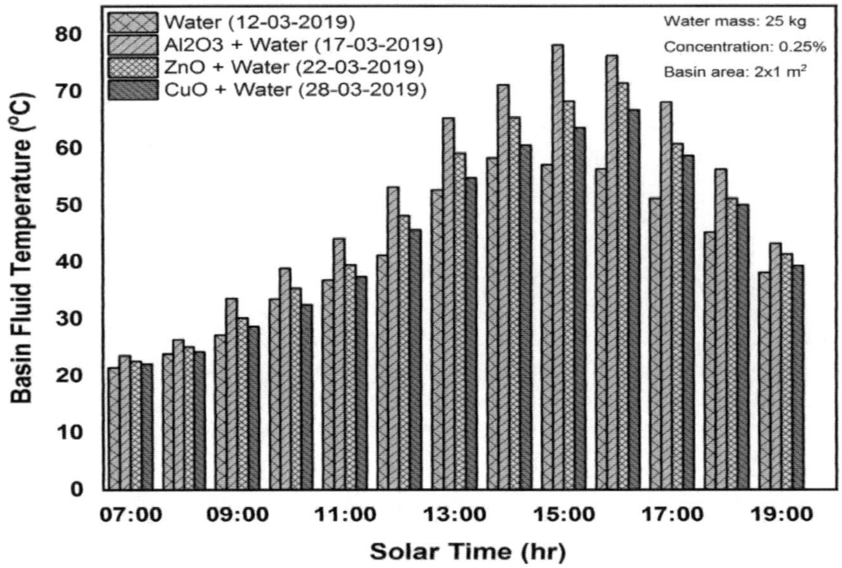

Figure 4. Effect of basin temperature with solar (with and without nanofluid)

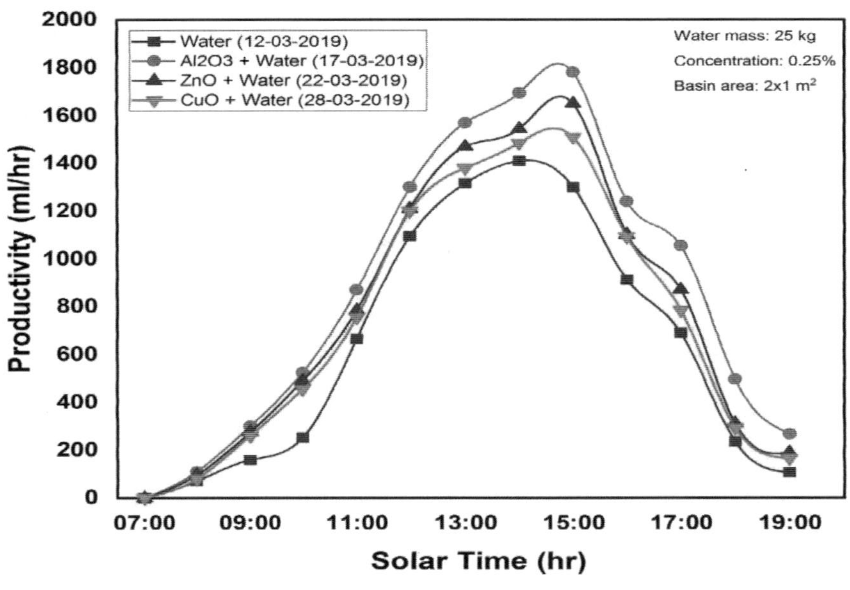

Figure 5. Effect of productivity with solar time (with and without nanofluid)

Due to increasing the rate of heat transfer, the evaporation rate was found higher which improved performance of the solar unit. Figure 6 depicts the cumulative productivity of different nanofluids at 0.25 concentrations along with base normal water. It was observed that the maximum productivity was found as 11209 ml/day through Aluminum Oxide (Al_2O_3) and as 9994 ml/day & 9461 ml/day through ZnO and CuO, respectively. It means productivity through all three nanofluids is better as compared with base water. The effect of production by nanofluids is far better than the conventional still with water (basin water). The production rate is about 36.55 % greater by Al_2O_3 as compared to convention solar still. Figure 7 depicts that, the comparative increment (% basis) of used nanofluids along with base water as taken basin water. The modified still gave better productivity as with Aluminum Oxide (Al_2O_3) 36.55% more followed by 21.75% for ZnO and 15.26 % for CuO. This study confirms that the vital role of thermal conductivity and it is a good indication for enhancing the yield of the system. Also analyzed various physicochemical parameters of water (before and after distillation) by using a water testing kit and found good agreement as per WHO standards.

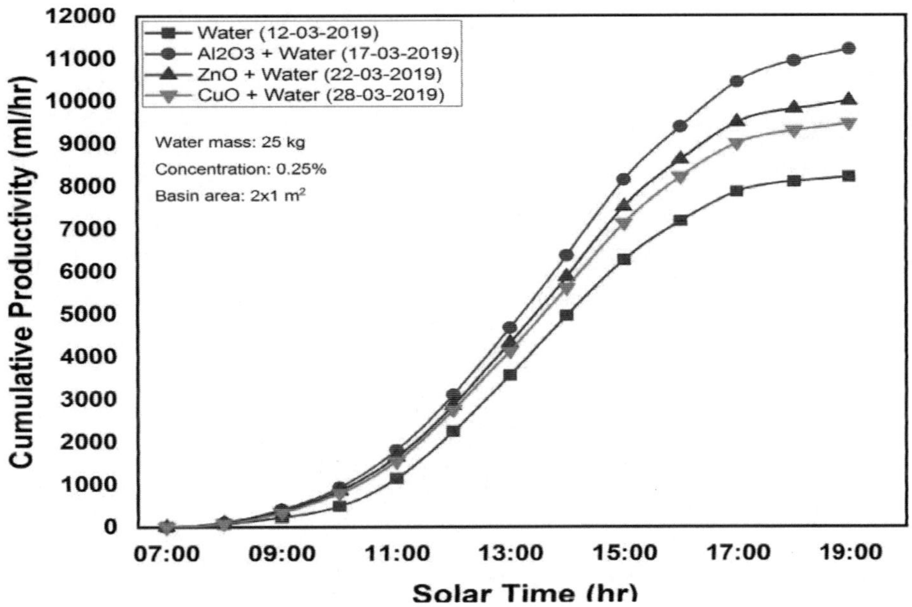

Figure 6. Cumulative productivity with different fluids.

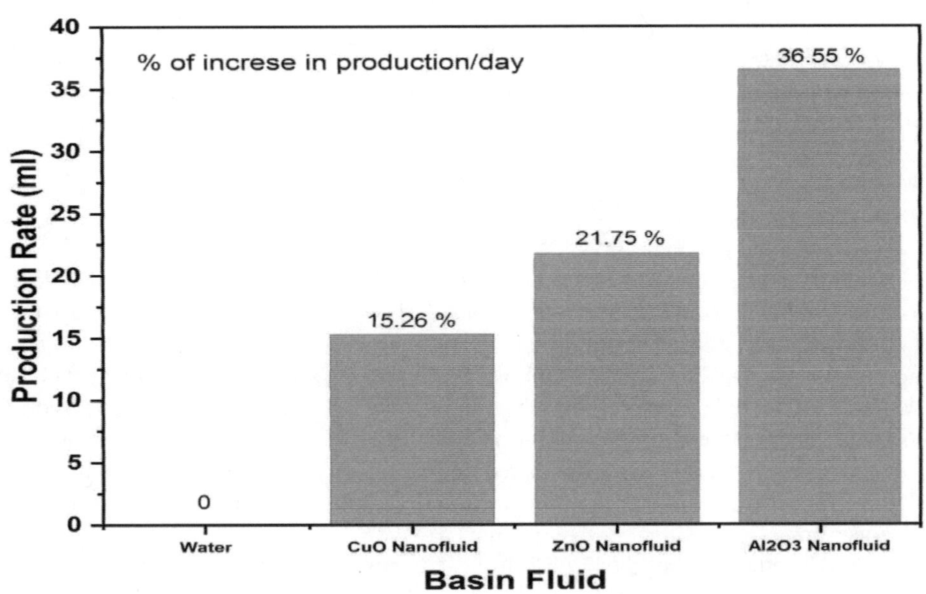

Figure 7. Increase (%) of production rate with nanofluid.

4. Thermal efficiency (daily efficiency) of MDSPSS

The overall thermal efficiency of the solar unit is represented by the ratio of total heat transfer rate (total distillate output) to the total heat input into the solar by incident solar thermal radiation and it is expressed as following equations [26, 5]:

$$\eta_0 = \frac{\sum \dot{m}_{ew} \times L_{vap}}{\sum(I(t) \times A_{ss} \times 3600)} \tag{4}$$

Similarly, the overall thermal efficiency of MDSSS:

$$\eta_0 = \frac{\sum_{i=1}^{i=12} \dot{m}_{ew} \times L_{vap}}{\sum_{i=1}^{i=12} I(t)_{input} \times dt} \times 100 \tag{5}$$

During the off sunshine hour, the value of $I(t)$ and $I(t)_{input}$ taken zero. Where, η_0 is the overall thermal efficiency, $\dot{m}_{ew}$ is the mass of evaporated water (ml/hr), L_{vap} is the latent heat vaporization (kJ/kg), $I(t)$ and $I(t)_{input}$ are the solar intensity (W/m^2) can be evaluated using expressions reported as:

$$L_{vap} = 3.1615x10^6[1 - (7.616x10^{-4}xT_v)]\, for\ T_v > 70°C \tag{6}$$

$$L_{vap} = 2.4935x10^6 \begin{bmatrix} 1 - (9.4779x10^{-4}xT_v) \\ +(1.3132x10^{-7}xT_v^2) \\ -(4.7974x10^{-9}xT_v^3) \end{bmatrix} for\ T_v < 70°C \tag{7}$$

The total input energy i.e. solar energy (instantaneous) to the solar unit by glass cover(s) and transparent walls is expressed by the following equation:

$$I(t)_{input} = I_{Eg}(t)xA_{Eg} + I_{Wg}(t)xA_{Wg} + I_E(t)xA_E + I_W(t)xA_W + I_s(t)xA_s \tag{8}$$

Where, A_{Eg} & A_{Wg} =East and Westside glass cover area (m^2), A_g=Glass cover area (m^2), A_S & W_E =South and West sidewall area (m^2), $\dot{m}_{ew}$=Total mass of distillate collected in a jar (kg/hr), $A_{SS=}$ Area of solar still in (m^2).

5. Economical Analysis

To check the viability of any process, the cost analysis is an important component. For this work, the evaluation of the payback period has been done considering the operating cost, maintenance cost, and capital cost of the entire system. The total manufacturing cost of modified solar still is considered in terms of capital cost (P_{cost}). The salvage value (15%), annual maintenance cost (15%), and rate of interest (12%) are considered for calculation. The daily (average value) production of the MDSPSS is found as 5.6 lit/m^2. From the economical point of view of MDSPSS, 320 days are considered as clear days, and the left over 45 days are considered as hazy/rainy days for the climatic condition of Lucknow (Uttar Pradesh), India. The procedure for calculation of payback time/period is mentioned by following equations:
Daily distilled water production per day= 11.2 lit/day (during the day time in summer season)
Indian market cost of distilled water (IMCDW) = Rs. 15/lit
The cost of the entire system has been increased by 25% by using nanofluids. Table 3 shows that the total cost analysis of the system. Although the total cost of the system is more, its productivity has

been increased up to 36.55%, which gave good reason for the high asset. The payback period is calculated for this MDSPSS system with nanofluids as 121 days, which is less than that of the still with ordinary water.

Table 3. Economical analysis based on one-day production data through MDSSS.

Materials and components	The total cost of the item (Rs.)	Cost type	Value (Rs.)
FRP sheet (5mm thick)	4750/-	CRF	0.147
Acrylic sheet (3 mm thick)	2440/-	SFF	0.027
Iron stand and pipes	580/-	Principle cost (P_{cost})	17600/-
Inlet/Outlet nozzle	80/-	Salvage value (*SV*)	2112/-
Black paint	120/-	Life of solar still (*n*)	15 years
Plastic clips	75/-	Annual first cost (*AFC*)	2587.2/-
Aero tape	85/-	Annual salvage value (*ASV*)	475.21/-
Silicon rubber (gasket)	360/-	Annual maintenance cost(*AMC*)	388.08/-
Glass putty	110/-	Total annual cost (*TAC*)	5000/-
Top glass cover (4mm thick)	1100/-	Annual yield (*AY*)	3264.0 lit
Nanofluids (Al_2O_3, ZnO, CuO)	4400/-	Annual useful energy (*AUE*)	2046.5 kWh
Fabrication cost	3500/-	Annual cost of potable water per lit (*ACPW*)	0.79Rs/lit
Total cost (Rs.)	**17,600/-**	Annual cost of potable water per kWh	2.44Rs./kWh
		Net profit (*NP*)	46381.4/-
		Payback period (*PP*)	121.4 days

* Costs are based on Lucknow (U.P), India market rate.

6. Conclusion

This study shows that the performance analysis of MDSPSS with base water, Aluminum Oxide (Al_2O_3 Copper Oxide (CuO)), and Zinc Oxide (ZnO) water-based nanofluids. It was observed that the potable water production rate was 36.55% more through Aluminum Oxide nanofluid for Zinc Oxide and Copper Oxide nanofluids than the solar still with base water i.e. (Al_2O_3>ZnO>CuO>water). It clearly says that the higher thermal conductive nanofluid absorbs higher sun rays than the base water. The production rate still increases due to the creation of maximum temperature difference between basin fluid and upper glass cover. It also observed that the thermal conductivity of nanofluids and weather conditions play a vital role in improving the yield of solar still. Also calculated average daily productivity is 5.6 lit/m^2, the annual produced water cost is 0.79 Rs. /lit, and the payback period with nanofluid is 121 days.

7. References

[1] Kalidasa K, Murugavel K and Srithar 2011 *Renew. Ener.* **36** pp 612-620.
[2] Patel S K, Singh D, Devnani G L, Sinha S and Singh D 2020 *Sustaina. Ener. Techno. and Asses.* **4** pp 100927.
[3] Tiwari G N and Mishra R K 2012 *Advanced Renewable Energy Sources,* RSC Publishing, Cambridge, UK.
[4] Patel S K, Singh D, Kumar B and Singh D 2020 *Int. J. of Recent. Tech. and Engg.* **8** pp 86-692.

[5] Patel S K, Kumar B, Pal P, Dev R and Singh D 2020 *Sol. Ener.* **209** (2020) 576-589.
[6] UN Environment, Global Environmental Outlook-GEO-6: 2019 *Healthy Planet, Healthy People*.
[7] Tiwari G N 2013 *Solar energy: fundamentals, design, modelling and applications,* Narosa Publishing House, New Delhi.
[8] Latteman S 2010 *Development of an environmental impact assessment and decision support system for seawater desalination plants,* CRC Press, London.
[9] Sarhaddi F, Tabrizi F F, Zoori H A and Mousavi S A 2017 *Ener. Conver. & Managt.* **133** pp 97-109.
[10] WHO-World Health Organization, Geneva, Guidelines for drinking-water quality: 2017 fourth edition incorporating the first addendum.
[11] Cheng W L, Huo Y K and Nian Y L 2019 *Desalina.* **455** pp 89-99.
[12] Sharshir S W, Peng G, Wu L, Essa F A, Kabeel A E and Yang N 2017 *Appl. Ener.* **191** pp 358-366.
[13] Kabeel A E, Omara Z M and Essa F A 2014 *Ener. Convers. Managt.* **78** pp 493-498.
[14] Sahota L, Shyam and Tiwar G N 2017 *Desalin.* **409** pp 66-79.
[15] Sharshir S W, Peng G, Wu L, Yang N, Essa F A, Mohamed S I T, Kabeel A E, Elsheikh A H 2017 *Appl. Therm. Engg.* **113** pp 684-693.
[16] Sahota L and Tiwari G N 2017 *Ener. Conver. and Managt.* **148** pp 413-430.
[17] Saidur R, Leong K Y and Mohammad H A 2011 *Sustain. Energy Rev.* **15** pp 1646-1668.
[18] Sahota L and Tiwari G N 2016 *Sol. Ener.* **130** pp 260-272.
[19] Hassan H, Yousef M S, Fathy M and Ahmed M S 2020 *Renew. Ener.* **155** pp 604-616.
[20] Sharshir S W, Peng G, Elsheikh A H, Edreis E M Eltawil M A, Abdelhamid T, Kabeel A E, Zang J and Yang N 2018 *Ener. Conver. and Managt.* **177** pp 363-375.
[21] Ahn H J, Lee J H, Jeong Y, Lee J H, Chi C S and Oh H J 2007 *Materi. Sci. and Engg. A* **449-451** pp 841-845.
[22] Mahian O, Kianifar A, Soteris, Kalogirou A, Pop I, Wongwises S 2013 *Int. J. Heat Mass Trans.* **57** pp 582-594.
[23] Zhu D, Li X, Wang N, Wang X, Gao J and Li H 2009 *Curr. Appl. Phys.* **9** pp 131-139.
[24] Haddad Z, Abid C, Oztop H F and Mataoui A 2014 *Int. J. Therm. Sci.* **76** pp 168-189.
[25] Elsheikh A H, Sharshir S W, Mostafa M E, Essa F A and Ali M K 2018 *Renew. and Sustaina. Ener. Revi.* **82** pp 3483-3502.
[26] Sahota L and Tiwari G N 2016 Desalina. **388** pp 9-21.

Acknowledgments

This research work was funded by Dr. A.P.J. Abdul Kalam Technical University, Lucknow under the Award of Grant “Visvesvaraya Research Promotion Scheme (VRPS)” for 2019-20. This is to provide financial assistance to create and update the general research capabilities of the faculty members.

Recovery of chromium using membrane containing charged material

Pallavi Mahajan-Tatpate[a,b], Supriya Dhume[a] and Yogesh Chendake[a*]
a: Department of Chemical Engineering, Bharati Vidyapeeth (Deemed to be) University, College of Engineering, Pune, India
b: School of Polymer Engineering, MIT World Peace University, Pune, India
*Email: yjchendake@bvucoep.edu.in

Abstract: Chromium (Cr) is one of the important materials of metal family with large applications in formation of ferrous alloys viz., steel. Due to its excellent chemical, mechanical and thermal properties it is largely used in mineral, leather tanning, dye, steel, and other alloy industries. Effluents of these industries containing traces of Cr are polluting water bodies and soil in surrounding. It would result in entering in food chain, where it has highly adverse effect on humans, animals, and environment. Hence its recovery from effluent and other streams is highly essential before disposal and exposure to surrounding. The polysulfone (PSF) based membranes were optimized using polyethylene glycol (PEG) as porogen, while ZnO as an additive either treated with or without acid for removal of Cr from water. A reduction in molecular weight of PEG resulted in decrease in water flux indicates reduction in pore size. Further incorporation of ZnO increases the Cr removal. The Cr removal increases further with incorporation of acid treated ZnO. Additionally, HCl treated ZnO showed higher rejection properties, while the effect is not that prominent in HNO_3 treated ZnO. This shows the importance of optimization of membrane surface charge for heavy metal removal.

Keywords: Polysulfone, polyethylene glycol, zinc oxide nanoparticle, chromium ion, rejection, permeation

1. Introduction:

Chromium (Cr) is one of the important metals due to its excellent chemical, mechanical and thermal properties. It is one of the most widely distributed metal ions in the earth's crust [Sivakumar S. et al 2005]. This wide availability and excellent properties lead to large applications of Cr in chemical and other industries for the formation of reactors and other systems. Though the stability and other properties are attractive, it poses some limitations on processibility and shaping. It can be used as an important component of alloys like stainless steel [Ray S. et. al. 2009]. Moreover, it is also used in plating on metal/ plastic materials, decorative or protective automotive, equipment coating, and in nuclear and high-temperature research [Bazrafshan E. et al. 2015, Fu F. et al. 2011, Shanker A. K. et al. 2005]. Hence the large quantity of Cr is extracted from minerals or recycled for the formation of these materials. During these extractions, recovery and recycle treatments, a large amount of Cr found its way to effluent and waste disposal. According to Ray et. al. (2009), 4500 kg/day of chromium was released in the environment due to different processes in the U.S.A.

Cr is present in nature in chromite ($FeOCr_2O_3$) form. It has three oxidative states Cr (II), Cr (III), and Cr (VI). Cr (III) is comparatively less harmful, approximately 500 times less than Cr (VI) as it has low water solubility [Alam N.E. et al. 2020, Ray S. et al. 2009]. The effects of industrialization and subsequent human activities like metal processing, electroplating, leather tanning, chemical production, paint, pigment, dye industries are releasing large quantities of Cr through effluent [Bazrafshan E. et al. 2015, Fu

F. et al. 2011]. This is resulting in large scale Cr contamination in environment-water and, soil. This contamination is adding to natural activities like weathered rocks, volcanic exhalations, biogeochemical processes, which is making the situation worse. These metal ions have tendency of accumulation and cause plant/ soil poisoning and disorder in liver, lung, kidney, brain, central nervous system functioning. Further, human exposure to Cr activates teratogenic processes, causes cancer, disturbs DNA synthesis, mutagens changes- malignant tumors, kidney, liver, gastric damage, nasal irritation/ ulcer, and fluctuates urinary level. It changes soil microbial populations and metabolism, affects the photosynthesis process, alters the germination process, affects the growth of roots, stems, and leaves [Ray S. et al. 2009, Bazrafshan E. et al. 2015, Singh H.P. et al. 2013]. Hence various limitations are defined by WHO for its permissible limit in the disposal stream. Cr maximum contaminated level (MCL) by USEPA is 0.05 mg/L [Gunatilake S. K. 2015].

Metal ions *viz.* arsenic, chromium, copper, iron, nickel, sodium, zinc, etc. are important at low concentration for normal growth/ functioning of human, animal, plant, and soil biota. Its consumption in higher quantities can cause serious health issues by interfering with normal body functioning. Hence it highly essential to control heavy metal pollution by human activity. Hence recovery or removal of Cr from the various process and effluent disposal streams is highly essential.

Various conventional methods are available for the separation of heavy metals such as chemical-precipitation, ion-exchange, adsorption, coagulation, flocculation, flotation, electrochemical, and advanced-membrane separation methods [Fu F. et al. 2011]. Conventional methods have good separation efficiency but have the issues of generation of a secondary pollutant, recovery issues restrict their applicability. Hence, there is a need for a reliable techno-economical, environment-friendly, sustainable separation, and recovery method. Membrane-based methods *viz.* reverse-osmosis (RO), nanofiltration (NF), electrodialysis (ED), ultrafiltration (UF) has ability to treat heavy metal containing water for component recovery without chemical contamination [Mahajan-Tatpate P. et al. 2021]. They have benefits of physical recovery of a component without addition of chemicals. Hence the recovered components can be recycled back/utilized in further processes. But RO, NF membrane operating pressure is high, and it can remove essential components such as monovalent Na^{+}, and divalent Ca^{+2} which are essential for the normal growth of human beings. Whereas ED process needs clean feed and regular cleaning and monitoring. UF membrane operates comparatively at low pressure and considering secondary treatment process involved in conventional process, it is cost comparable. Hence UF membranes are utilized for heavy metal separation. But there is a need for optimizing the membrane properties for the desired separation with a high transport rate to make the process techno-economically beneficial along with ecological benefits. For this purpose, the present study focuses on optimizing the membrane transport and Cr rejection properties with the use of PEG as pore forming agent and ZnO as rejection improving additive. The effect of dope solution composition on transport rate and Cr rejection is investigated.

2. Materials and Methods:

2.1. Materials:

Reagents used in the present study are all synthesis-grade chemicals. Polysulfone (PSF) was procured from Otto Chemie Pvt. Ltd., N, N'-Dimethyl acetamide (DMAc), and Polyethylene glycol (PEG) with molecular weight 1500 (PEG-1500) was purchased from Loba Chemie Pvt. Ltd. India. PEG with molecular weight 400 (PEG-400), 600 (PEG-600) were procured from High Purity Lab. Pvt. Ltd. India, and 6000 (PEG-6000) was procured from Sisco Research Lab Pvt. Ltd. Zinc oxide (ZnO) nano powder of size 80-100 nm was procured from Nanoshel LLC. Hydrochloric acid (HCl) and Nitric acid (HNO_3) were procured from Merck Ltd. India. Non-woven polyester backing 3324 was procured from Ahlstrom Hollytex. Potassium dichromate was procured from Sisco Research Laboratory.

2.2. Method of Membrane Preparation:
PSF based porous membranes were prepared by the phase inversion method. The casting solution was prepared by dissolving 29 % PSF (w/v) and 8% PEG (w/w of PSF) in DMAc as discussed earlier [Dhume S. et al. 2020]. It is followed by dispersing 0.8% ZnO nano powder optionally treated with acid (HCl or HNO_3) in the formed solution. The formed solution is degassed to remove dissolved air and used for casting on non-woven polyester backing 3324. Water is used as a non-solvent, at ambient temperature. An automatic membrane casting machine (Shivom membrane casting system) was used to prepare membrane. The formed membrane samples were stored and maintained at 4°C before use in flux, and rejection property analysis.

3. Characterization:

3.1. Water Flux Analysis:
The water flux of the prepared membrane was measured at 2 bar pressure using a dead-end cell, at room temperature. The water transport rate/ flux was measured using Eq. 1.

$$J = \frac{V}{A \times \Delta t} \quad (1)$$

Where, V is a volume (Liter) of water transported across the membrane in time (Δt, hour) through the membrane of cross-sectional area (A).

3.2. Chromium ion Rejection Analysis:
For chromium ion rejection analysis, a 1000 ppm solution of potassium dichromate was prepared. This solution was passed through the membrane and permeate was collected. Chromium ion retained on the membrane was determined by collecting permeate and its analysis was done by double beam UV visible spectrophotometer (UV 3000$^+$ from Lab India) by measuring absorbance at 313 nm. Percent rejection (% R) was calculated by

$$\%R = \left[1 - \frac{C_p}{C_f}\right] \times 100 \quad (2)$$

Where C_p and C_f is the concentration of permeate and feed respectively.

4. Result and Discussions

4.1. Effect of molecular weight of PEG on Pure Water Flux:
PSF based membranes prepared using the phase inversion method showed a large effect of dope solution concentration on transport properties. An increase in dope solution concentration leads to a reduction in pore size [Phale J. S. et al. 2015]. It resulted in increased polyethylene glycol (MW 6000 Da) (PEG 6000) rejection, desired property for heavy metal (Cr) removal. It also leads to a reduction in water flux which would affect the separation economy.
Further use of PEG is reported as pore forming agent for PSF based membranes, which affects the hydrophilicity and transport properties of formed membranes [Dhume S. et al. 2020]. It is highly soluble

in water, which can be leached by water used as a gelation agent during membrane formation. Such leaching would result in the formation of pores at the place of PEG in membranes. This increases the porosity of the membrane by maintaining pore size, which results in increase of pure water flux while maintaining selectivity or rejection properties for PEG 6000 [Dhume S. et al. 2020]. The increase in transport rate and maintaining rejection properties are attributed to the formation of the pore by removal of PEG [Chakrabarty B. et al. 2008]. It is reported that with decrease in molecular weight of PEG, water flux decreases, which attributes to a decrease in pore size of the membrane, which is desired property for current work to improve rejection of metal ion. Hence, it was thought that a reduction in PEG molecular weight would help to improve the transport rate while improving the rejection properties.

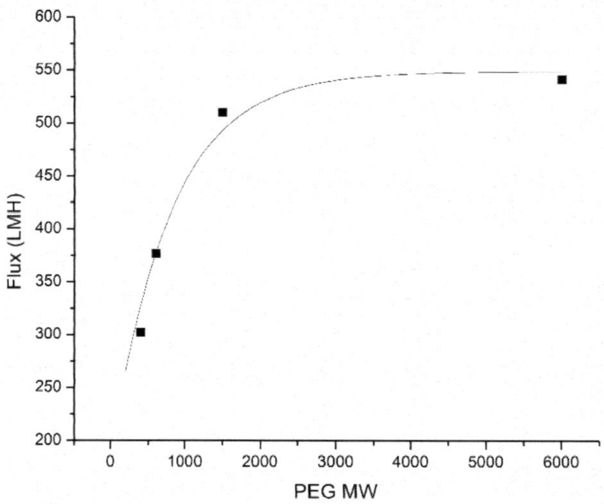

Figure 1. Effect of PEG molecular weight on Pure Water flux

PEG with different molecular weights ranging from 400 to 6000 was used to prepare the membrane. Formed membranes were analyzed for water flux to determine the effect of PEG molecular weight on transport properties of porous membranes [Fig.1]. The membrane flux (302 to 541 LMH) increases logarithmically with an increase in PEG molecular weight. This could be attributed to a reduction in pore size which increases resistance for transport of water across the membranes. It was observed that with a decrease in molecular weight of PEG, pure water flux has decreased. It attributes to a decrease in pore size with a decrease in molecular weight. [Kim J.H. et al. 1998, Chakrabarty B. et al. 2008].

4.2. Effect of acid treated ZnO nanoparticle on chromium ion rejection:

It was well reported by Chung Y.T. et al. 2015 that nano sized ZnO is effectively used to remove heavy metals from water. ZnO nanoparticles improve water flux, which means it forms a more porous structure of the membrane and it also helps to improve rejection. The equal distribution of these metal nanoparticles in dope solution improves PSF membrane hydrophilicity and decreases its fouling [Chung Y. T. et al. 2015]. Polysulfone membrane performance has improved by the addition of ZnO nanoparticle even for CO_2, CH_4 separation [Moradihamedani P. et al. 2013]. Leo C.P. 2012 stated that the hydrophobic nature of PSF membrane has changed to hydrophilic by addition of ZnO nanoparticles which results in reduction in the fouling of membrane. Also, it increases the thermal stability of the membrane. [Leo C.P. et al. 2012].

Membrane prepared with PEG-400 has the minimum water flux as compared with high molecular weight PEG. The rejection properties for chromium ions were studied on a membrane prepared with PSF/PEG 400/ZnO. The addition of ZnO has shown improved membrane properties. Here in this study, ZnO was modified with acid treatment.

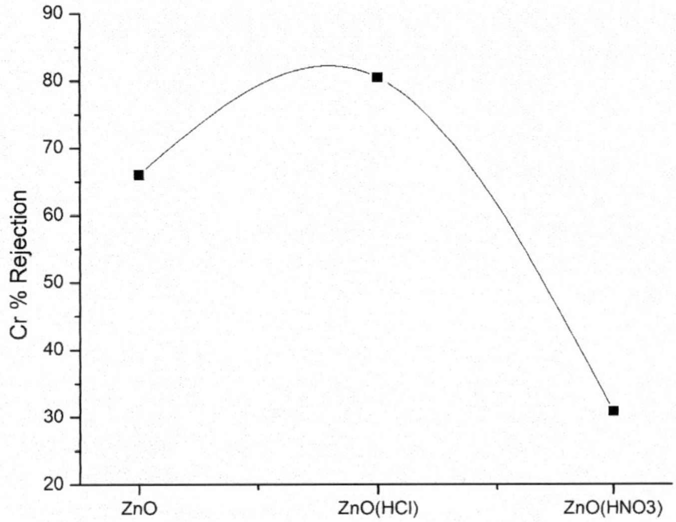

Figure 2. Effect of Modified/ unmodified ZnO incorporated membrane on % Rejection

ZnO nanoparticle reacts with HCl and produces $ZnCl_2$. This $ZnCl_2$ reacts with polysulfone, this interaction gives surface charge to the membrane [Mukherjee R. et al. 2015]. The addition of $ZnCl_2$ in dope solution decreases membrane roughness and membrane become smoother, results in a decrease in pore size of membrane [Mondal M. et al. 2015, Panda S.R. 2014]. Here the $ZnCl_2$ binds with the PSF by weak interaction between strong chloride ions and PSF molecules. Thus, it will provide a charge to membranes. The potassium dichromate solution has a negatively charged dichromate anion. According to the Donnan exclusion principle, there is repulsion between similar charge ions. So, there is repulsion between membrane surface and chromium ions. The presence of Zn would provide the repulsive interactions with metal ions enhancing the rejection properties. Thus, the rejection property of membrane cast with ZnO modified with HCl increases from 66 to 80%. This can be attributed to increase in Zn ionization in chloride format which would improve the Donnan rejection properties for Cr.

A curious case was observed when the ZnO was modified with HNO_3, the rejection of Cr was reduced from 66 to 38 % for the membrane containing virgin ZnO and ZnO modified with HNO_3, respectively. This can be attributed to formation of zinc nitrate during the interaction between ZnO and HNO_3. Here the negative charge of N and membrane is masked by the presence of O_3. It would reduce the interactions and repulsion of similarly charge Cr and in turn the rejection properties.

5. Process Implementation and Economy:

Chromium is one of the major pollutants in various industries including cement where it is found in different oxidation states like 3, 4.6, 5, 6, etc. [Sinyoung S. et al. 2011, Estokova A. 2018]. The leachate material is one of the major sources of heavy metal pollution in cement industry. Further, a tannery industry releases a large amount of chromium in the effluent stream used for leather tanning [Alam N. E. et al. 2018, Alam N. E. et al. 2020]. It is also a part of effluent generated from steel/metal plating, automobile, battery manufacturing, refinery, pharmaceutical, as observed from the industrial zones of Balanagar, Hyderabad. Here sediments from conventional effluent treatment units contain chromium up to 96.2 to 439.6 mg/kg [Machender G. et al. 2014].
For these effluent treatment conventional effluent treatment methods *viz.* chemical precipitation, ion-exchange, adsorption, coagulation, electrocoagulation, flocculation, flotation, electrochemical, etc. used, require chemical addition for heavy metal separation. During process it can generate toxic intermediate compound. This hazardous sludge if not treated further, can cause secondary pollution, and has disposable issues. Formed complexes or separated phase needs further treatment for regeneration and recovery of metal ions. This generated secondary sludge, which affects the economy and real-life applicability of those processes.
Further, some of the advanced technologies like membrane-based methods *viz.* RO, NF, ED, UF is available to treat the wastewater from those industries [Alam N. E. et al. 2020]. Some of these methods show good separation efficiency based upon physical separation and recovery methods without any addition of chemicals. They have limitations towards design of membranes and energy intensive processing due to high pressure applications (RO, NF), electricity driven processes (ED), etc.
Ultrafiltration based membrane separation possesses benefits of higher transport rates and low energy requirements. They can be used at moderate process conditions and combined with conventional process streams for continuous separation of metal ions, its recovery and recycling of other desired streams. They need the membranes to be designed carefully for obtaining desired selectivity properties while maintaining transport rate. This would provide techno-economic benefits along with continuous recovery and recycle of heavy metals. This would enhance the benefits further. Considering all limitations of conventional separation methods and RO. NF, ED membrane-based methods, use of surface-modified UF membranes provide multiple benefits about its commercial implementation and process enhancement for recovery of metal ions in pure form.

6. Conclusion:

Chromium is an important material in different industrial application along with part of various processes in the form of chemical compositions. During these processes and material formation large amount of Cr is released to environment through effluent and waste streams. This along with the natural activities enhances its concentration in water and soil biota. Where it can prove to be detrimental to human, animal, soil, and environment if its concentration is above standard permissible limit. It possesses the properties of bioaccumulation which makes it more dangerous upon entry into food chain. Hence it is highly important to control its release to environment through human activity. Though various processes have been reported for its treatment conventional separation processes have limitations of addition of chemicals and generation of intermediate compound which needs disposal or secondary treatment. Membrane based processes physically separate metal ions in nascent form without any change in chemical property. RO and NF have high operational cost and removes the desired components from water. UF membrane is techno-economical and sustainable considering all aspects, but it requires careful membrane optimization for desired transport and selectivity properties. Current work shows PSF based membranes modified with PEG as porogenic agent and ZnO (alternatively treated with acid) shows excellent (80 %) recovery of Cr ions, with excellent transport properties at low operational pressure (2 bar). This makes the membranes techno-economically feasible and environmentally attractive choice for treatment of process streams for Cr removal before their release or disposal to the atmosphere.

7. Acknowledgment:

It gives great pleasure in presenting the work for chromium ions separation from water. The authors are thankful to Bharati Vidyapeeth (Deemed to be University) College of Engineering, Pune, and MIT-World Peace University, Pune for providing us the useful resources. The authors also thank DST- Nano Mission for the financial support in the format of project Sanction No. SR/MN/NT/-1029/2015.

8. References:

1. Sivakumar S A, and Subbhuraam C V 2005 *Ecotoxicology and Environmental Safety* **62** 93–98
2. Ray S A, Ray M K 2009 *Al Ameen J. Med. Sci.* **2** 57–63
3. Bazrafshan E, Mohammadi L, Ansari-Moghaddam A, Mahvi A H 2015 *J. Environ. Health Sci. Eng.* **13** (1) 1-16
4. Fu F and Wang Q 2011 *J. Environ. Manage.* **92** (3) 407–418
5. Shanker A K, Cervantes C, Loza-tavera H, and Avudainayagam S 2005 *Environment International* ***31*** 739–753
6. Alam N E, Mia A S, Ahmad F, and Rahman M 2020 *Appl. Water Sci.* **10** (9) 1–22
7. Singh H P, and Priyanka S. 2013 *Environ. Chem. Lett.* **11** 229–254
8. Gunatilake S K 2015 *J. Multidiscip. Eng. Sci. Stud.* **1**(1) 12–18
9. Mahajan-Tatpate P, Dhume S, Chendake Y 2021 *International Journal of Membrane Science and Technology* **8** 1-21
10. Phale J S, Chendake Y J 2016 *International Journal of Science and Research* **5**(6), 2569–2573
11. Dhume S, Mahajan-Tatpate P, Chendake Y 2020 *J. Appl. Membr. Sci. Technol.* **24** (3) 57–73
12. Chakrabarty B, Ghoshal A K, and Purkait M K 2008 *J. Membr. Sci.* **309** (1–2) 209–221
13. Kim J H, Lee K H 1998 *J. Membr. Sci.* **138** 153–163
14. Chung Y T, Ba-abbad M M, and Mohammad A W 2016 *Desalin. Water Treat.* **57** 7801-7811
15. Moradihamedani P, Ibrahim N A, Ramimoghadam D, Zin W, Yunus W, and Yusof N A 2013 *J. Appl. Polym. Sci.* 1–9
16. Leo C P, Lee W P C, Ahmad A L, and Mohammad A W 2012 *Sep. Purif. Technol.* **89** 51–56
17. Mukherjee R, Mondal M, Sinha A, Sarkar S, De S 2015 *J. Environ. Chem. Eng.* **15**
18. Mondal M, De S 2015 *Polym. Adv. Technol.* **26** 1155–1167
19. Panda S R, De S 2014 *Desal.* **347** 52–65
20. Sinyoung S, Songsiriritthigul P, Asavapisit S, Kajitvichyanukul P 2011 *J. Hazar. Mater.* **191** 296–305
21. Estokova A 2018 *Int. J. Environ. Research and Public Health Study* **824** 1-13
22. Alam N E, Mia A. S., Ahmad F., Rahman M. 2018 *Applied Water Science* **8** 1–7
23. Alam N E, Mia A S, Ahmad F, Rahman M 2020 *Applied Water Science* **10** 1–22
24. Machender G, Dhakate R 2014 *Arab J Geosci.* **7** 513–525

Study in the Changes on the Functional Groups Present in Biomass during Pyrolysis Process

S Suman[1], AM Yadav[2], T Jain[1], A Ali SK[1]

[1]Department of Mechanical Engineering, Meerut Institute of Engineering and Technology, Meerut-250005, Uttar Pradesh
[2]Department of Chemical Engineering, Meerut Institute of Engineering and Technology, Meerut-250005, Uttar Pradesh
Email: swapan.suman@miet.ac.in

Abstract. In this study, investigate the changes occurs on the functional groups of biochar materials. The biomass materials subjected to the pyrolysis, which is a thermochemical conversion process to obtain biochar material. It was produced by slow-pyrolysis method by subjected in different pyrolysis temperatures (400, 600, 800 and 1000°C). Effect of the treatment procedures was determined by the modifications in the functional groups of the obtained bio-char during the pyrolysis treatment was determined with the Fourier transform infrared spectroscopy (FT-IR). FTIR methods deliver fast, low-cost and non-destructive analysis and have shown qualitative and quantitative outcomes. The analysis was applied to some agricultural biomass materials such as saw dust, rice hull, wheat hull, tea wastes, and eucalyptus shell. Results are associated with literature, and species potentially computable by Fourier Transform Infrared Spectroscopy are reviewed. The result showed that the aliphatic, hydroxyl and aromatic groups gradually diminishes by increasing the temperature from 400 to 1000°C.

Keywords: Biomass; Pyrolysis; Biochar; FTIR Spectroscopy

Graphical Abstract

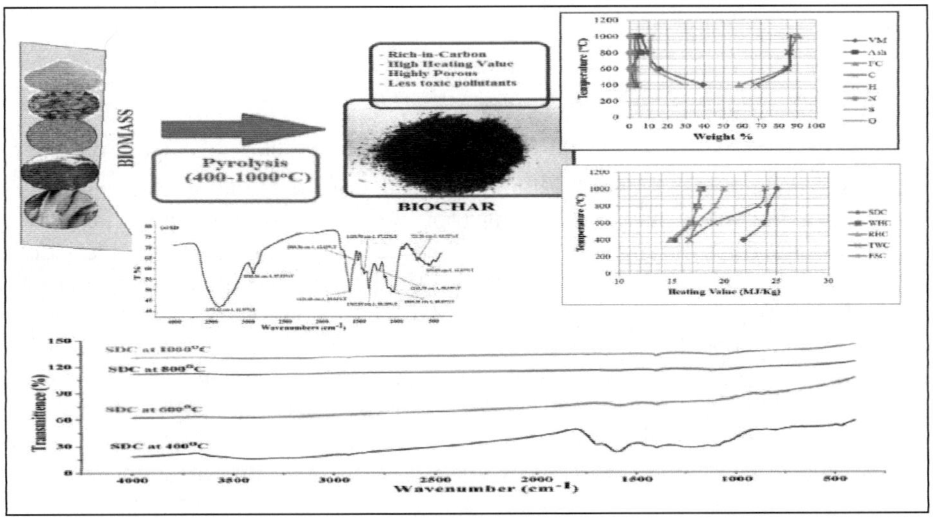

Statement of Novelty

This research demonstrates the optimization and changes in functional groups of selected biomass during pyrolysis process at different pyrolysis temperature. Throughout the pyrolysis method, oxygen and hydrogen contents were reduces and enrich carbon rest remains in form hydroxyl (C–H stretching) and aromatic carbon bonds and aromatic groups are gradually diminishes. During the pyrolysis chemical modification enriches the functional groups of biochar. This study has shown to high potential of biochar to get a deeper understanding of structural relationships with their functional groups.

1. Introduction

Recently, energy needed in the world has increased dramatically. Therefore, it is likely that fossil fuels cannot meet the energy demand in the next some decade. Besides, the fossil fuels and its derivatives have caused significant environmental risks due to the formation of harmful oxides likes nitrogen, carbon, and sulphur. These reasons have moved our attention towards alternative energy source that is renewable and environmentally friendly. Solid bio-waste is one of the demandable source of energy for future energy demand due to its sustainable nature.

Biomass is the leading energy source in the world which comes after fossil fuels and it providing about 14% of the world's energy demand. Biomass is providing 35% of their energy in developing nations [1].

Biomass has likely to provide energy with almost zero-emissions [2]. Biomass and its application provides an effort to utilize the energy stored in it as an alternate fuel in order to conserve the other energy sources like coal, oil etc.

Pyrolysis is one of the thermo-chemical processes that are conducted at high temperatures (400-1100°C) in the oxygen-less condition. During the pyrolysis process, biomass is converted into solid, liquid and gaseous products based on heating rates and time of pyrolysis [3].

Hence we can say that energy generated from biomass has a high capacity to bridge the global energy shortage gap caused by the depletion of fossil fuel resources while keeping environmental sustainability.

This study includes the influence of temperature on different biomass feedstocks and their physicochemical properties. The different agricultural biomass were characterized by proximate & ultimate analysis, higher heating value (HHV) and field-emission Fourier transform infrared (FTIR) spectroscopy.

2. Materials and Methodology

2.1. Biomass materials

The conditions in which pyrolysis process is accomplished can dramatically change the final properties and composition (solid, liquid and gaseous) of the products obtained from the process. The parameters like feedstock type and heating profile (i.e. pyrolysis temperature) have been frequently recognised as important variables for determining the composition and characteristics of produced bio-char [4-7]. Besides, the significance of pyrolysis time during biochar production is less conclusive due to a limited number of studies varying these conditions. Therefore detailed investigations would aid in reaching a definitive conclusion on the importance and preferred value for these variables.

Biomass samples used were mainly woody waste, crop residues and shells, this includes Saw Dust (SD), Rice Hull (RH), Wheat Hull (WH), Tea Wastes (TW), and Eucalyptus Shell (ES).

The pyrolysis temperature range studied throughout the experimental work covered a range from 400°C to 1000°C with 600°C and 800°C values serving as intermediate values. This temperature choice covered the main regions of biomass degradation as well as the respective upper and lower limits of temperatures associated with slow pyrolysis at slow heating rate (15-25°C/min). A low heating rate was selected to provide longer heating time in an attempt to provide tolerable time for sufficient heat transfer and heat penetration into the biomass particles. Based on realistic times seen in industrial sized units to generate fast conversion of feedstock to biochar, the residence times chosen were therefore 60 minutes.

2.2. Preparation of Biochar

Biochar is a solid and carbonaceous by-product of biomass. To obtain biochar, selected biomass subjected to reactor at different pyrolysis temperature at a low heating rate for characterization [8]. After pyrolysis, products were determined for biochar and yields of product were calculated as a proportion of feedstock weight on a dry weight basis.

2.3. Methodology

2.3.1. Proximate and ultimate analyses and higher heating value (HHV)

"Proximate analysis" was employed on selected and prepared materials for the estimation of physiochemical properties with ensuing standards "(ASTM E871-82, E1755-01, and E872-82)" [9]. And the proximate analysis for obtained biochar from studied samples was evaluated according to "ASTM D1762-84" standard method [10].

"ASTM E777, E778 and E775" standard method was ensued for "ultimate analysis" [11] to calculate the elemental components of the selected samples and their corresponding biochar using "CHNS analyser" (Vario EL III). In the ultimate analysis we determine the Carbon, Hydrogen, Nitrogen and Sulphur in weight percentage.

Higher Heating Value (HHV) of all samples were estimated by "bomb calorimeter" (Model: AC-350 LECO) according to "ASTM D4809" standard [12].

2.3.2. Fourier-transform infrared (FT-IR) spectroscopy

FT-IR spectroscopy has been used as an influential methodical tool for the swift description of different types of biomass. FTIR spectroscopy tracks the existence of 'molecular vibrations' that are representative of a chemical compound. This technique has broadly been used for the qualitative and quantitative interpretation of modifications in biomass structure during the process [13]. For FT-IR study, FT-IR spectrometer (Perkin Elmer Spectra 2, USA) was exploited by the 'pellet method' by blend with dried bio-char and pulverized KBr in the ratio proportion of 1:200 (Potassium Bromide).

3. Results and Discussion

3.1. Proximate, ultimate analyses and heating values of studied biomass samples

The physicochemical characteristics of the selected biomass samples are tabulated in Table 1. As per the literature, the parametric values in case of a variety of biomass vary in the range: ash (1.45-18.20%), VM (67.38-81.33%), FC (13.11-17.22%), C (39.40-50.11%), O (37.06-41.78%), H (3.59-5.50%), N (0.27-0.75%), and S (0.12-0.18%). These results are verified with literature values [14]. Their ash depends on the sample types and their geographic region.

The HHV of selected biomass vary in the range of 14.77-19.61 MJ/Kg. These values are also within the range of literature value [15] and that is comparable with primary and other renewable energy source. Heating value directly affected by ash, its high or less weight percentage depends on the calorific value which shows their heating energy.

From above evident, it can be concluded that high ash containing samples makes it less desirable as a 'fuel' [16-17].

Table 1. Proximate analysis (dry basis), Ultimate analysis and higher heating value of studied biomass samples

Samples	VM%	ASH%	FC%	C%	H%	N%	S%	O%	HHV(MJ/Kg)
Saw Dust (SD)	81.33	1.45	17.22	51.11	5.50	0.27	0.12	41.67	19.61
Wheat Hull (WH)	71.24	14.56	14.20	40.32	5.34	0.43	0.15	39.20	15.43
Rice Hull (RH)	69.62	17.27	13.11	39.40	4.23	0.75	0.18	39.17	14.77
Tea Waste (TW)	73.33	11.47	15.20	42.35	3.59	0.67	0.14	41.78	16.11
Eucalyptus Shell (ES)	67.38	18.20	14.42	39.46	4.84	0.31	0.13	37.06	15.68

Agricultural biomass contains much more ash-forming components than some woody biomass [18]. In particular, the straw, grasses, and fast growing crops mostly have higher content of both the ash and high mobile elements (like Cl, K, Mg, S, Na, and Si) than wood biomass; and reverse is true for C content [18-19]. The higher HHV of Saw Dust and Coconut Shell is because of lower content of ash i.e. an incombustible component and higher amount of combustible components such as volatile matter, fixed carbon, carbon and hydrogen than in other biomass materials.

3.2. Fourier transforms infrared spectroscopy of studied biomass samples

Figure 1(a)-(e) demonstrate the FTIR spectra plots of selected biomass samples. These graphs are the plot versus wavenumbers (cm^{-1}) and transmittance %, which shows the aliphatic losses in their functional groups. The all selected biomass have mutual bands i.e., O-H stretching occurs at 3390-3420 cm^{-1}, CH, CH_2 stretching occurs at 2900-2930 cm^{-1}, C=C stretching occurs at 1615-1640 cm^{-1}, benzene ring occurs at 1500-1520 cm^{-1}, aromatic skeletal vibrations occurs at 1405-1465 cm^{-1}, Phenolic OH and aliphatic C-H deformation occurs at 1315-1380 cm^{-1}, C-O-C stretching occurs at 1235-1245 cm^{-1} in alkyl aromatic, C-O-C asymmetry stretching occurs at 1155-1165 cm^{-1}, C-O stretching occurs at 1030-1060 cm^{-1}. Other peaks seemed at 1720 cm^{-1} i.e., C=O stretching and at 605cm^{-1} i.e., OH bending modes in WC [20-21]. In Addition the peak at 440, 805 and 1045-1108 cm^{-1} seemed in Wheat Hull and Rice Hull, i.e., for SiO_2 [22].

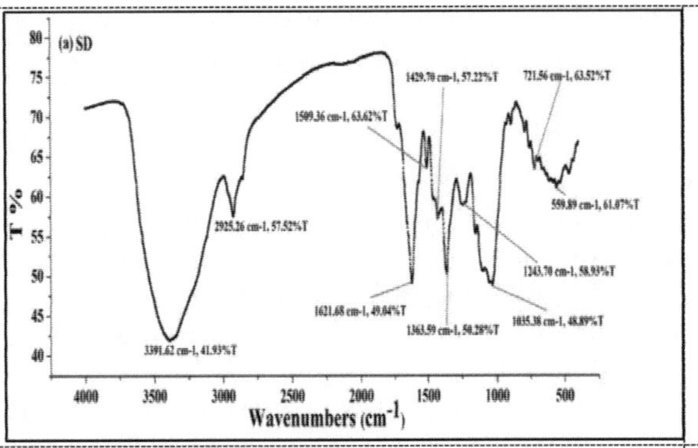

Figure 1(a). FTIR analysis of Saw Dust

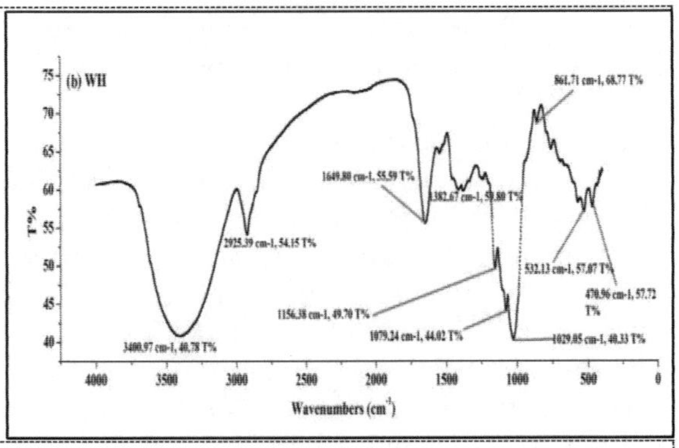

Figure 1(b). FTIR analysis of Wheat Hull.

During FTIR analysis (Figure 1(a)-(e)), the covalent bonds of the functional groups absorbed from infrared radiation causing the bonds to be stretched, vibrated, bended, rocked or twisted. The main application of this method is to observe the cellulosic contents which can be clearly observed by the O-H bonds. The O-H bonds after the absorption of energy were stretched resulting in enhancement of the peak's intensity. The peak intensity is indicated by either the peak's sharpness or broadness and also the percentage transmittance (%T). The transmittance percentage indicates the how much lights been transmitted through the molecules which means the absorption energy by the O-H bonds is the deduction of the transmittance percentage by hundred per cent.

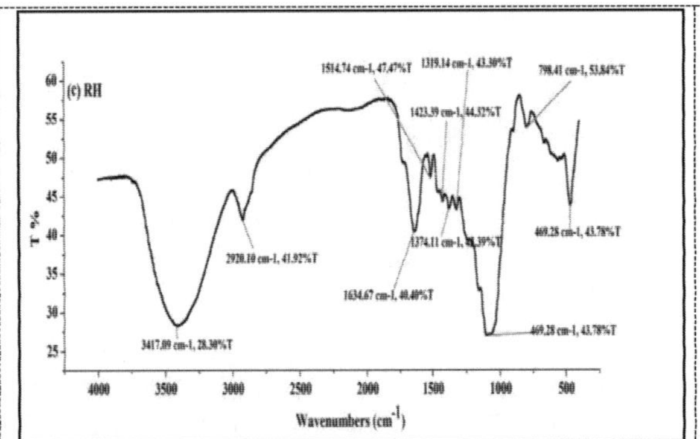

Figure 1(c). FTIR analysis of Rice Hull

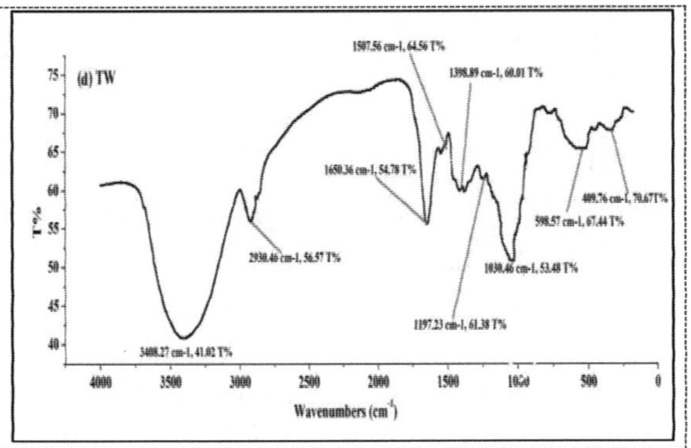

Figure 1(d). FTIR analysis of Tea Wastes

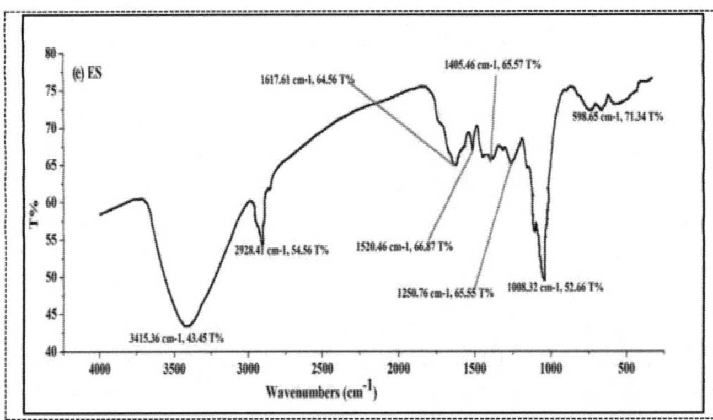

Figure 1(e). FTIR analysis of Eucalyptus Shell

IOP Conf. Series: Materials Science and Engineering 1146 (2021) 012023 doi:10.1088/1757-899X/1146/1/012023

3.3. Proximate, ultimate analyses and higher heating value of obtained biochars

The 'proximate', 'ultimate analyses' and 'HHV' of the carbonized products of studied biomass samples at different temperature are shown in Figure 2 (a)-(f). Biochar is mainly formed from the thermal breakdown of biomass. During pyrolysis the volatile matter is distorted into the gas and residue left as ashes [23]. From the Figure 2 (a)-(e), it can be seen that volatile matter gradually decreases with rise in temperature from 400 to 1000°C during the pyrolysis process. Biochar derived at low temperature contains high VM because the existence of lignin in the biomass sample. Lignin cannot decompose at low temperature i.e. 400°C [24]. When the temperature increases upto 800°C, a rapid loss seen in volatile matter content [25-26] because volatile organic compounds are breakdown at high temperature in organic cyclic and aromatic molecules. The observations of higher content of ash in WHC, RHC, and ESC than SDC, and TWC are similar to the observations of others [27-28]. Enriched content of silica, K_2O, and other ash forming elements presents the in straw and husk materials, may be this reason of higher content of ash than in woody materials [25, 27]. The content of inorganic salts influences the property of volatiles and reduction in the yield of the formation of biochar [28]. The 'fixed carbon' in SDC, WHC, RHC, TWC, and ESC increases with increase in temperature. The related range values have presented in form of figures as we can see.

From ultimate analysis data, it can be observed that the principal components [22] such as, C, H, N, S, and O, differed in both cases (biomass and its corresponding biochars). The higher the content of VM and FC in SDC are reason for the higher the content of C and H in SDC than others. The carbon contents of obtained biochars are 86.57, 59.82, 52.52, 71.82 and 70.05% for SDC, WHC, RHC, TWC, and ESC, respectively. Among all these elements, the contribution of carbon was highest in most of the biochars [23], which was followed by oxygen, then hydrogen and then the rest of the components, although the compositions of nitrogen and sulphur content in biochar materials are low or approximately same. Presence of these C, H and O elements in biomass leads to more char formation as well as to the higher heating value of the product. Among C, H, N, S and O, the contribution of carbon (C) is highest in most of the biochars. Although, nitrogen and sulphur contents are low or approximately same.

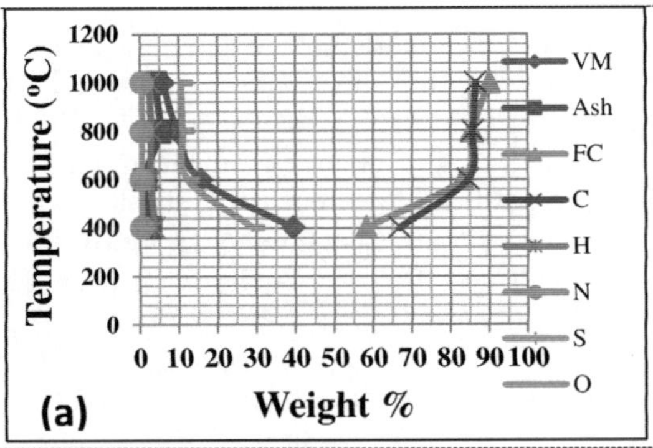

Figure 2. Physiochemical analysis of (a) SDC of all studied biochar samples

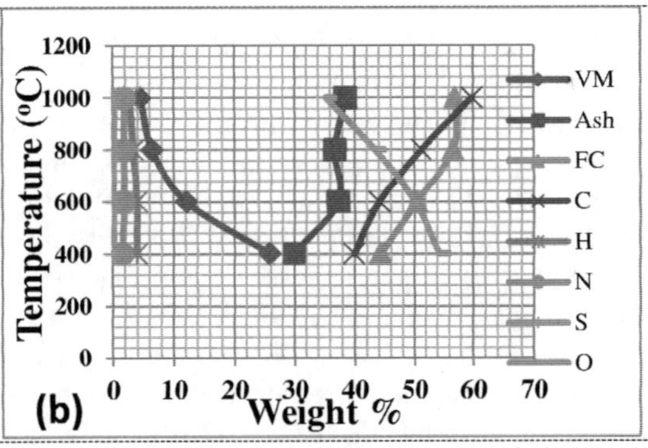

Figure 2. Physiochemical analysis of (b) WHC of all studied biochar samples

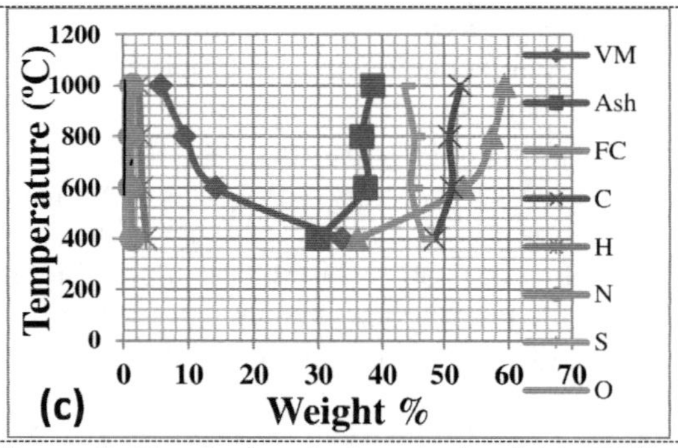

Figure 2. Physiochemical analysis of (c) RHC

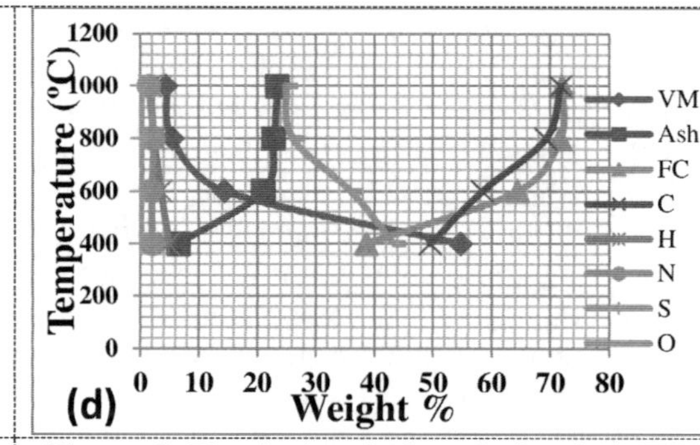

Figure 2. Physiochemical analysis of (d) TWC

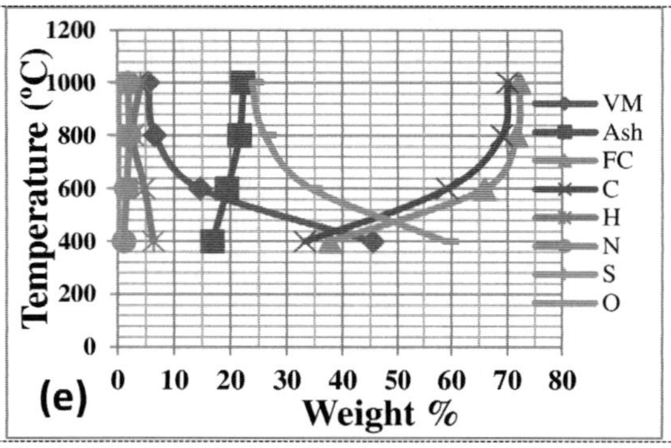

Figure 2. Physiochemical analysis of (e) ESC

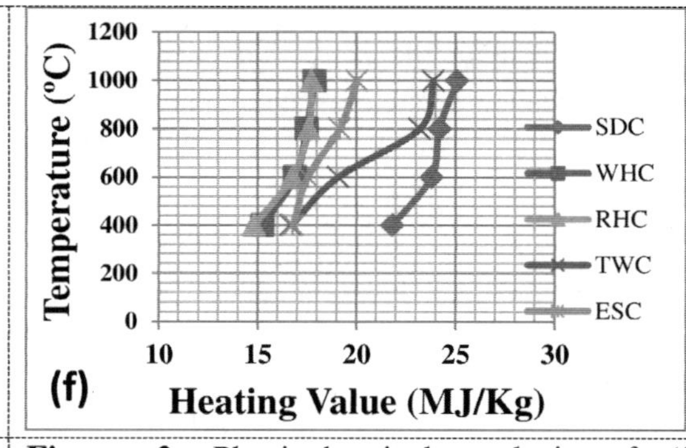

Figure 2. Physiochemical analysis of (f) Heating value

3.4. Fourier transforms infrared spectroscopy analysis of obtained biochar samples

During the process of biomass carbonization/pyrolysis, many reactions takes place in parallel and series as well, including 'aromatisation', 'isomerization', 'dehydration', and 'charring' [29-30]. Pyrolysis process comprises an increase in aromatic structures in the obtained biochars structure. Figure 3 (a)-(e) demonstrate the FTIR spectra of biochar derived from different agricultural biomass samples at different temperatures. From Figures, it was seen that with increasing temperature from 400 to 1000°C, the aliphatic losses occurs at 2944–2845 cm^{-1} band i.e. aliphatic C–H stretch [31]. The peaks appeared at (753–908 and 3040–3005 cm^{-1}), (1375–1445 cm^{-1}) and (1581–1708 cm^{-1}) for aromatic carbon include C–H stretching, C=C & C–C and C–O Carbonyl stretch respectively [32-33]. This is obviously because during the carbonization process modification takes place in the functional group and with this aliphatic C-H band decreases but aromatic C-H band increases [34]. But, at 800-1000°C, the strength of the hydroxyl groups and aromatic groups [31, 35], gradually diminishes, i.e. band occurs at (3204–3411 cm^{-1}) and (1570–1610 and 3045–3005 cm^{-1}) respectively. At lower temperature (300 and 500°C) the presence of number of bands representing functional groups in biochars but it disappeared in the temperature 800-1000°C derived bio-chars. The shoulder around 1610 cm^{-1} in the WHC and RHC bio-chars, allocated to the 'aromatic compounds', is still present at 800°C temperature during the pyrolysis.

At 400°C, there was greater transmittance at band of 1700 to 1000 cm^{-1}, stretching of the carboxyl group of C=O and the molecular bonds of C=C of the aromatic compounds compared to the biochar produced at higher temperatures.

In general, the presence of organic components with chemical responses and distinct structures is observed in the biochar samples, which affects on the reactivity of these materials. The biochars of

IOP Conf. Series: Materials Science and Engineering 1146 (2021) 012023 doi:10.1088/1757-899X/1146/1/012023

SD, WH, RH, TW and EC are composed of the carboxylic acid, ketone, alcohol, ester and aromatic compounds were identified. The persistence of char in the environment for a long time is due to its polycyclic aromatic structure [36] (GLASER et al., 2001). The stability and recalcitrance of biochar can contribute over the long period to retention of water in the soil [37] (OH et al. 2012).
The chemical interactions between biochar and the environment are directly related to its surface structure. The presence of the carboxyl and hydroxyl groups propose that all biochar samples present in this paper could have utilized as a soil amendment for improving of the cation exchange capacity in soil [37] (OH et al. 2012).

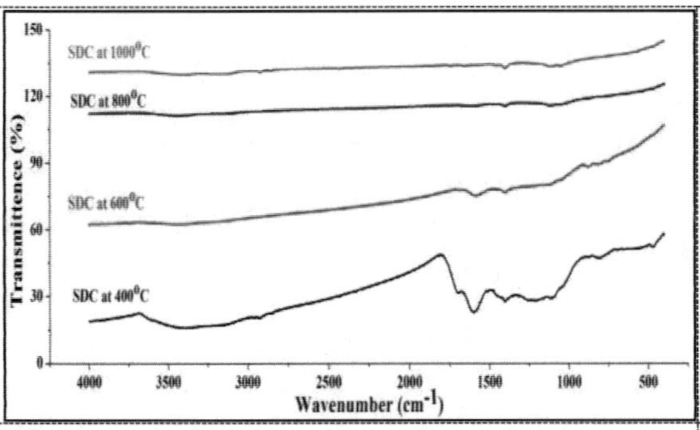

Figure 3(a). FTIR analysis of saw dust char

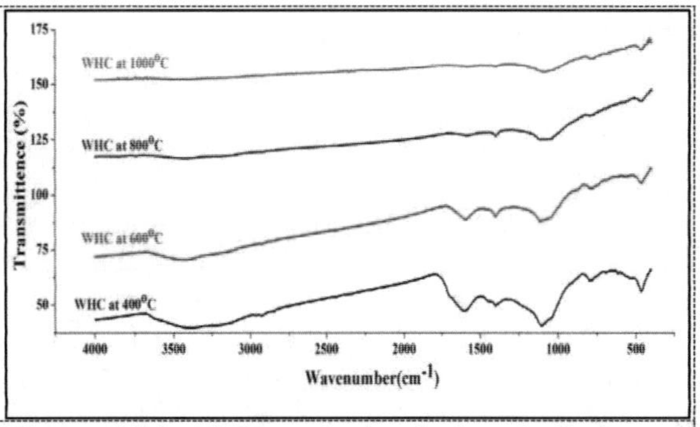

Figure 3(b). FTIR analysis of wheat hull char

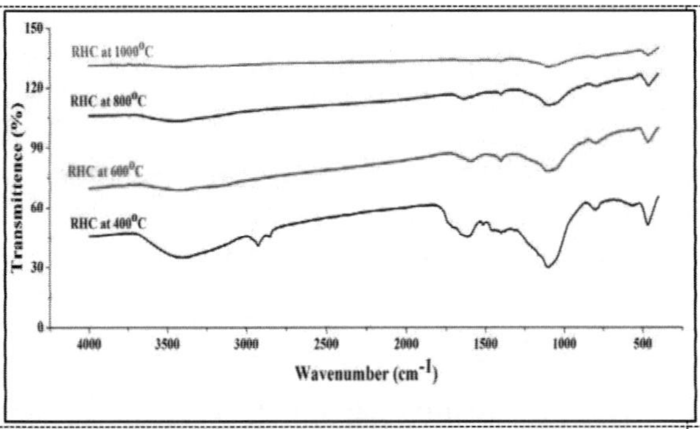

Figure 3(c). FTIR of rice hull char

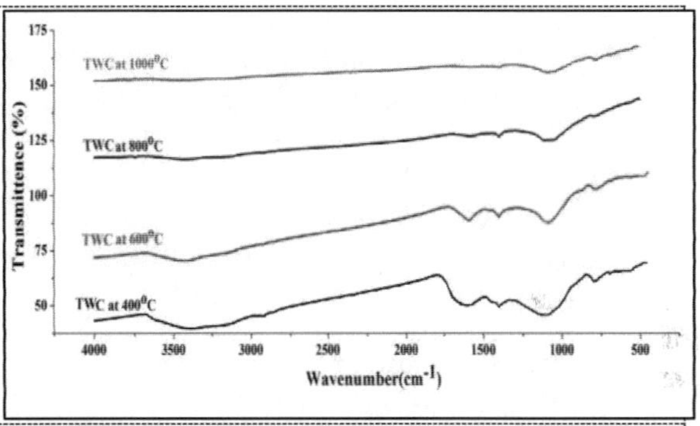

Figure 3(d). FTIR analysis of tea wastes char

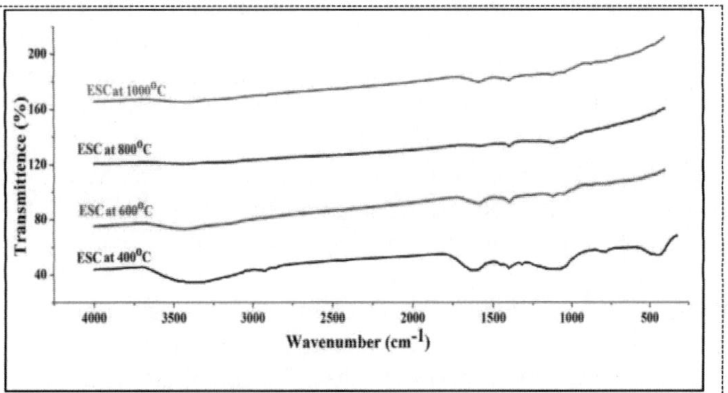

Figure 3(e). FTIR analysis of eucalyptus shell char

4. Conclusion

This study demonstrated the pyrolysis temperature effect on the physical and chemical characteristics of prepared biochars. The influence of the thermal modification on the Saw Dust, Rice Hull, Wheat Hull, Tea Wastes, and Eucalyptus Shell was investigated. With increasing in pyrolysis temperature, the yield%, VM and hydroxyl and aromatic functional groups of biochar significantly decreases. The degree of carbonization for biochars was enhanced with increasing temperature from 400°C to 1000°C. The aromaticity was considerably improved in biochar at high temperatures because a phase transition occurred in studied biomass samples.

FT-IR analyses have shown to high potential to get a deeper understanding of structural relationship with their functional groups.

References

[1] McGowan F 1991. Controlling the greenhouse effect: the role of renewables. Energy Policy, 111–8.
[2] Demirbas A 2005. Potential applications of renewable energy sources, biomass combustion problems in boiler power systems and combustion related environmental issues. Progress in Energy and Combustion Science, 31, 171–192.
[3] Chew JJ and Doshi V 2011. Recent advances in biomass pretreatment—Torrefaction fundamentals and technology. Renewable and Sustainable Energy Reviews, 15, 4212–22.
[4] Williams PT and Besler S 1996. The influence of temperature and heating rate on the slow pyrolysis of biomass, Renewable Energy 1481, 6–7.
[5] Antal MJJ and Grønli M 2003. The Art, Science, and Technology of Charcoal Production, Industrial & Engineering Chemistry Research, 42, 1619–1640.
[6] Downie A and Crosky A 2009. Munroe, P. Physical properties of biochar. In: Biochar for Environmental Management: Science and Technology. 1st edn (eds Lehmann J, Joseph S), Earthscan, London, 13–29.
[7] Angin D 2013. Effect of Pyrolysis Temperature and Heating Rate on Biochar Obtained from Pyrolysis of Safflower Seed Press Cake. Bioresource Technology, 128, 593–597.
[8] Yaman S 2004. Pyrolysis of biomass to produce fuels and chemical feedstock. Energy Conversion and Management, 45 (5), 651-671.
[9] Demirbas A 2004a. Effects of temperature and particle size on bio-char yield from pyrolysis of agricultural residues. J. Ana. and App. Pyro., 72, 243–248.
[10] Demirbas A 2004b. Combustion Characteristics of different biomass fuels. Prog. in En. and Comb. Sci., 30, 219–230.
[11] Mena HLE, Pecora AAB, Beraldo AL 2014. Slow Pyrolysis of Bamboo biomass: Analysis of Biochar properties. Chemical Engineering Transactions, 37, 115-120.
[12] Movasaghi Z, Rehman S, ur Rehman I 2008. Fourier Transform Infrared (FTIR) Spectroscopy of Biological Tissues. Applied Spectroscopy Reviews, 43, 134-179.
[13] Vassilev SV, Baxter D, Andersen LK, Vassileva CG 2010. An overview of the chemical composition of biomass. Fuel, 89, 913–933.
[14] Demirbas A 2002a. Relationships between heating value and lignin, moisture, ash and extractive contents of biomass fuels. Energy Explor. Exploit., 20, 105–11.
[15] Demirbas A 2002b. Fuel characteristics of olive husk and walnut, hazelnut, sunflower and almond shells. Energy Sources, 24, 213–9.
[16] Titiloye JO 2013. Abu Bakar, M.S. Odetoye, T.E. Thermochemical cahracteristic of agricultural wastes from West Africa. Industrial Crops and Products, 47, 199–203.
[17] Cuiping L, Yanyongjie CW, Haitao H 2004. Chemical elemental characteristics of biomass fuels in China. Biomass Bioenerg., 27, 119–30.
[18] Baxter LL, Miles TR, Jr Miles TR, Jenkins BM, Milne T, Dayton D 1998. The behaviour of inorganic material in biomass-fired power boilers: field and laboratory experiences. Fuel Process Technol., 54, 47–78.

[19] Qian Y, Zuo C, Tan J, He J 2007. Structural analysis of bio-oils from sub and supercritical water liquefaction of woody biomass. Energy, 32, 196-202.
[20] Ghani WAWAK 2014. Sawdust-derived Biochar: Characterization and CO_2 adsorption/desorption study. J. of App. Sci., 14(13), 1450-1454.
[21] Guo J and Chen B 2014. Insights on the molecular mechanism for the recalcitrance of biochar: interactive effects of carbon and silicon components. Environ. Sci. Technol., 48, 9103–9101.
[22] Sukiran MA, Kheang LS, Bakar NA, May CY 2011. Production and characterization of bio-char from the pyrolysis of empty fruit bunches. J. Appl. Sci., 8, 984–988.
[23] Jindo K, Mizumoto H, Sawada Y, Sanchez-Monedero MA, Sonoki T 2014. Physical and chemical characterization of biochars derived from different agricultural residues. Biogeosciences, 11, 6613–6621.
[24] Keiluweit M, Nico PS, Johnson MG, Kleber M 2010. Dynamic molecular structure of plant biomass-derived black carbon (Biochar). Environmental Science & Technology, 44, 1247–1253.
[25] Spokas KA 2010. Review of the stability of biochar in soils: predictability of O:C molar ratios. Carbon Manag, 1(2), 289-303.
[26] Bi YY, Guan JP, Wang DL 2008. Comprehensive Utilization Technology of Straw Resources in China (in Chinese), Beijing: China Agricultural Science and Technology Press.
[27] Jensen A, Dam-Johansen K, Wojtowicz MA, Serio MA 1998. TG-FTIR Study of the Influence of Potassium Chloride on Wheat Straw Pyrolysis, Energy & Fuels, 12, 929–938.
[28] Collard FX and Blin J 2014. A review on pyrolysis of biomass constituents: Mechanisms and composition of the products obtained from the conversion of cellulose, hemicelluloses and lignin. Renew. Sustain. Energy Rev., 38, 594–608.
[29] Kan T, Strezov V, Evans TJ 2016. Lignocellulosic biomass pyrolysis: A review of product properties and effects of pyrolysis parameters. Renewable and Sustainable Energy Reviews, 57, 1126-1140.
[30] Haslinawati MM, Matori KA, Wahab ZA, Sidek HAA, Zainal AT 2009. Effects of temperature on the ceramic from rice husk ash. Int. J. Basic Appl. Sci., 9, 111–116.
[31] Wu W, Yang M, Feng Q, McGrouther K, Wang H, Lu H, Chen Y 2012. Chemical characterization of rice straw-derived biochar for soil amendment. Biomass & Bioenergy, 47, 268–276.
[32] Chen B and Chen Z 2009. Sorption of naphthalene and 1-naphthol by biochars of orange peels with different pyrolytic temperatures. Chemosphere, 76, 127-133.
[33] Lee JW 2010. Characterization of of biochars produced from corn-stovers for soil amendment. Environmental Science and Technology, 7970–7974.
[34] Peng X, Ye LL, Wang CH, Zhou H, Sun B 2011. Temperature and duration-depend rice straw-derived biochar: characteristics and its effects on soil properties of an Untisol in southern China. Soil Tillage Res., 112, 159–166.
[35] Yuan JH, Xu RK, Zhang H 2011. The forms of alkalis in the biochar produced from crop residues at different temperatures. Bioresource. Technol., 102, 3488–3497.
[36] GLASER, B. et al. The Terra Preta phenomenon - a model for sustainable agriculture in the humid tropics. Naturwissens chaften. v. 88, p. 37-41, 2001.
[37] Oh TK, Choi BS, Shinogi Y, Chikushi J (2012) Characterization of biochar derived from three types of biomass. J Fac Agric, Kyushu Univ 57(1):61–66.

Study of non – Newtonian flow, heat transfer and entropy generation characteristics in trapezoidal corrugated channel

D P Singh[1*], S K Rathore[1] and S Chaudhuri[2]

[1]Department of Mechanical Engineering, National Institute of Technology Rourkela, Rourkela - 769008 (Odisha), India
[2]School of Mechanical Engineering, KIIT, Deemed to be University, Bhubaneswar, Odisha, 751024, India

*corresponding author's e-mail address : dps.12jhs@gmail.com

Abstract. This study is focused on the thermo-hydraulic and entropy generation characteristics of a non-Newtonian fluid in a trapezoidal shaped wavy channel. The study has been conducted for different Reynolds number ranging from 25 – 100 and power law index varying from 0.5 – 1.5. The variation of average Nusselt number, pressure drop, thermal entropy generation, viscous entropy generation are being calculated. For the discretization of the governing equations, finite volume method is adopted in Ansys Fluent. Average Nusselt number value increases with an increase in the value of Reynolds number and decreases with an increase in power law index. On the other hand, pressure drop decreases on increasing the value of Reynolds number, whereas on increasing the value of power law index, pressure drop value showing a decreasing trend. Thermal entropy generation and viscous entropy generation follows almost the same trend but both having different order of magnitude. The trend shows that on increasing the value of *Re* both the entropy generations increases whereas on increasing the power law index, it decreases.

1. Introduction

The technological developments in the past shows us that more emphasis is given on to increase the output by supplying the same amount of input. Many technological modifications have been carried out in the industries related to process engineering, air conditioning, refrigeration, beverages industries, where the heat transfer is increased between the two fluids exchanging heat from each other or only between one fluid and heat source. Generally, two methods are adopted for increasing the heat transfer namely active and passive method. In active method changes in input is made whereas in passive method changes in the geometry of the flow passage is made in order to achieve an increased output.

Many studies have been reported on the performance of the wavy channels using numerical and experimental techniques. Rush *et al.* [1] experimentally concluded that an enhancement in the heat flow in corrugated channel is seen because of the formation of recirculation areas and these area depends upon geometry of channel and flow variables like *Re*. Yang and Chen [2] numerically found out thermo-hydraulic characteristics for a Newtonian fluid in V shaped wavy plate, *Re* ranging between 2000 – 5500, they observed that heat transfer increment is directly proportional to the included angle in the V shaped wavy plate.

Nomenclature

Re	Reynolds number
Nu	Nusselt number
Pr	Prandtl number
Be	Bejan number
P	Pressure($N\ m^{-2}$)
T	Temperature (K)
C_p	Specific heat at constant pressure ($J\ kg^{-1}\ K^{-1}$)
x, y	Cartesian coordinate directions
u, v	Velocities in axial and transverse directions, respectively ($m\ s^{-1}$)
k_f	Thermal conductivity of fluid($W\ m^{-1}\ K^{-1}$)
a	Amplitude of corrugated profile (m)
E	Starting point of wavy profile
F	Ending point of wavy profile
2L	Average gap between corrugated profile (m)
s	Profile of the corrugation part
k_c	Flow consistency coefficient
Δp	Dimensionless pressure drop
m	Power law index
h	Coefficient of heat transfer($W\ m^{-2}\ K^{-1}$)
$s_{gen,thermal}$	Entropy generation per unit volume due to temperature gradient ($W\ K^{-1}\ m^3$)
$s_{gen,viscous}$	Entropy generation per unit volume due to viscous effects ($W\ K^{-1}\ m^3$)
s_{gen}	Total Entropy generation per unit volume ($W\ K^{-1} m^3$)

Greek symbols

Φ	Length of corrugation surface from Point E to x
μ	Fluid dynamic viscosity ($N\ s\ m^{-2}$)
$\dot{\gamma}$	Rate of shear strain
τ	shear stress
μ_a	Apparent viscosity
ρ	Fluid density ($kg\ m^{-3}$)
λ	Wavelength of wavy part (m)

subscripts

i	inlet
W	wavy wall
avg	average

Wang and Chen [3] investigated numerically, the changes in Nu_{avg} and Nu_{Local} and coefficient of skin friction in sinusoidal channel using Newtonian fluid, and their variations on changing the parameters like sinusoidal wave amplitude, wavelength, Prandtl number and Reynolds number. Wang and Vanka [4] simulated numerically the steady Newtonian flow in a wavy passage for *Re* upto 180 and they reported that the value of Nu_{avg} is more for corrugated channel as compared to that of a straight channel and it increases as the state of flow changes from laminar to transition region. Bahaidarah *et al.* [5] numerically measured the changes in momentum and heat flow in a corrugated channel and compared the trend with straight channel for a Newtonian fluid having $Pr = 0.7$ and *Re* in the range 25 – 400. They noticed that heat flow rate is same in both the channels at low *Re* but at high *Re* wavy channel shows more heat transfer. Mehta and Pati [6] numerically studied characteristics of heat flow in sinusoidal channel which is imposed to the sinusoidal heat flux for *Re* ranging from 5 to 300, and they found out that for the lower amplitude of sinusoidal heat flux, average Nusselt number is highest. Mohammed *et al.* [7,8] made studies about variation of heat flow and other quantities, nearby entrance length of a corrugated channel for a forced convection in laminar regime. Pati *et al.* [9] made numerical studies and concluded that heat flow performance factor of the raccoon profile is always lower than serpentine wavy profile.

Metzner and Reed [10] performed studies on steady non-Newtonian fluids and provided correlation for the general Reynolds number in turbulent, laminar and transient regimes. Akbarzadeh *et al.* [11] made studies numerically to find the variations in heat flow, entropy generations, performance parameter and drop in pressure in three different wavy channels for a Newtonian fluid for *Re* in the range 400 – 1400. Guha and Pradhan [12] conducted numerical simulations on non-Newtonian fluid to study the forced convection on a horizontal plate. They concluded that *Pr* and index of power law are the two main factors that contributes to the change in pressure, temperature and velocity. Shubham *et al.* [13] studied the characteristics of heat flow for a non-Newtonian flow flowing in a sinusoidal corrugated profile channel and they reported that minimum pressure drop and maximum heat transfer occurs in case of pseudo-plastic fluids. Esfahani and Shahabi [14] studied entropy generation in a pipe flow using high Prandtl number fluid and the pipe wall subjected to different combinations of wall fluxes. Their studies show that maximum entropy generations are in the situation where heat flux is of decreasing nature along the length of the flow. Bejan [15,16] have given entropy generation minimization method for the convective heat flow which deals with the calculation to find the entropy generated because of heat transfer through finite temperature difference and viscous irreversibilities present in the fluid flow. It helps in making the process more efficient as one can reduce the factors that are mainly responsible for entropy generation.

As non-Newtonian fluids are majorly being used in the industries and less literature and studies are reported to illustrate the heat transfer and entropy generation characteristics. This study is an effort to fill this gap. The profiles of Nu_{avg}, dimensionless pressure drop, thermal and viscous entropy generations for non-Newtonian fluid comprising of both the pseudoplastic ($m < 1$) and dilatant fluid ($m > 1$), for Re = 25 to 100, Pr = 6.93 and the index of power law varying from 0.5 to 1.5 are reported in this study.

2. Problem definition

A 2D trapezoidal domain is considered for study, depicted in figure 1. Average distance between the two trapezoidal surface is 2L, starting and ending length of 3L and 5L, respectively are smooth adiabatic wall in nature. This starting and ending length are made in such a way that full development of flow can be experienced at entry of the wavy profile and no recirculation is formed at the exit of wavy section [3,7,8]. Length of 12L between the start and end length are isothermal in nature and trapezoidal in geometry. Amplitude of trapezoidal wavy profile is 0.4L, wavelength is 4L.

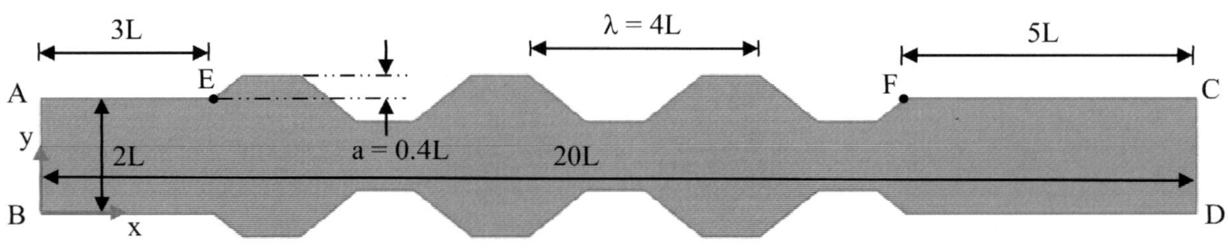

Figure 1. Schematic geometry of a trapezoidal channel.

In the above figure AB is the inlet, CD is the outlet, point E is the starting point of the wavy channel and point F is the ending point of it. Assumptions made for this study are: -

I. Ostwald-de waele model (power law model) is used.
II. Flow is considered as incompressible, 2D and steady in nature.
III. Transverse heat conduction and viscous dissipation terms are neglected.
IV. No internal heat generation.

Ostwald-de waele model states that, shear stress and rate of shear strain are directly proportional for a given range of rate of shear strain. Mathematically, it can be written as:

$$\tau_{xy} = \mu_a \left(\dot{\gamma}_{xy}\right) \tag{1}$$

Where, μ_a is the apparent viscosity of the non-Newtonian fluid, mathematically it is expressed as:

$$\mu_a = k_c\left(\dot{\gamma}_{xy}\right)^{m-1} \tag{2}$$

In the above equation (2), by varying the value of *m* from less than 1 to greater than 1, one can move from shear thinning fluid (pseudoplastic) to shear thickening fluid (dilatant).

3. Mathematical equations

The governing equations are mentioned below:

- Mass conservation

$$\frac{\partial u}{\partial x} + \frac{\partial v}{\partial y} = 0 \tag{3}$$

- Momentum conservation equations in x and y coordinates respectively

$$\rho\left(u\frac{\partial u}{\partial x} + v\frac{\partial u}{\partial y}\right) = -\frac{\partial p}{\partial x} + \mu\left(\frac{\partial^2 u}{\partial x^2} + \frac{\partial^2 u}{\partial y^2}\right) \tag{4}$$

$$\rho\left(u\frac{\partial v}{\partial x} + v\frac{\partial v}{\partial y}\right) = -\frac{\partial p}{\partial y} + \mu\left(\frac{\partial^2 v}{\partial x^2} + \frac{\partial^2 v}{\partial y^2}\right) \tag{5}$$

- Energy conservation equation

$$\rho C_p\left(u\frac{\partial T}{\partial x} + v\frac{\partial T}{\partial y}\right) = k_f\left(\frac{\partial^2 T}{\partial x^2} + \frac{\partial^2 T}{\partial y^2}\right) \tag{6}$$

4. Boundary condition

Above mathematical equations are solved using the below boundary conditions:

- Inlet of the channel.

$$\frac{u}{u_{avg}} = \frac{2m+1}{m+1}\left(1 - \left|1 - \frac{y}{L}\right|^{\frac{m+1}{m}}\right), v = 0, T = T_i \tag{7}$$

- Channel wall.

$$\frac{\partial T}{\partial y} = 0 \tag{8}$$

$$T = T_W \text{ (such that } T_W > T_i) \tag{9}$$

Equation (8) is valid for straight wall and equation (9) is valid for corrugated part.

- Channel outlet

$$\frac{\partial u}{\partial x} = 0\,;\ \frac{\partial v}{\partial x} = 0\,;\ \frac{\partial T}{\partial x} = 0 \tag{10}$$

5. Solution methodology

Governing equations are solved using the method of finite volume in Ansys Fluent using SIMPLE algorithm [17]. Second order upwind scheme is used for the discretization. The convergence criteria are taken as 10^{-06} for mass and momentum conservation, and 10^{-09} for energy conservation.

- Thermo-hydraulic transport characteristics analysis.
  Local and average Nusselt number can be calculated using below equations:

$$Nu_{Local} = \frac{-\left(\frac{\partial T}{\partial \sigma}\right)L}{T_W - T_i} \; ; \; Nu_{avg} = \frac{h_{avg}\, L\, \phi}{K\, x} \quad (11)$$

$$\text{where, } h_{avg} = \frac{1}{E - F}\int_F^E h_x ds \quad (12)$$

- Entropy generation analysis.
  By this analysis one can know about the losses occuring while the process is taking place. Thermal and viscous entropy generations are calculated using entropy generation minimization technique [16].

$$S_{gen,thermal} = \frac{k_f}{T^2}\left[\left(\frac{\partial T}{\partial x}\right)^2 + \left(\frac{\partial T}{\partial y}\right)^2\right] \quad (13)$$

$$S_{gen,viscous} = \frac{k_c}{T}\left(\frac{u_{avg}}{L}\right)^{m-1}\left[2\left\{\left(\frac{\partial u}{\partial x}\right)^2 + \left(\frac{\partial v}{\partial y}\right)^2\right\} + \left(\frac{\partial u}{\partial y} + \frac{\partial v}{\partial x}\right)^2\right] \quad (14)$$

$$S_{gen} = S_{gen,thermal} + S_{gen,viscous} \quad (15)$$

$$Be = \frac{S_{gen,thermal}}{S_{gen,thermal} + S_{gen,viscous}} \quad (16)$$

Equation (16) shows the relation of Bejan number, which helps in finding the dominance of entropy generation because of temperature gradient in the total entropy generation. Dimensionless form of pressure difference, entropy generations are given below:

$$\Delta p = \frac{P}{\rho u_{avg}^2} \; ; \; N_{gen,thermal} = \frac{S_{gen,thermal}\, L^2}{k_f} \; ; \; N_{gen,viscous} = \frac{S_{gen,viscous}\, L^2}{k_f} \quad (17)$$

6. Grid independence test and meshing

Four grids are considered for performing grid independence test. The properties of the fluid used is such that Prandtl number is 6.93, *Re* = 100 and *m* = 1.6. Based on the percentage difference in average Nusselt number (Table 1) for the considered grid sizes, a grid size of 140×1400 is taken for the present study. Figure 2 shows the meshed computational domain. The mesh is created in such a way that it is fine nearby surface and coarse in the core region because the gradients values of temperature, pressure and velocity are more near the walls.

Table 1. Average Nusselt number value and percentage difference at different grid size.

Grid size	Nu_{avg}	Percentage difference
80 × 800	4.310725	-
100 × 1000	4.290502	0.4691
120 × 1200	4.282599	0.1841
140 × 1400	4.278881	0.0868

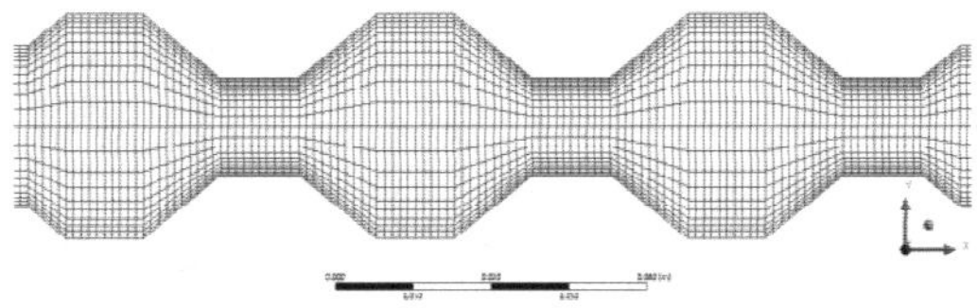

Figure 2. A sample meshing used in the present study.

7. Validation study

Numerical techniques used in the present study are validated using the value of Nu_{Local} obtained by Wang and Chen [3] for Prandtl number 6.93, Reynolds number 100 and index of power law 1 using sinusoidal wavy channel whose dimensionless amplitude is 0.2. Another validation is done to compare the entropy generations values with the findings of Esfahani and Shahabi [14] as shown in figure 3(b).

Esfahani and Shahabi [14] have conducted the study for a high *Pr* fluid (*Pr* = 13400) for a straight pipe of length 1m, diameter 0.025m and the velocity at inlet 0.02 m/s.
Figure 3 shows the graphs obtained in the validation study, a close match can be seen in both the cases. Figure 3(a) depicts that Nu_{Local} is maximum for wave one and decrement is seen on moving in flow direction and this decrement is significant upto fourth wave and after that this decrement becomes almost nil as formation of thermal boundary layer takes place.

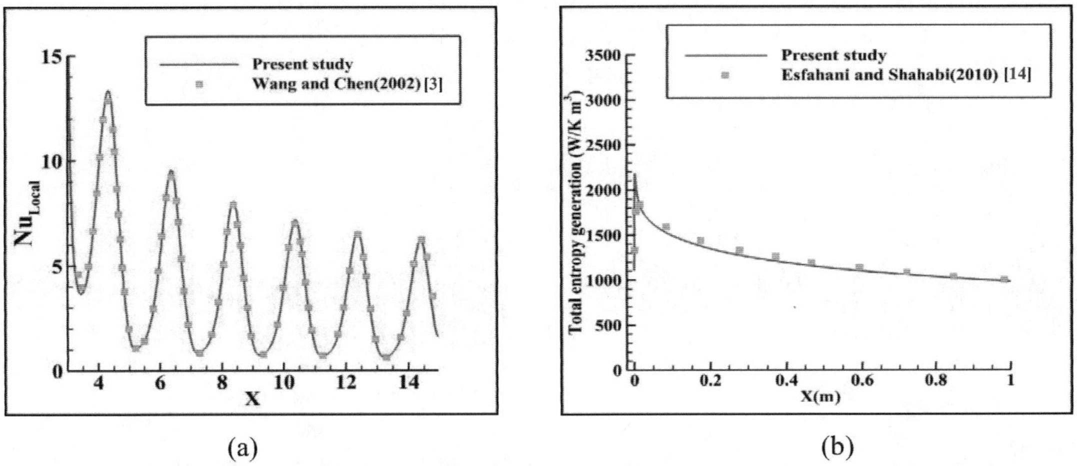

Figure 3. Comparison of present study with (a) Wang and Chen [3] results (b) Esfahani and Shahabi [14] results.

8. Results and discussion

The velocity streamline contours are displayed in figure 4 for *Re=25* and *Re=100* for the values of *m=0.5* and *m=1.5*. It can be observed that flow reversal is taking place in the diverging wall because of which a recirculation zone is created and the size of recirculation zone is directly proportional to *Re*. For pseudoplastic fluids (*m*<1), velocity is higher inside the channel as compared to dilatant fluids (*m*>1) because the apparent viscosity value is higher in dilatant fluids, offering higher resistance to the flow.

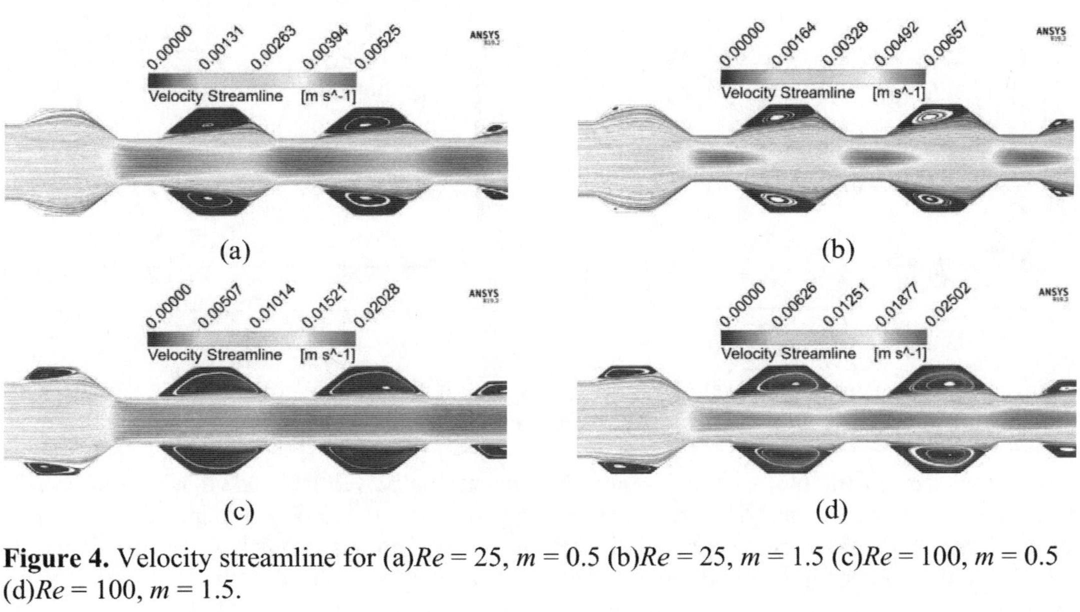

Figure 4. Velocity streamline for (a)*Re* = 25, *m* = 0.5 (b)*Re* = 25, *m* = 1.5 (c)*Re* = 100, *m* = 0.5 (d)*Re* = 100, *m* = 1.5.

As the velocity in the region near the corrugated surface is lower compared to the velocity in the core region, therefore, the convection is dominant at the central region. Therefore, convective transport of energy is higher in the core, resulting in a lower temperature region in the central region. Because of lower velocity near the walls, convective transport of energy is lower, causing higher temperature in that region. Temperature contours are depicted in figure 5 for *Re=25* and *Re=100* for the values of *m=0.5* and *m=1.5*. It can be observed that as the nature of the fluids changes from shear thinning to shear thickening category, waviness of the contour displayed in blue colour, increases in the core region. The increase in waviness is seen because of the velocity pattern followed by the shear thickening fluids as a result of increase in the apparent viscosity.

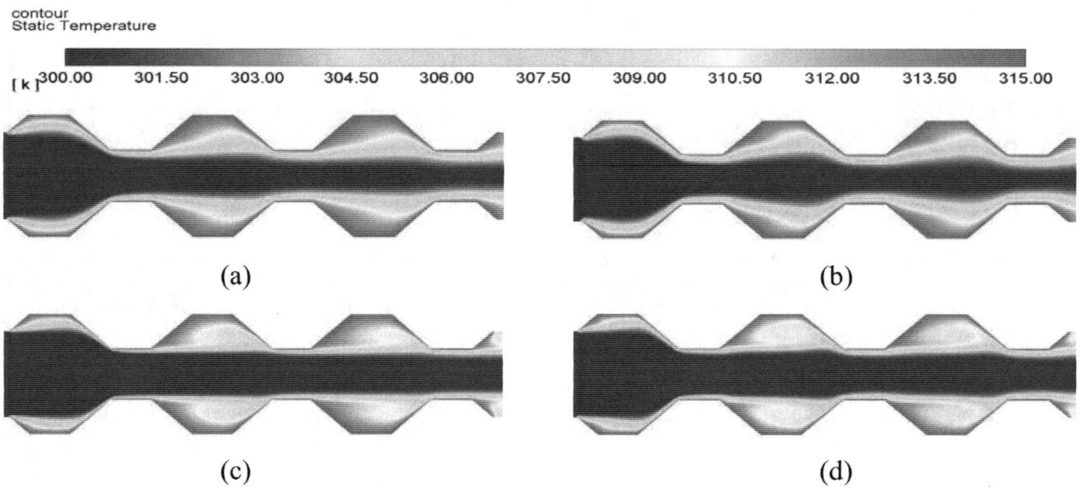

Figure 5. Contour of temperature for (a)Re = 25, m = 0.5 (b)Re = 25, m = 1.5 (c)Re = 100, m = 0.5 (d)Re = 100, m = 1.5.

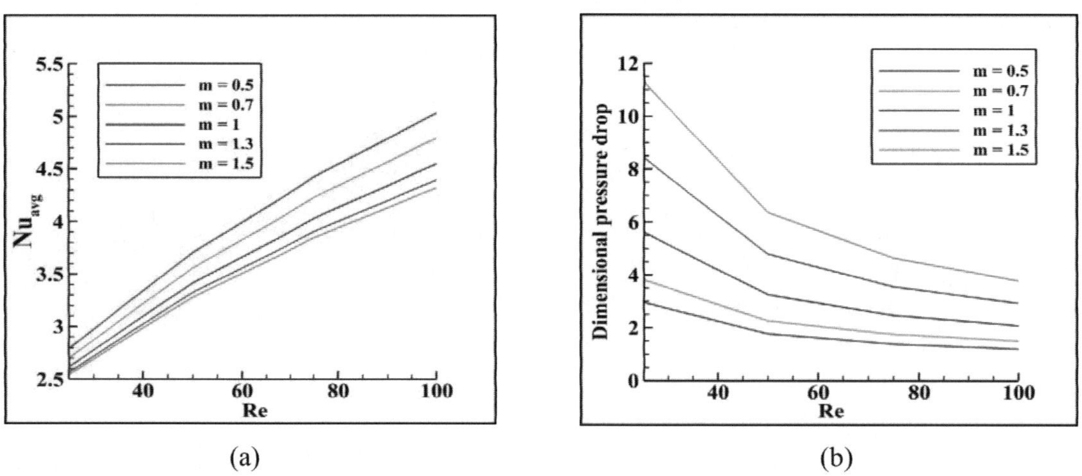

Figure 6. (a) Average Nusselt number variation with respect to *Re* (b) variation of dimensionless pressure drop versus *Re* for different values of power law index.

Figure 6(a) depicts the plot of Nu_{avg} versus *Re* for various power law index, whereas figure 6(b) displays the dimensionless pressure drop. It is observed, as the value of *Re* increases, flow velocity increases which enhances convection and therefore Nu_{avg} increases. Also, as we move from pseudoplastic to dilatent fluids at the same *Re*, Nu_{avg} decreases because of the increase in the value of apparent viscosity. Pressure drop shows a decreasing trend as value of Reynolds number is increased.

As the value of Reynolds number increases, it results in an increase in the velocity which helps in pushing the fluid forward. However, the viscosity is higher for $m > 1$ therefore, a higher-pressure drop is observed for shear thickening fluids. Entropy generation value shows an increment when an increase in temperature gradient is seen, since the temperature gradient is high for the peudoplastic fluids therefore, its thermal entropy generation is higher. However, for higher value of Reynolds number, the size of recirculation region is increased which increases the mixing of fluid and decreases the temperature gradient as depicted in figure 7. When index of power law is greater than unity, fluid viscosity increases. Therefore, the fluid gets sufficient time to transfer heat and this results in the decrement of temperature gradient. Figure 8 shows the contours of viscous entropy generation, where it is evident that by increasing the Reynolds number, viscous entropy generation increases as increase in the size of recirculation zone offers a kind of viscous resistance.

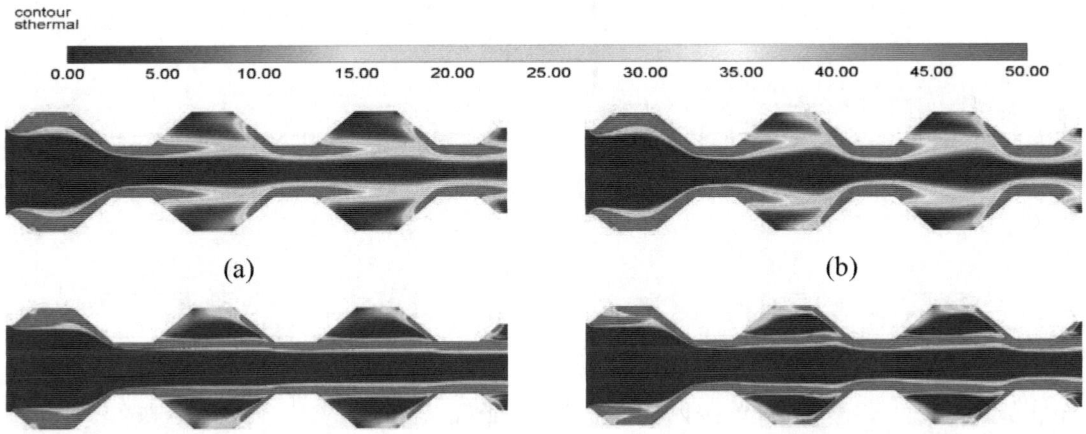

Figure 7. Thermal entropy generation contours for (a)$Re = 25$, $m = 0.5$ (b)$Re = 25$, $m = 1.5$ (c)$Re = 100$, $m = 0.5$ (d)$Re = 100$, $m = 1.5$.

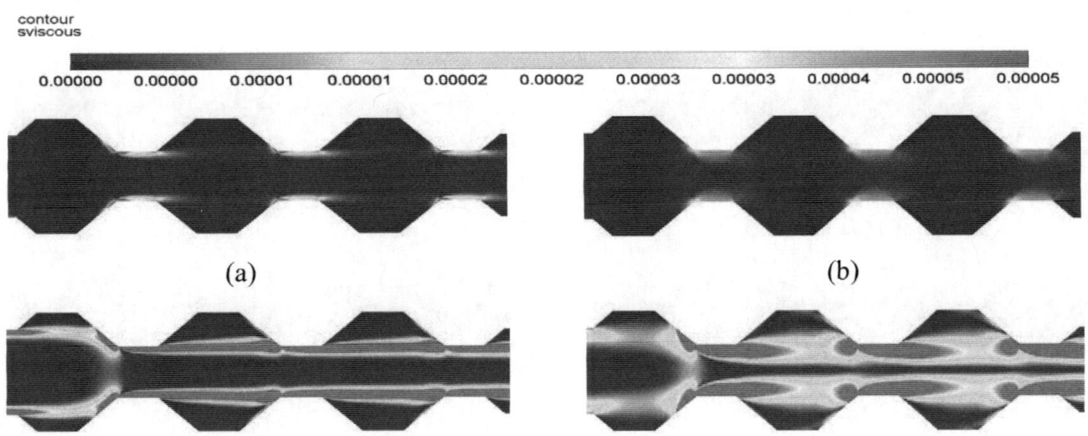

Figure 8. Viscous entropy generation contours for (a)$Re = 25$, $m = 0.5$ (b)$Re = 25$, $m = 1.5$ (c)$Re = 100$, $m = 0.5$ (d)$Re = 100$, $m = 1.5$.

Figure 9 depicts the plot of dimensionless entropy generations, where figure 9(a) shows the plot of thermal entropy generation versus Reynolds number, whereas figure 9(b) shows the viscous entropy generation trend with Reynolds number for different values of *m*. Both the graphs depict almost the same trend where an increment in the value of entropy generations can be seen on increasing the Reynolds number. For a constant Reynolds number, entropy generation value decreases due to an increase in the value of *m*. Bejan number contours shown in figure 10 help in determining the portion of thermal entropy generation in the total entropy generated and its value lies between 0 and 1. Where 0 means that viscous entropy generation is dominating; whereas a value of Bejan number equal to unity indicates thermal entropy generation is dominating. From the above contour, it is visible that with increment in the value of Reynolds number and power law index, thermal entropy generation dominance shows a decrement in the computational domain.

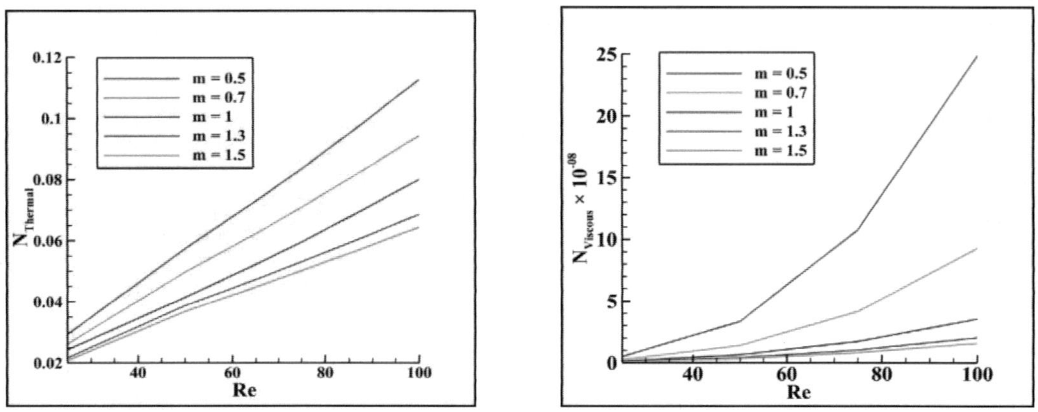

(a) (b)

Figure 9. (a) Dimensionless thermal entropy generation variation with respect to Re (b) variation of dimensionless viscous entropy generation with respect to the *Re* for different values of index of power law.

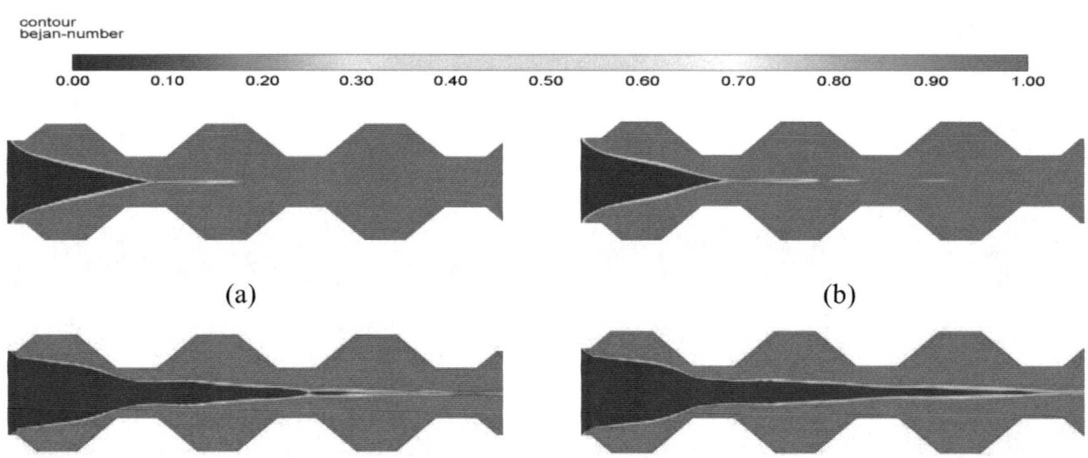

Figure 10. Bejan number contours for (a)$Re = 25$, $m = 0.5$ (b)$Re = 25$, $m = 1.5$ (c)$Re = 100$, $m = 0.5$ (d)$Re = 100$, $m = 1.5$.

9. Conclusions

Some of the important findings from this study are listed below:

i. Average Nusselt number will be maximum for shear thinning fluids and for high Reynolds numbers.
ii. As Reynolds number increases, pressure drop decreases, shear thinning fluids have higher pressure drop compared to shear thickening fluids.
iii. Both thermal and viscous entropy generation increases on increasing the *Re* and is higher for shear thinning fluids.
iv. Bejan number shows that as Reynolds number and index of power law are increased, dominance of thermal entropy generation starts decreasing in the computational domain.

References

[1] Rush TA, Newell TA, Jacobi AM 1999 An experimental study of flow and heat transfer in sinusoidal wavy passages. *Int J Heat Mass Transf*; **42**:1541–53.
[2] Yang Y-T, Chen P-J 2010 Numerical simulation of fluid flow and heat transfer characteristics in channel with V corrugated plates. *Heat Mass Transf*; **46**:437–45
[3] Wang C-C, Chen C-K 2002 Forced convection in a wavy-wall channel. *Int J Heat Mass Transf*; **45**:2587–95. https://doi.org/10.1016/S0017-9310(01)00335-0
[4] Wang GV, Vanka SP 1995 Convective heat transfer in periodic wavy passages. *Int J Heat Mass Transf*; **38**:3219–30. https://doi.org/10.1016/0017-9310(95)00051-A
[5] Bahaidarah HMS, Anand NK, Chen HC 2005 Numerical Study of Heat and Momentum Transfer in Channels with Wavy Walls. *Numer Heat Transf Part Appl*; **47**:417–39.
[6] Mehta SK, Pati S 2018 Effect on non-uniform heating on heat transfer characteristics in wavy channel. *Fifth Int. Conf. On Compuational Methods for Thermal Probems.* ThermaComp-18, July 9-11, 2018. Bangalore, India; p. 498–501
[7] Mohamed N, Khedidja B, Belkacem Z, Michel D 2007 Numerical study of laminar forced convection in entrance region of a wavy wall channel. *Numer Heat Transf Part Appl*; **53**:35
[8] Mohamed N, Wided BR, Mohamed EM, karim MA el, Mohamed B 2013 Numerical Investigation on the Fluid Flow and Heat Transfer in the Entrance Region of Wavy Channel. *Energy Procedia*; **36**:76–85. https://doi.org/10.1016/j.egypro.2013.07.010
[9] Pati S, Mehta SK, Borah A 2017 Numerical investigation of thermo-hydraulic transport characteristics in wavy channels: comparison between raccoon and serpentine channels. *Int Commun Heat Mass Transf*; **88**:171–6
[10] Metzner AB, Reed JC 1955 Flow of non-newtonian fluids—correlation of the laminar, transition, and turbulent-flow regions. *AIChE J*; **1**:434–40.
[11] Akbarzadeh M, Rashidi S, Esfahani JA 2017 Influences of corrugation profiles on entropy generation, heat transfer, pressure drop, and performance in a wavy channel. *Appl Therm Eng*; **116**:278–91. https://doi.org/10.1016/j.applthermaleng.2017.01.076
[12] Guha A, Pradhan K 2014 Natural convection of non-Newtonian power-law fluids on a horizontal plate. *Int J Heat Mass Transf*; **70**:930–8.
[13] Shubham, Saikia A, Dalal A, Pati S 2018 Thermo-hydraulic transport characteristics of non-Newtonian fluid flows through corrugated channels. *Int J Therm Sci*; **129**:201–8. https://doi.org/10.1016/j.ijthermalsci.2018.02.005
[14] Esfahani JA, Shahabi PB 2010 Effect of non-uniform heating on entropy generation for the laminar developing pipe flow of a high Prandtl number fluid. *Energy Convers Manag*; **51**:2087–97. https://doi.org/10.1016/j.enconman.2010.02.022
[15] Bejan A 2013 *Entropy Generation Minimization: The Method of Thermodynamic Optimization of Finite-Size Systems and Finite-Time Processes*; CRC Press
[16] Bejan A 1979 A study of entropy generation in fundamental convective heat transfer. *J. Heat Transfer;* **101(4)**: 718-25. https://doi.org/10.1115/1.3451063
[17] Patankar SV, Minkowycz WJ, Sparrow EM 1980 *Series in computational methods in mechanics and thermal sciences*. McGraw-Hill Book Co NY:1–197

Study on PCM Assisted Constant Temperature Water Heating System

Pragnan Lad[1], Rajat Saxena[1, 2,*] and Jatin Patel[1]

[1]Department of Mechanical Engineering, School of Technology, Pandit Deendayal Energy University, Gandhinagar (Gujarat), India
[2]Department of Chemical Engineering, School of Technology, Pandit Deendayal Energy University, Gandhinagar (Gujarat), India

[*]Corresponding Author`s E-mail: Rajat.Saxena@sot.pdpu.ac.in

Abstract. Spread of novel coronavirus across the globe creates panic situations internationally. Major challenge of this pandemic situation is the absence of a specific treatment or vaccine for curing the disease. As per published report of WHO, temperature above 55 ^{0} C is potentially kill Corona virus. This study focuses on design a water heating system that could provide water above the required temperature with minimal input power even during the off-sunshine hours. The thermal performance of Phase Change Material (PCM) integrated solar water heating system providing water temperature above 56 ^{0} C both during the day and the off sunshine hours has been discussed in this present study.

The phase change behaviour of paraffin-wax-6035 confined within a copper tube of 100 mm diameter and 1000 mm length and 1 mm thickness, has been studied using CFD analysis. The analysis is carried out for three different cases viz. conventional solar water heating system, modified solar water heating system integrated with bare PCM tube and finned PCM tube of 1 mm fin thickness. The thermal storage capacity of solar water heater is compared based on the thermal storage capacity factor which is calculated to be 1.1149 & 1.1586 for bare PCM tube and finned PCM tube case, respectively with respect to the conventional solar water heater.

Keywords: Phase Change Material; COVID-19; energy efficiency; CFD; solar water heater

Table 1. Indexes & Abbreviation used in present article

Index	Description	Abbreviation	Description
Cu	Copper domain	CFD	Computational Fluid Dynamics
Fin	Fin	HTF	Heat Transfer Fluid
PCM	Phase Change Material	PCM	Phase Change Material
Steel	Storage tank wall	WHO	World Health Organization
Wall	Isothermal inner wall	SWH	Solar Water Heater
Water	Water filled in storage tank	FCC	Face Centre Cubic-structure

Table 2. Symbols used in present article

sign	Description	sign	Description	sign	Description
T_B	Boiling temperature (K)	h_{mol}	Molar enthalpy (kJ/mol)	$\hat{j}$	Unit vector in Y direction
t	Duration (s)	M_o	Molecular weight (g/mol)	u	Velocity component in X direction (m/s)
b	Correlation coefficient	T_{wall}	Temperature of considered wall (K)	v	Velocity component in X direction (m/s)
z	Carman-Kozeny constant for the source term (kg/m^{3}s)	$\mathcal{t}$	Thermal status maintenance time (s)	$\vec{V}$	Velocity field of fluid (m/s)
$\mathcal{t}_{\mathcal{R}}$	Duration for water to touch solidus temperature of PCM (s)	r	Small supportive constant in source term for solid domain	T_{ref}	Reference temperature (K)
$\mathcal{t}_S$	Duration required for complete solidification (s)	T_S	Solidus temperature (K)	ρ	Density (kg/m^3)
T_f	Final temperature (K)	C	Specific heat (J/kgK)	λ	Liquid fraction
$\vec{g}$	Gravity (m/s^2)	A_s	Surface area (m^2)	σ	Root mean square deviation (%)
Q	Heat (J)	ΔT	Temperature difference (K)	ϕ	Source term
$\hbar$	Heat transfer coefficient (W/m^{2}K)	T	Temperature field (K)	ξ	Thermal storage capacity factor
T_i	Initial temperature (K)	P	Pressure (pa)	μ	Viscosity (Pas)
L	Latent heat (J/kgK)	k	Thermal conductivity (W/mK)	β	Volumetric coefficient of thermal expansion (K^{-1})
T_l	Liquidus temperature (K)	h	Total enthalpy (J)	ψ	Melt fraction enrichment factor
T_m	Melting temperature(K)	$\hat{\imath}$	Unit vector in X direction		

1. Introduction

Nowadays, the world is witnessing global pandemic situation due to novel Coronavirus-COVID-19. Almost all countries are affected by this virus since January 2020. Several actions including social distancing, use of mask and frequent sanitization, self-quarantine, taking immunity booster drink (kadha, hot water & herbal tea, etc.) have been prescribed by Government worldwide [1]. As a precaution use of warm water has also been prescribed. Studies show that viruses and bacteria are neutralized at exposer of temperature range (50 ^{0} C to 70 ^{0} C) [2]. For thermal disintegration of Corona virus, 20 min duration to expose at temperature range between 50 ^{0} C to 55 ^{0} C is required [3]. PCM integrated solar water heater system can be an energy efficient solution to provide a constant water temperature above 55 ^{0} C.

PCM solar water heater is an attractive solution as it increase the heat storage capacity and thus, maintains an isothermal temperatures for longer duration [4], [5], [6]. Thus, it can be an energy efficient solution that can be implemented in hospitals, public places, restaurants, schools and house-holds [7], [8]. This study is mainly dedicated to modified solar water heater with PCM tube. A noticeable amount of heat can be stored by PCM during sun shine hours which can be further used for maintaining required temperature of water during off sun shine hours [9], [10]. Paraffin wax offers excellent characteristics among all different types of PCMs for solar water heating purpose. But it is suffered from low thermal conductivity [11], [10].

Solar water heating systems are principally classified as flat plat solar water heating system and evacuated solar water heating system. Solar water heating systems are secondary classified based on manner of water circulation, as active solar water system (external pump assisted) and passive solar water heater system (governed by natural circulation principle) [12]. However, passive system is more dominated over active system due to better thermal efficiency and no external power requirement [4], [13].

In-depth analysis on the solar water heating system and resisted heat transfer by suitable ceiling isolation materials inside the indoor construction is carried out by D Prakash et al. [14]. They gain the results as by improving the roof system, overall production of 25 L of hot water per day with 60 ^{0} C temperature rises is gained in the summer season. Also ceiling temperature maintained at 27 ^{0} C is throughout the day in summer.

The exhaustive study done by Manoj Kumar & Mylsamy [15] related to solar water heater included with a Nano fused phase alteration material with three various cases viz. with PCM ,without PCM and with Nano fused PCM. They gain the result as, exergy efficiencies of the system were calculated as 19.6 %, 22.0 %, and 24.6 %, for without PCM, with PCM, and with Nano fused PCM, respectively and energy efficiencies for the three cases were calculated to be 58.74 %, 69.62 %, and 74.79 %, respectively.

2. Proposed model

As shown into the Figure 1, the PCM paraffin wax encapsulated copper tube is held at the core portion of the water storage tank. Thermopsyphonic convection characterized by circulation of fluid is mainly responsible for heating of water during day time. At each rotation, the warm water up surges in evacuated tubes of solar water heater and stored in to insulated storage tank. And cold water down surges for absorbing the heat.

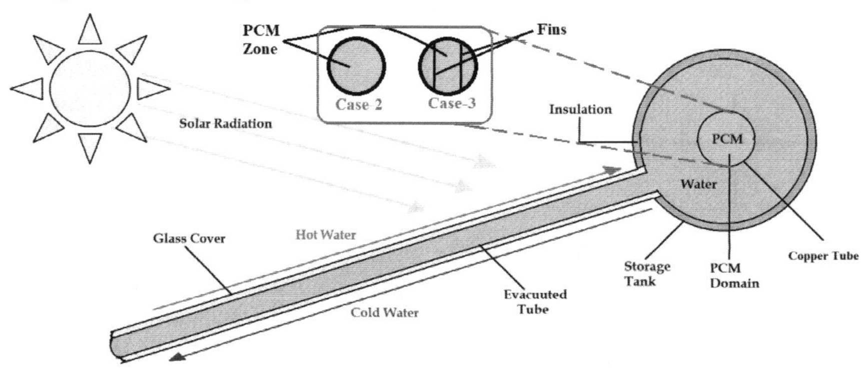

Figure 1. Schematic diagram of evacuated tube solar water heater with PCM encapsulated tube

Table 3. Thermo-physical properties

Properties	Water	PCM	Copper
h_{mol}	285820 (kJ/mol)	14800 (kJ/mol)	---
M_o	18.01528 (g/mol)	374 (g/mol)	29 (g/mol)
T_B	373 (K)	595 (K)	---
T_S	273 (K)	331 (K)	---
T_l	273 (K)	333 (K)	---
C	4186 (J/kgK)	2100 (J/kgK)	381 (J/kgK)
L	2260 (J/kgK)	189000 (J/kgK)	---
k	0.6 (W/mK)	0.21 (W/mK)	387.6 (W/mK)
β	0.21×10^{-3} (K^{-1})	0.75×10^{-3}	---
μ	0.0017 (Pas) at 273 K to 0.00025 (Pas) at 373 K	0.00689 (Pas) at 333 K to 0.00370 (Pas) at 363 K	---
ρ	1000 (kg/m^3) at 273 K to 959 (kg/m^3) at 373 K	920 (kg/m^3) at 293 K to 745 (kg/m^3) at 363 K	8978 (kg/m^3)
Chemical formula	H_2O	$C_{20\sim40}H_{42\sim82}$	Cu
Structure	Triangular (polar molecule)	iso chain (soft in nature)	FCC (Crystalline Solid)

Three different cases are considered for CFD analysis viz. (1) Conventional solar water heater (2) Modified solar water heater with integration of PCM encapsulated bare tube and (3) Modified solar water heater with integration of PCM encapsulated finned tube. As soon as, the warm water gets in touch with the PCM encapsulated tube, it transfers thermal energy to PCM in sensible way. So, absorption of heat is done by PCM via sensible mode up to melting point and then via phase alteration process at isothermal condition. This produces localized drop in temperature of hot water. This low temperature water down surges towards evacuated tube for gaining the heat and fresh charge of heated water gets in touched with PCM zone. And next circulation initiated. Thus, storage of thermal energy is done in PCM encapsulated copper tube. This stored energy can be further utilized at night time for water heating.

Table 4. Geometric Specifications

Component	Material	Dimensions
Storage Tank (insulated)	Steel	570 mm diameter,1020 mm length
PCM encapsulation tube	Copper	100 mm diameter, 1000 mm length & 1 mm thickness
Fins	Copper	86 mm height, 1000 mm length & 1 mm thickness

3. Fluid Dynamics & Thermal Equations

Certain relevant assumptions which considered for CFD analysis are (1) Transportation of heat from the hot water to the PCM through tube wall and fin walls is done in two dimensional manners and also temperature field associated with PCM zone has two space co-ordinates. So, whole problem is considered as two dimensional [16]. (2) Copper has thermal conductivity considerably better enough to nullify the effect self-conductive resistance of 1 mm thick tube wall and fin walls. So, all copper walls are considered as diathermic in nature. (3) Molecular diffusion thermal energy transformation occurs in both solid and liquid phase of PCM. This is strictly true with thermo-physical properties that may differ. (4) The evacuated tube wall is diathermic in nature because wall is painted with good absorptivity material. So, glass wall is considered as diathermic wall. (5) Water is heated via natural convection and temperature-gradients are considerably small enough to neglect the radiation effects. (6) The liquid generated from its solid part is characterized as Newtonian, laminar, compressible, and viscous in nature. (7) All properties of paraffin wax are varying with respect to temperature and also the melting pattern displays less discontinuity. So, PCM is considered as pure, homogeneous and isotropic substance for thermodynamic calculations. (8) Mushy zone is saturated porous zone by the fluid portion of PCM and porosity of this porous zone is equal to fluid fraction of that particular instant.

Based on above assumptions following are the governing equations for all considered computational domains.

Water Domain (Water portion of the system) & Liquid Domain (Liquid portion of PCM)

Continuity Equation

$$\left[\nabla(\rho\vec{V}) + \left(\frac{\partial\rho}{\partial t}\right)\right] = 0 \quad (1)$$

Navier stokes equation

The momentum equation is given for the liquid domain of PCM is following

$$\left[\frac{\partial(\rho\vec{V})}{\partial t} + \vec{V}\nabla(\rho\vec{V})\right] = -\nabla P + \nabla(\mu\nabla\vec{v}) - \rho\vec{g}\beta\Delta T + \phi\vec{V} \quad (2)$$

The melt front is formed as porous media between solid and fluid domain with porosity equal to liquid fraction for the saturated porous zone termed as mushy zone. So, whatever fluid dynamics present in this zone is handled by enthalpy-porosity approach by providing the mushy zone parameter ϕ in the momentum equation. In order to achieve this particular fluid dynamics, an appropriate definition of ϕ is given as:

$$\phi = \left\{ x/x = \left(\frac{z(1-\lambda)^2}{\lambda^2 + r} \right) ; \lambda = f(T) ; \forall \lambda \in (0,1) \ \& \ \forall T \in (T_s, T_l) \right\} \tag{3}$$

This is known as *Carman–Kozeny relation*. The constant z has enough large value which can be able to suppress velocity and r is a small constant used to avoid a division by zero when the two dimensional computational domain is under action of solid domain. The choice of the constants is arbitrary. In this work, $z = 161110 \, \mathrm{kg/m^3s}$ and r=0.0003 are used [17]. And for the water domain there is no phase transformation is done so, $\phi = 0$ for water.

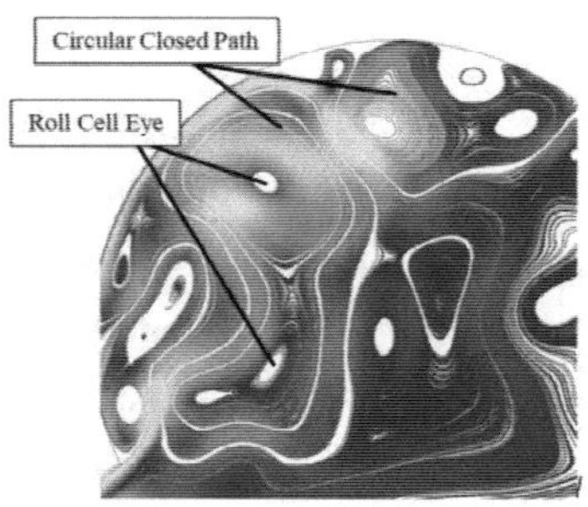

Figure 2. Circulating flow field in water domain

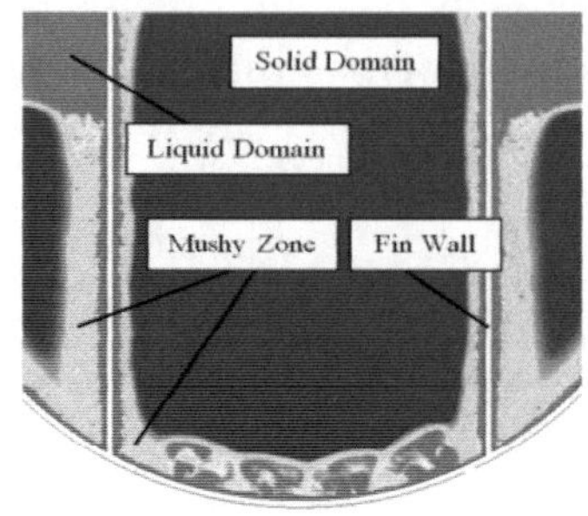

Figure 3. Various Domains in PCM zone

Energy Equation

$$C\left[\frac{\partial(\rho h)}{\partial t} + \nabla(\rho \vec{V} h)\right] = k\nabla^2 T \tag{4}$$

Where h is *total enthalpy* for both domains is defined as:

$$h = \left(\int_{T_{ref}}^{T} C dT \right)_{Water} \quad \& \quad h = \left(\int_{T_{ref}}^{T} C dT + \lambda L \right)_{PCM} \tag{5}$$

Solid Domain (Solid portion of PCM) & Copper Domain (Tube Wall and Fin portion)

$$C\left[\frac{\partial(\rho h)}{\partial t}\right] = k(\nabla^2 T) \tag{6}$$

Equivalent temperature of water & Heat transfer rate equation

For providing isothermal conditions, equivalent temperature of water gets defined as:

$$T_f = \left(\sum_{p=0}^{n} \sum_{q=0}^{n} T_q \cdot \Delta t_p \Big/ \sum_{p=0}^{n} \Delta t_p \right) \tag{7}$$

Where T_q the temperature of water at particular time and Δt_p is time duration between two sequential observations. And also $\forall T_q \geq T_m$ and Δt_p is always taken after melting started.

$$\left(\frac{\partial Q}{\partial t}\right) = \left\{ x/x = \left[\frac{\partial(\rho V)}{\partial t} \int_{T_i}^{T_f} C dT \right]_{Water} + \delta \left[\frac{\partial(\rho V)}{\partial t} \left(\int_{T_i}^{T} C dT + \lambda L \right) \right]_{PCM} ; \delta = \begin{cases} 0; \lambda = 0 \\ 1; \lambda \neq 0 \end{cases} ; \lambda = f(T) ; \forall \lambda \in (0,1) \ \& \ \forall T \in (T_s, T_l) \right\} \tag{8}$$

Melt fraction enrichment factor (ψ) *& Thermal Storage Capacity Factor* (ξ)

Both are the performance parameters for comparing case-2 and case-3 configurations of water heater.

$$\psi = \left(\frac{\tau_{Bare\ Tube}}{\tau_{Double\ Finned\ Tube}} \right) \ \& \ \xi = \left[\left(\frac{\partial Q}{\partial t}\right)_{Case-2\ or\ 3} \Big/ \left(\frac{\partial Q}{\partial t}\right)_{Case-1} \right] \tag{9}$$

Thermal Status Maintenance Time
It is duration up to which solar water heater can maintain the thermal status of water above 56^0 C. So it is defined as following ($\hbar \cong 30\ \mathrm{W/m^2K}$ for PCM [18]).

$$t = t_{\mathcal{R}} + \left(\frac{Q}{\hbar A_s \Delta T}\right)_{PCM} \tag{10}$$

4. Boundary conditions & Computational Methodology

Initial boundary conditions for the CFD analysis of conventional solar water heating system are (1) At the hot surface of Glass tube, wall temperature $T_{wall} = T_f = 70$ °C ; (2) Initial temperature of the water is considered as $T_i = 27$ °C ; (3) At the cold wall (back side of glass tube) and side wall (end portion of glass tube) of Glass tube, $\left[k_{glass}\left(\partial T_{glass}/\partial t\right)\right]_{wall} = 0$; (4) At the tank wall $[k_{steel}(\partial T_{steel}/\partial t)]_{wall} = 0$. And initial boundary conditions for the CFD analysis of remaining both cases are (1) At the outer surface of Copper tube, wall temperature $T_{wall} = T_f = 70$ °C ; (2) Initial temperature of the PCM is considered as $T_i = 27$ °C ; (3) At the PCM and fin surface interface conjugate heat transfer is done so $[k_{fin}(\partial T_{fin}/\partial t)]_{wall} = [k_{PCM}(\partial T_{PCM}/\partial t)]$; $T_{fin} = T_{PCM}$; (4) At the outer wall of Copper tube is held at constant temperature so, at the wall $[k_{Cu}(\partial T_{Cu}/\partial t)]_{wall} = 0$. (5) At the tank wall $[k_{steel}(\partial T_{Steel}/\partial t)]_{wall} = 0$.

Complete simulation of phase alteration process and the fluid dynamics incorporated with the melting process is carried out by ANSYS-FLUIENT-16.0. Second order upwind differencing scheme is used for solving both momentum and energy equations and SIMPLE scheme is considered for pressure–velocity coupling with 0.1 second is considered as time step for calculations. The grid size with 31293, 36258 and 38355 cells is checked to evaluate grid independence case of bare tube. Finally grid size with 36258 cells is selected, and further refinement doesn't show any remarkable improvement in accuracy of liquid fraction. While, the grid size with 785293, 805424 and 925683 cells is checked to evaluate grid independence case of finned tube. Finally grid size with 805424 cells is selected, and further refinement doesn't show any remarkable improvement in accuracy of liquid fraction.

5. Validation

CFD simulation result of bare Copper tube is validate against experimental data taken of average temperature of PCM after phase alteration event started at the day time from thermally equivalent model for phase alteration process presented by Manoj Kumar & Mylsamy, [15]. Root mean square σ (%) deviation and correlation coefficient (b) are calculated to validate the CFD simulations according to following expressions [8] respectively.

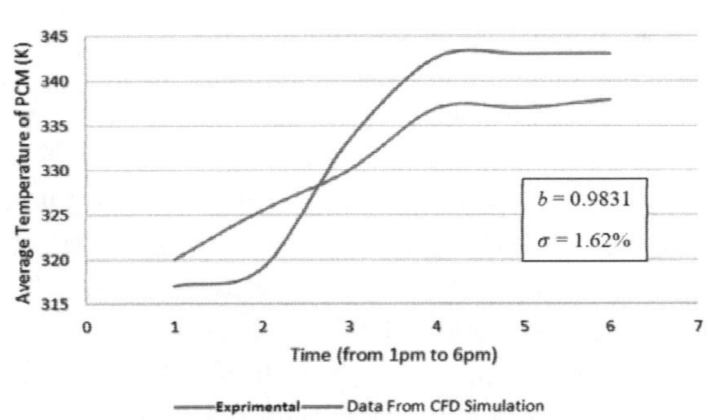

Figure 4. Validation curve

Here $b > 0$ indicates positive linear relationship, $b < 0$ indicates negative linear relationship and $b = 0$ indicates no relationship between two variables. Also X_i is CFD simulation data and Y_i is experimental data.

$$b = \left(\frac{N\sum X_iY_i - (\sum X_i)(\sum Y_i)}{\sqrt{N\sum X_i^2 - (\sum X_i)^2}\sqrt{N\sum Y_i^2 - (\sum Y_i)^2}}\right) \ \&\ \sigma(\%) = \sqrt{\frac{\sum\left[\left(\frac{X_i - Y_i}{X_i}\right)\times 100\right]^2}{N}} \tag{11}$$

6. Results and Discussion

Consider the Figure 5, which indicates the simulated profile of the average water temperature of the water with respect to time which approximately near to the experimental profile obtained by Manoj Kumar & Mylsamy, [15] for same time interval. Water gets heated from gaining the heat from the evacuated tube. Then, this heated water up surges due to density gradient in to the evacuated tube towards the storage tank. So, overall clockwise circulation cycles are generated in the system.

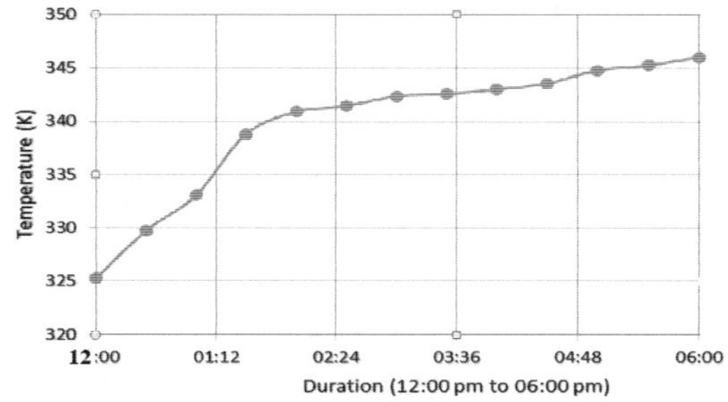

Figure 5. Change in average temperature of water with respect to time

As shown in to Figure 6, 7 & 11, the growth of liquid fraction is higher in case of the finned tube as compare to bare tube but physical nature of both curves get restored. This restoration of physical nature of curves is the indication of same fluid field gets generated in to fluid domain. But thermal response of the finned tube higher compare to bare tube because two fins provide extra nucleating sites for localized circulating child roll cell growth.

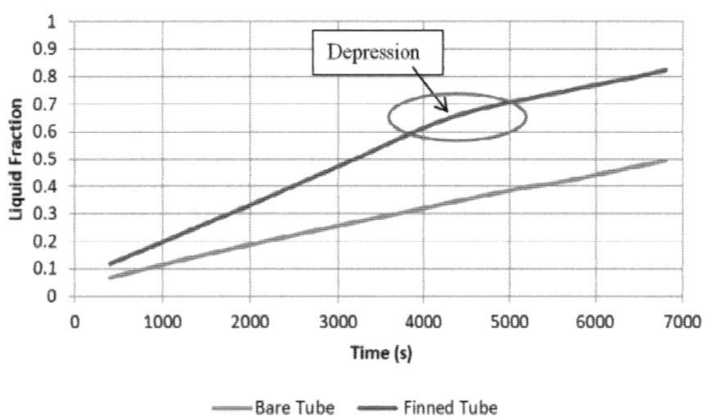

Figure 6. Growth of liquid fraction with respect to time

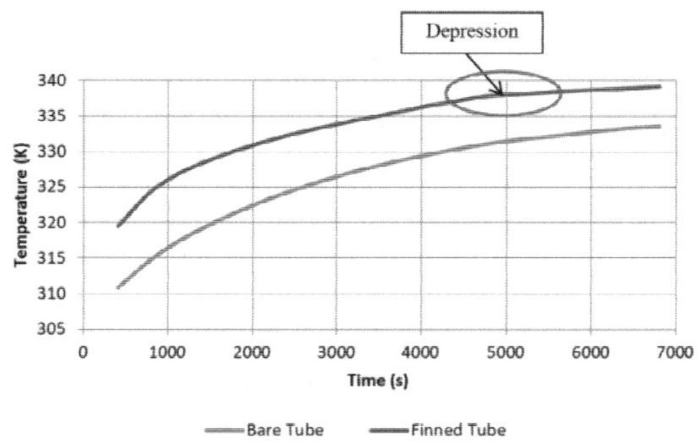

Figure 7. Growth of average temperature of PCM zone with respect to time

While in case of bare tube, only tube wall is participated in this process. As time progresses, these new born circulating roll cells became matured. So their size get increased and due to that the tangential velocity gets increased as admiring more fluid particles in to self for circulation around the eye of self. So, most of the fluid domain is dominated by advection heat transfer mode instead of conduction heat transfer mode. But still some regions of the fluid domain such as tube & fin walls and mushy zone get dominated by this molecular conduction heat transfer mode due to activation of no slip condition and self-nature as saturated porous media with porosity equal to liquid fraction respectively. As shown in to Figure 8, 9 & 11, in fully developed fluid domain whatever child roll cells generated became more matured as participating more fluid particles and during this process, their localized temperature increases, velocity also gets increases and density gets decreases as $\rho \propto 1/T$. So, due to density difference considered local circulating roll cell tries to up surge by overcoming solid domain. And get settle down at the top portion of the tube, resulting in two large circulating roll cells observed at the top portion of fluid domain in to tube. But during this up surging process, fins act as the resistance by providing small space between self and solid domain. Also due to presence of various active nucleating sites on the fin wall, child roll cells born and the fluid particles join these child roll cells' circulating field by leaving the matured roll cell circulating field. In addition to this, also activation of no slip condition and diffusion fluid films supports molecular thermal excitations instead of advection which creates zone which termed as diffusion fluid film.

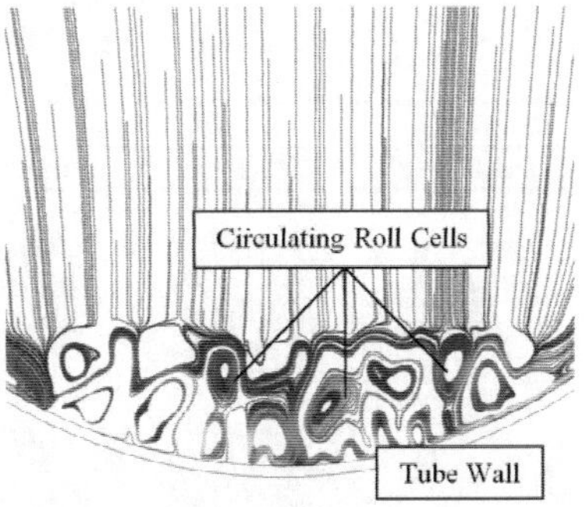

Figure 8. Matured fluid domain in case-2

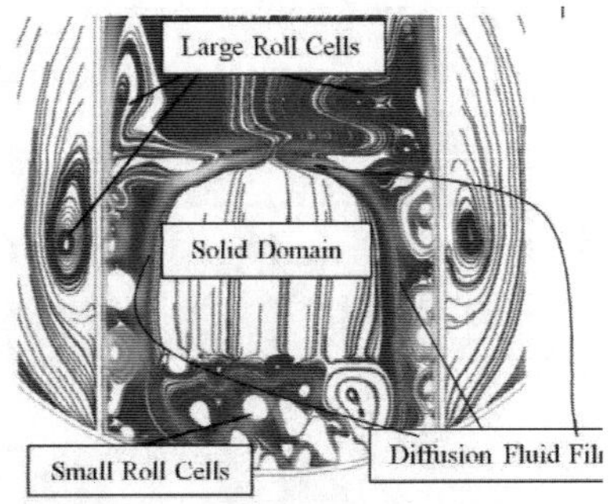

Figure 9. Matured fluid domain in case-3

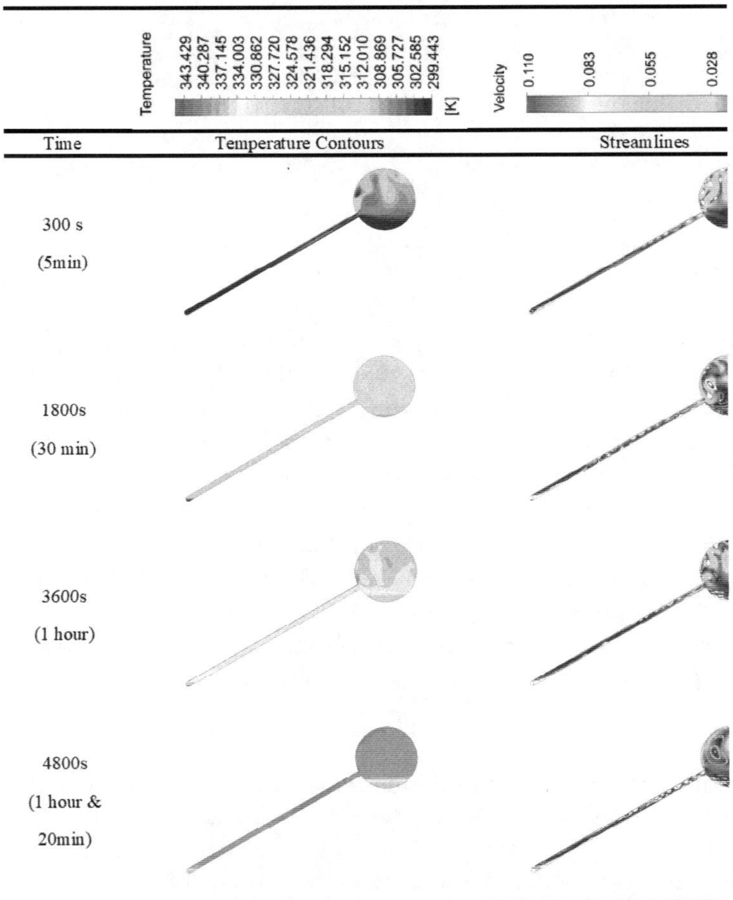

Figure 10. CFD Results of conventional solar water heater

Table 5. Thermal Status Maintenance Time of various cases

Case	Thermal Status Maintenance Time
Case 1	Up to midnight
Case 2	Up to 8:00 am
Case 3	Up to 4:30 am

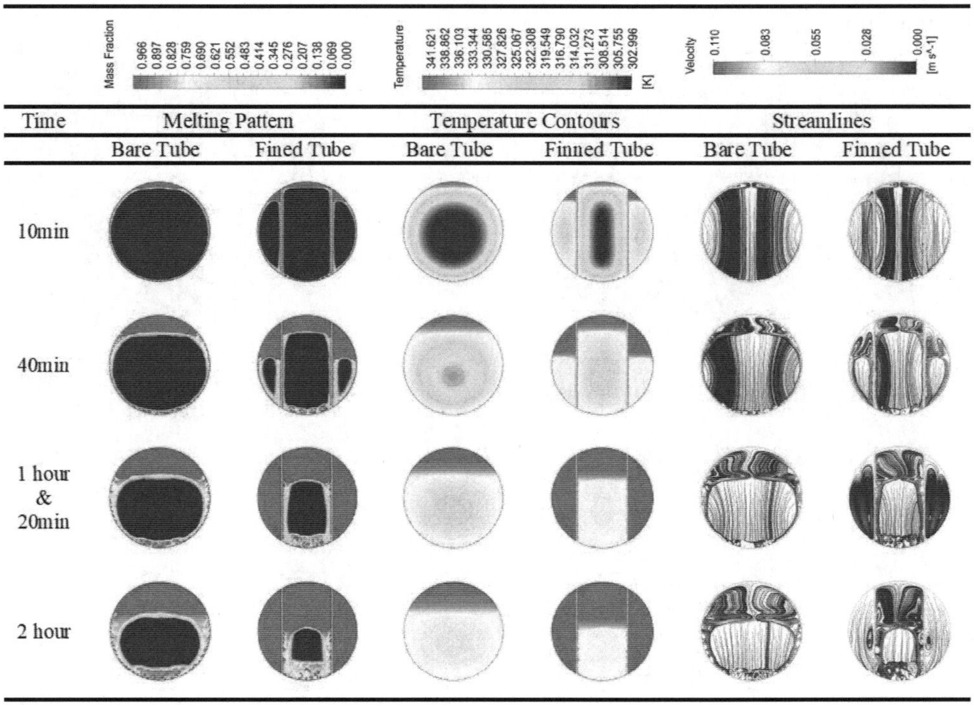

Figure 11. CFD Results

So, overall fluid field of matured roll cell gets destroyed up to certain considerable extent by combine effects provided by fin walls, solid domain & diffusion fluid film. So, destroyed large roll cells are observed in Figure 9. This phenomenon gets reflated in all three profiles of the finned tube as slight depression. This is not happen in bare tube due to more space available in fluid domain, and roll cells get perfect pattern in bare tube fluid domain. So, all three profiles of bare tube case are not depressed. According to Figure 6 & 7, it is concluded that the physical nature of fluid domain is remain same in both cases but thermal response of the finned tube is batter as compared to bare tube because of thermal supports provided by the fins to the fluid domain. But due to depression in to the thermal profiles of finned tube, this advantage of fins is became slight darker. As in view of reverse temperature profile of water throughout off sun shining hours [14], water touches 55 ^{0} C at middle of the night. t is calculated for both cases by using equation (26).

7. Conclusions

Present study illustrates Phase alteration pattern for three different cases viz. conventional solar water heating system, PCM assisted solar water heater under different configurations. Following inferences are drawn from this study:

(1) Thermal storage capacity of solar water heater gets increased due to integration of PCM tube inside the water storage tank. Thermal Storage Capacity Factor is calculated to be 1.1586 and 1.1149, for PCM SWH with and without fin case, respectively. (2) PCM encapsulated bare tube failed in completion of stored thermal energy delivery which causes lacking of post sensible heat transfer is observed in bare tube case which is not the scenario in finned tube case. (3) Overall performance of PCM encapsulated bare tube is better than finned tube among all considered physical and thermal constrains. (4) Modified evacuated solar water heating system integrated with finned tube can able to provide hot water above 56 ^{0} C throughout the day and night time. Thus, it can be treated as better compassionate choice for defending present COVID-19 pandemic situation.

In this present article, two configurations of PCM integrated tube incorporated inside conventional water heater are studied. Another configuration of PCM integrated tubes viz. single vertical finned tube, crossed finned tube, triple finned tube will be studied.

8. References

[1] A. Khandelwal, A. Agrawal, and A. Kumar, "An Outbreak of Coronavirus (COVID-19) Epidemic in India: Challenges and Preventions," vol. 8, no. 2, pp. 1–7, 2020.

[2] A. Soltani and T. R. Sinclair, "Soil water.," *Model. Physiol. Crop Dev. growth yield*, pp. 161–169, 2012, doi: 10.1079/9781845939700.0161.

[3] J. P. Abraham, B. D. Plourde, and L. Cheng, "Using heat to kill SARS-CoV-2," *Rev. Med. Virol.*, vol. 30, no. 5, pp. 8–10, 2020, doi: 10.1002/rmv.2115.

[4] A. Shukla, D. Buddhi, and R. L. Sawhney, "Solar water heaters with phase change material thermal energy storage medium: A review," *Renew. Sustain. Energy Rev.*, vol. 13, no. 8, pp. 2119–2125, 2009, doi: 10.1016/j.rser.2009.01.024.

[5] B. Zalba, J. M. Marín, L. F. Cabeza, and H. Mehling, "Review on thermal energy storage with phase change: Materials, heat transfer analysis and applications," *Applied Thermal Engineering*. 2003, doi: 10.1016/S1359-4311(02)00192-8.

[6] R. Saxena, D. Rakshit, and S. C. Kaushik, "Experimental assessment of Phase Change Material (PCM) embedded bricks for passive conditioning in buildings," *Renew. Energy*, vol. 149, pp. 587–599, 2020, doi: 10.1016/j.renene.2019.12.081.

[7] R. Saxena, D. Rakshit, and S. C. Kaushik, "Phase change material (PCM) incorporated bricks for energy conservation in composite climate: A sustainable building solution," *Sol. Energy*, vol. 183, no. January, pp. 276–284, 2019, doi: 10.1016/j.solener.2019.03.035.

[8] R. Saxena, K. Biplab, and D. Rakshit, "Quantitative Assessment of Phase Change Material Utilization for Building Cooling Load Abatement in Composite Climatic Condition," *J. Sol. Energy Eng. Trans. ASME*, vol. 140, no. 1, 2018, doi: 10.1115/1.4038047.

[9] F. Agyenim, N. Hewitt, P. Eames, and M. Smyth, "A review of materials, heat transfer and phase change problem formulation for latent heat thermal energy storage systems (LHTESS)," *Renewable and Sustainable Energy Reviews*. 2010, doi: 10.1016/j.rser.2009.10.015.

[10] R. Saxena, C. Dwivedi, V. Dutta, S. C. Kaushik, and D. Rakshit, "Nano-enhanced PCMs for low-temperature thermal energy storage systems and passive conditioning applications," *Clean Technol. Environ. Policy*, no. 0123456789, 2020, doi: 10.1007/s10098-020-01854-7.

[11] E. M. Anghel, A. Georgiev, S. Petrescu, R. Popov, and M. Constantinescu, "Thermo-physical characterization of some paraffins used as phase change materials for thermal energy storage," *J. Therm. Anal. Calorim.*, vol. 117, no. 2, pp. 557–566, 2014, doi: 10.1007/s10973-014-3775-6.

[12] L. M. Ayompe, A. Duffy, M. Mc Keever, M. Conlon, and S. J. McCormack, "Comparative field performance study of flat plate and heat pipe evacuated tube collectors (ETCs) for domestic water heating systems in a temperate climate," *Energy*, vol. 36, no. 5, pp. 3370–3378, 2011, doi: 10.1016/j.energy.2011.03.034.

[13] D. A. G. Redpath, "Thermosyphon heat-pipe evacuated tube solar water heaters for northern maritime climates," *Sol. Energy*, vol. 86, no. 2, pp. 705–715, 2012, doi: 10.1016/j.solener.2011.11.015.

[14] D. Prakash, "Thermal analysis of building roof assisted with water heater and insulation material," *Sadhana - Acad. Proc. Eng. Sci.*, vol. 43, no. 3, pp. 1–10, 2018, doi: 10.1007/s12046-017-0781-y.

[15] P. Manoj Kumar and K. Mylsamy, "Experimental investigation of solar water heater integrated with a nanocomposite phase change material: Energetic and exergetic approach," *J. Therm. Anal. Calorim.*, vol. 136, no. 1, pp. 121–132, 2019, doi: 10.1007/s10973-018-7937-9.

[16] S. W. Webb, M. T. Itamura, N. D. Francis, and D. L. James, "CFD calculation of internal natural convection in the annulus between horizontal concentric cylinders," *Proc. ASME Summer Heat Transf. Conf.*, vol. 2003, no. October, pp. 911–916, 2003, doi: 10.1115/ht2003-47515.

[17] A. Joulin, Z. Younsi, L. Zalewski, D. R. Rousse, and S. Lassue, "A numerical study of the melting of phase change material heated from a vertical wall of a rectangular enclosure," *Int. J. Comut. Fluid Dyn.*, vol. 23, no. 7, pp. 553–566, 2009, doi: 10.1080/10618560903203723.

[18] S. P. Jesumathy, M. Udayakumar, S. Suresh, and S. Jegadheeswaran, "An experimental study on heat transfer characteristics of paraffin wax in horizontal double pipe heat latent heat storage unit," *J. Taiwan Inst. Chem. Eng.*, vol. 45, no. 4, pp. 1298–1306, 2014, doi: 10.1016/j.jtice.2014.03.007.

Theoretical design of solar-powered vapor absorption refrigeration system coupled with latent heat energy storage

Devendra Raut [1]**, *Arunendra Kumar Tiwari*** [2] ***and Vilas R Kalamkar*** [1*]

[1] Department of Mechanical Engineering, Visvesvaraya National Institute of Technology, Nagpur, India,
[2] Solar Energy Division, Sardar Patel Renewable Energy Research Institute (SPRERI), Anand, India,

*Email address for correspondence: vilas.kalamkar@rediffmail.com
Abstract. In this work, the vapor absorption refrigeration system (VARS) with a cooling capacity of 1kW is designed. VARS is designed to be driven by hot water available from the solar thermal collector during the day. In the Sun's absence, the system is designed to run by a coupled latent heat energy storage system. The unit has three major components: evacuated tube collector, latent heat energy storage, and VARS. The system is designed for a cooling capacity of 1 kW. The size of other components has been calculated by adopting the heat exchanger design approach and the relevant equations presented in the published works. The corresponding size of absorber, generator & condenser comes out to be 1.412kW, 1.46 kW & 0.96 kW, respectively for the cooling capacity of 1 kW. The evacuated tube collector of 100 LPD capacity is sufficient to provide the required heat to run the VARS and to charge the thermal energy storage. Latent heat energy storage of 2.25 kW charging and discharging time of 3 hours kW is coupled with the system. The designed system gives the COP of 0.875.
Keywords*: Solar vapor absorption refrigeration system, solar thermal collector, latent heat energy storage, evacuated tube collector.*

1. Introduction

Refrigeration has become an essential part of human life in one or another way. The process which cools down or maintains the temperature of the substance, lower than surrounding is called refrigeration. The most conventional system to achieve the refrigeration is vapor absorption refrigeration system (VARS). It is a low-grade heat-driven system. Conventionally, electric energy is the prime source to operate it, but limited electricity availability provokes to think of an alternate source of energy for the energy-consuming system. In this work, the author presented a theoretical design of water LiBr VARS, operated on solar energy during sunny days, and latent heat energy storage system during non-sunny hours.

Solar energy emerges as a great alternative to conventional energy. It is a clean and eco-friendly source of energy. It is abundant and reported that if only 5% of incident solar radiations are captured and utilized efficiently, the energy demand will be fulfilled worldwide. Though solar is clean and eco-friendly, it is non-continuous during the night and sun-off days[1]. The energy storage could have answered this issue of discontinuity and will make the system run without hindrance. The current research shows that many researchers did a numerical and experimental study on either solar refrigeration or latent heat energy storage[2–6]. Lie et al.[2] presented phase change material (PCM) TRNSYS model portable refrigeration. An off-vehicle refrigeration unit charged the system, and cooling is provided when PCM discharges. In such a system, PCM is directly used for cooling. Other solar-assisted refrigeration works are reported in references [5,7].

The theoretical work on solar-driven water- LiBr VARS unit with latent heat energy storage is not available. This urges the authors to present a theoretical design of a 1kW (0.25 TR) VARS system driven by hot water supplied by the solar collector during daytime and by PCM-based thermal energy storage system Sun-off days or night time. The layout of the solar-powered vapor absorption refrigeration system coupled with latent heat energy storage is shown in Figure 1.

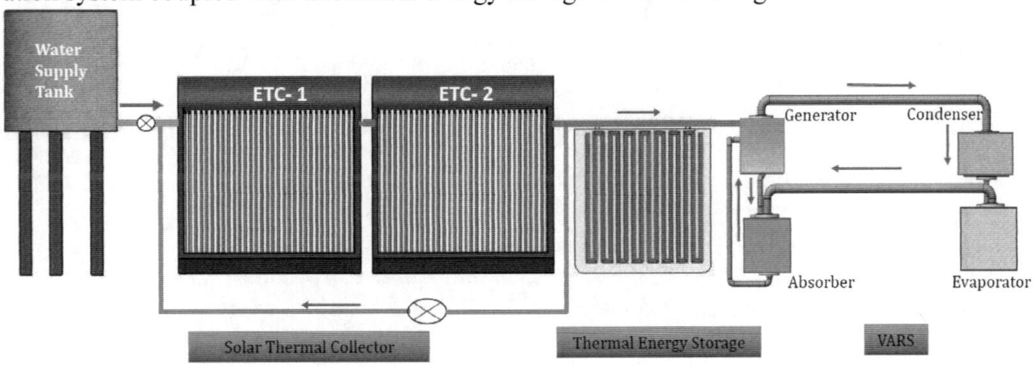

Figure 1. The layout of solar powered vapor absorption refrigeration with heat Storage.

In the given layout of the solar VARS, the water is supplied from the water source. It passes through the evacuated tube collector. There, water gets heated and passed to the VARS generator unit. The design of each component is based on the standard method of thermal analysis and heat exchanger design. The system's main component is explored for temperature, pressure, the working fluid's mass flow rate, and component size for required capacity in the following section. The cooling capacity of refrigeration is considered is 1kW at the evaporator. Further, estimations are made accordingly.

2. Vapour Absorption Refrigeration System

The water- lithium bromide refrigeration system has become popular in the last decades for space cooling application because of low-grade energy operation. In this system, water is used as a refrigerant and a homogenous solution of LiBr as absorbent. When the water-LiBr solution is heated in a generator, LiBr does not evaporate, and water gets evaporated thus, both can be separated without the use of any additional unit of rectification. The single effect VARS considered for calculation is presented in Figure 2.

2.1. Thermal Analysis

Each component's thermal analysis evaluates the operating parameters like temperature, pressure, solution strength, mass flow rates, and heat exchanging capacity. The design calculations are based on the following requirements:

Required Refrigeration Capacity: **1kW**
Desired Evaporator Temperature: **5°C**
Condenser Temperature: **45°C.**
LiBr Strength X_{LiBr} : **0.55**

At first, the saturation pressure and temperature are estimated at each component from the Pressure-Temperature- Concentration characteristics of Water- LiBr solution.

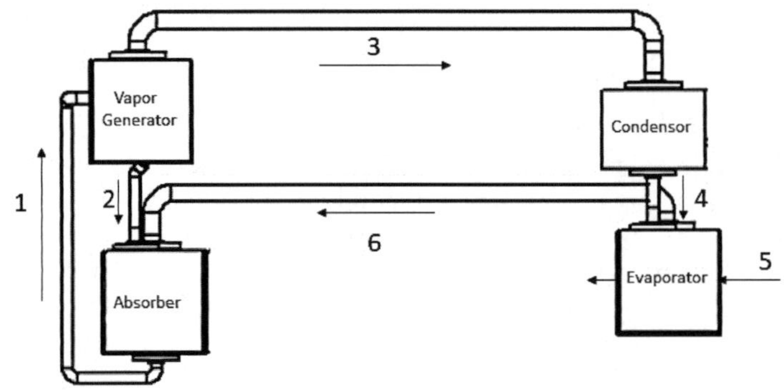

Figure 2. Schematic of Single Effect VARS

The following values are obtained corresponding to H_20 (refrigerant) at 5˚C:
Saturation Pressure = 8.7 mbar = 870Pa and solution strength is 60%.
Thus, Evaporator Pressure (P_E) = Absorber Pressure (P_C) = 870 Pa.
Since condenser temperature is 45 ˚C, the saturation temperature of water = 9.58 kPa
Thus, vapor generator pressure P_G =9.58 kPa
For P_G = 9.58 kPa and T_G = 80 ˚C.
Required H_20-LiBr strength X_{LiBr} = 55%

Since the hot water temperature is limited to 85 ˚C, the generator's solution temperature is restricted to approx. 80 ˚C only. The next step is calculating the heat potential of working fluids in enthalpy at the different locations within the unit.

2.1.1. Enthalpy Calculation. The enthalpy of the water and water vapour can be estimated using the general expression as follow:

For water, h = 4.1867t kJ/kg t in ˚C. (1)

For water vapour, h = 2501 +1.88t kJ/kg, t in ˚C. (2)

At generator exit, phase of fluid is vapour so, enthalpy of water vapour (h_3) at 80 ˚C is given by Eqn. (1)
At Absorber, T_{abs} = 45 ˚C & Pressure = 8.52 kPa at solution concentration X_1 = 0.58.
enthalpy h_1 corresponding to inlet to generator = -145 kJ/kg
While the enthalpy h_2 corresponding to solution at X_2 = 0.55 T_G = 80 ˚C is -70 kJ/kg.
At the evaporator exit, enthalpy of water vapour leaving the evaporator at 5 ˚C is given by Eqn. (2)
At condenser exit enthalpy of water h_4 =146.5 kJ/kg
In enthalpy calculation, h_5 is not considered as it not the integral part of the calculation. The next step is to calculate the required mass flow rate of working fluids.

2.1.2. Mass Flow Rate Calculation. The following steps are adopted to calculate the mass flow rate at various points of the refrigeration system.

At Evaporator; $m_6 = 1/(h_6-h_7) = 0.00043$ kg/s (3)

At Absorber; $\dot{m}_2 x_2 + \dot{m}_6 x_6 = \dot{m}_1 x_1$ & $\dot{m}_2 + \dot{m}_6 = \dot{m}_1$ gives $\dot{m}_2 = 0.44 \times 10^{-2}\, kg/s$ (4)

At Generator; $x_1 \dot{m}_1 = x_2 \dot{m}_2 + x_3 \dot{m}_3$ and $\dot{m}_1 = \dot{m}_2 + \dot{m}_3$ gives $\dot{m}_3 = 0.004$ kg/s. (5)

2.1.3. Heat Exchange Capacity

Absorber Capacity: $Q_A = \dot{m}_2 h_2 + \dot{m}_6 h_6 - \dot{m}_1 h_1$ gives $Q_A = 1.416$ kW (6)

Generator Capacity $Q_G = \dot{m}_2 h_2 + \dot{m}_3 h_3 - \dot{m}_1 h_1$ gives $Q_G = 1.46$ kW (7)

Condenser Capacity $Q_c = \dot{m}_3 (h_3 - h_4)$ gives $Q_c = 0.965$ kW (8)

In this section capacity estimation of each component is done and the summarized capacity is presented in
Table 1 as below:

Table 1. The capacity of VARS Component

Component	Capacity[kW]
Generator	1.462
Condenser	0.965
Absorber	1.412
Evaporator	1

And, theoretical COP comes out to be 0.875 with a circulation factor of 10.232.
The next step in the calculation is calculating the dimension of the component and heat transfer arrangement.

2.2. Sizing of VARS Component

The sizing of each component is based on the heat exchanger design approach. The standard assumptions and approximation are made as per requirement. The standard equations used for each component are:
Heat exchange capacity,

$$Q = UA\Delta T_{LMTD} \quad (9)$$

The overall heat transfer coefficient,

$$\frac{1}{U} = \frac{1}{h_i} + \frac{r_i \ln\left(\frac{r_i}{r_o}\right)}{k_{wall}} + \frac{r_i}{r_o . h_{sol}} \quad (10)$$

Reynolds Number is given by $\mathrm{Re} = \frac{4m}{\pi D_i \mu}$ (11)

Logarithmic Temperature Difference is $LMTD(\Delta T_m) = \frac{\left(T_{h_i} - T_{h_{out}}\right) - \left(T_{c_i} - T_{C_{out}}\right)}{\ln\left(\frac{T_{h_i} - T_{h_{out}}}{T_{c_i} - T_{C_{out}}}\right)}$ (12)

2.2.1. Generator. Since the generator's theoretical heat exchange capacity is 1.5 kW, all the calculations are done with respect to that capacity.

The calculation for heat transfer coefficient of single-phase flow of water inside the tube h_i:

The heat transfer coefficient of water inside the tube (h_i) is given by Gnielinski [8]:

$$h_i = \frac{\frac{f}{2}(\mathrm{Re}-1000)\mathrm{Pr}}{1+2.7(\frac{f}{2})^{\frac{1}{2}}(\mathrm{Pr}^{2/3}-1)} \cdot \frac{k}{D}$$ Where $f = [1.82\log(\mathrm{Re}) - 1.64d]^{-2}$

(13)

The calculation for heat transfer coefficient of H_2O-LiBr solution outside the tube h_{sol}

The following equation calculates the heat transfer coefficient of H_2O-LiBr solution outside the tube:

For flooded generator, Cornwell and Houston relation[9].

$$h_{sol} = 9.7P_c^{0.5}(1.8\mathrm{Pr}^{0.17}\ 4Pr^{1.2}\ 10\mathrm{Pr}^{10})(\frac{qD}{\mu_f H_{fg}})^{0.67}(\frac{C_p \mu_f f}{k_f})^{0.4}\frac{k_f}{D} \quad (14)$$

Substitute the value of h_i and h_{sol} in the Eqn. 10 of U and solve for area A in Eqn. 9.

$A = \pi d_o l$, where l= length of the tube

On solving the above Eqn 9-14, we get the characteristics listed in Table 2.

Table 2. Characteristics of Vapor Generator

Parameter	Value	Unit
h_i	6756.6	W/m^2K
h_w	2102.40	W/m^2K
U	1535.70	W/m^2K
LMTD	14.427	K
Length of Tube(l) for a tube diameter of 12.5 mm	1.72	m

2.2.2. Condenser

To estimate the overall heat transfer coefficient U as per Eqn 10, we need to solve for heat transfer coefficient inside tube(h_i)and outside the tube (h_o). If Reynolds number (estimated from Eqn. 11) lies in $10^4 < \mathrm{Re} < 5X10^6$ and Pr = 0.5 < Pr <2000. Then the Nusselt number equation given by Petukhov-Popov Equation is used [9] :

$$Nu = \frac{(\frac{f}{8})\mathrm{Re.Pr}}{K_1 + K_2(\frac{f}{8})^{0.8}(\mathrm{Pr}^{\frac{2}{3}}-1)}$$, f is friction factor and given by $f = [1.82\log(\mathrm{Re}) - 1.64d]^{-2}$ (15)

Constants; $K_1 = 1.04$ *and* $K_2 = 11.7 + \frac{1.8}{Pr^{\frac{2}{3}}}$ and $Nu = \frac{h_i D_i}{k}$ (16)

Eqn. 16 will give h_i.

The calculation for heat transfer coefficient of water outside the tube h_o

The heat transfer coefficient of water outside the tube h_o (h_m) is estimated by Nusselt Theory for average heat transfer coefficient [9] and equation is given as

$$\frac{h_m D}{k_l} = 0.728[\frac{\rho_l(\rho_l - \rho_g)gh_{fg}d_0^3}{\mu_l(T_{sat} - T_w)k_l}]^{1/4} \quad (17)$$

D and d_0 are shell diameter and outer diameter of tube respectively.

In the case of forced convection, the most conservative relation

$$\frac{h_m d_o}{k_l} = 0.59\tilde{\mathrm{Re}}^{1/2} \quad (18)$$

3. When both velocity and gravity is considered

$$\frac{Nu_m}{\tilde{\mathrm{Re}}^{1/2}} = 0.64(1+(1+1.69F)^{1/2})^{1/2} \tag{19}$$

$$F = \frac{g d_o \mu_l h_{fg}}{u_g^2 k_l \Delta T} \tag{20}$$

Substitute the value of h_i and $h_m = h_o$ in the Eqn.10 of U and solve for area A.

$A = \pi d_o l$, l= length of the tube (21)

Table 3. Characteristics of Condenser

Parameter	Value	Unit
h_i	2432	W/m^2K
h_o	905	W/m^2K
U	531.5	W/m^2K
LMTD	13.8	K
Length of Tube(L) for tube diameter of 12.5 mm	3.4	m

2.2.3. Absorber. The absorption process of an aqueous lithium bromide refrigeration system usually occurs in the vertical tube as shown in fig. 3. The control variables of such absorption are pressure, temperature, the mass flow rate of solution, and the solution's concentration. The absorber's performance assessment parameters are mass absorption flux, the cooling of the outlet solution, and the falling film heat transfer coefficient.

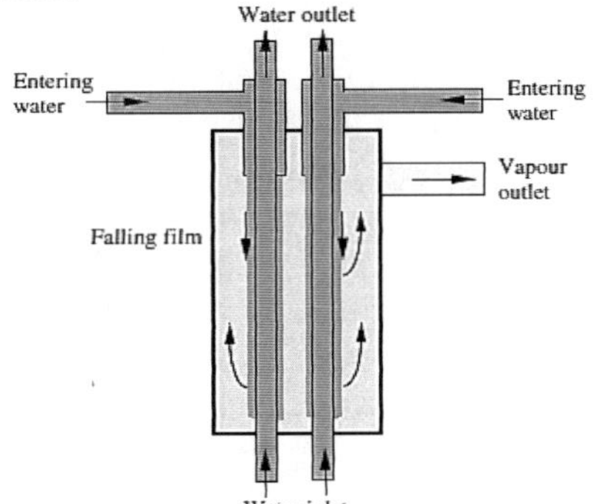

Figure 3. Schematic of Absorber[10]

The variables that affect absorber performance are pressure, solution inlet concentration, solution mass flow rate, and absorber temperature. The correlation for mass flow rate and length of the absorber is:

$$L = a\dot{m}b \tag{22}$$

Where $a = -132(\ln(100 - A_p)/86.0)$, $b = 1.33$ & $\overset{*}{m}$ = mass flow rate per width of plate (kg/m s);

A_p= "absorption percentage"

$$A_p = \frac{C_{IN} - C_{OUT}}{C_{IN} - C_{EQ}} \times 100 \qquad \text{Where,} \quad C_{EQ} = \frac{X_{EQ}}{100} \tag{23}$$

X_{EQ} is estimated from the iterative method presented in ref [10]. L's assumed value in Eqn. (24) calculates the mass flow rate since the external area of tube per unit length is known, so the required tube number can be estimated.

The next step to ensure whether the calculated surface area is enough for absorption or not. Eqn. 9 and 10 in coordination with the following relationships, are used to calculate the heat exchange area:

The heat transfer coefficient h_s can be evaluated using Wilke's relation

$$h_s = \frac{k_s}{\delta}(0.029(\text{Re}_s)^{0.53}\,\text{Pr}_s^{0.344}) \text{ where, } \quad \delta = (\frac{3\mu\Gamma}{\rho^2 g})^{1/3} \text{ (m)} \tag{24}$$

and the Reynolds number (Re_s) of solution flow in the vertical tube is $\text{Re}_s = \frac{4\Gamma}{\mu}$ (25)

The heat transfer coefficient h_i for cooling water can be estimated from $Nu = \frac{h_i D_i}{k}$.

The values of h_s and h_i will be a substitute in Eqn. 10 and U overall heat transfer will be available. And, the length per tube would be obtained. This length should be equal to the initial length assumed. The above calculation steps may be repeated until the length of the pipe matches.

2.2.4. Evaporator. The evaporator can be designed similarly to that of the absorber. The falling film of water around the evaporator tubes (through which water flows) extract heat to vaporize. The heat transfer coefficient is continuously increasing with distance. It is difficult to calculate all of the vaporisation characteristics theoretically because of the involvement of a large number of variables and the complexity of the multiphase flow patterns that occur as the quality of the vapour-liquid mixture increases during vaporization [10]. So, the mean heat transfer coefficient could be to be determined by experimental investigation.

3. Latent Heat Energy Storage

The another component of the proposed system is latent heat energy storage, in which, heat is stored during the daytime and could be used during the night or cloudy days. In this section, specifications of LHES are estimated, which would be sufficient to run the VARS in the absence of solar thermal energy. The phase change material is used to store heat in latent form, this section presents the initial guess of desired properties to match the heat requirement for VARS. The present calculation is done with known properties of paraffin wax with a melting temperature of 56°C-58°C and exercised for other PCMs also. The properties of Paraffin wax are listed in Table 4.

The design of the unit is the same as that of the standard heat exchanger. The following set of equations are used to calculate the mass of the PCM, size of the container, and heat transfer area. As the generator's capacity is 1.5 kW, considering a design factor of 1.5, the design capacity is 2.25 kW. The energy storage should be able to run the refrigeration unit for 3 hours. Thus, the amount of energy to be store in PCM will be 24300 kJ.

The amount of PCM required to store Q_{stored} will be estimated by equation [11]:

$$Q_s = m[C_{Ps}(t_m - t_i) + f\nabla q + C_{Pl}(t_f - t_m)] \tag{26}$$

Q_s = Storage capacity in kJ, m=Mass of the storing medium in kg., t_m= The melting point of the storing medium in °C., t_i = The initial temperature of the storing medium in °C, t_f =The final temperature of the storing medium in °C, C_{Pl} = Latent specific heat capacity, C_{Ps} = Sensible specific heat capacity.

f = Melt fraction; ∇q =Latent heat of fusion, in kJ/kg.

Thus, the required mass of the PCM is approx. 95 kg.

Volume of the PCM Container = (mass/density) = 0.120 m^3.

Liquid density is considered to account for maximum volume.

For, cylindrical container with height = 1000 mm.

The inner diameter of the container = 400 mm

The effectiveness of an energy storage tank can be evaluated the same as that of the heat exchanger. For this case, shell and tube arrangement is considered and the LMTD approach is adopted.

Table 4 Properties of the PCM

Stored Heat Q_{stored}	24300 kJ
Melting Temp. t_m	56 °C
Initial Temp., t_i	30 °C
Final Temp. t_f	80 °C
C_{Pl}	2.88 kJ/Kg K
C_{Ps}	2.93 kJ/Kg K
f	1
∇q	115 kJ/kg
Density liquid (ρ)	780 kg/m^3

For inlet and outlet condition: T_{h_i} = 85 °C, $T_{h_{out}}$ =70°C, T_{c_i} = 30°C, $T_{c_{out}}$ = 80°C;

From Eqn. 12, LMTD = 56°C

The inner and outer heat transfer coefficient is calculated and substitute in Eqn. 10 to get U.

$$h_{water} = \frac{\frac{f}{2}\left(\mathrm{Re}-1000\right)\mathrm{Pr}}{1+2.7(\frac{f}{2})^{\frac{1}{2}}(\mathrm{Pr}^{2/3}-1)}.\frac{k}{D} \quad (27)$$

D is the outer diameter of the water tube = 12.5 mm

Other parameters in the equation can be estimated the same as in Eqn 16 and 17.

Thus, $_{water}$ = 9052 W/m^{2}k

And h_{pcm} can be calculated by using the Nusselt equation in terms of Rayleigh number as given below[12]

$$\mathrm{Nu} = 0.133\mathrm{Ra}^{0.326}(\frac{l}{\Delta r_m})^{-0.0686} \quad (28)$$

Δr_m is the thickness of heat storage material $\frac{d_o - d_i}{2}$; l is the heat exchanger length.

$$Ra = \frac{g\beta(T_{wall} - T_{pcm})D_i^3}{\upsilon^2}.\mathrm{Pr} \quad (29)$$

All the thermophysical properties are at the mean temperature.

h_{pcm} can be calculated from the following equation:

$Nu = \frac{h_{pcm}.D_{eff}}{k_{pcm}}$, where D_{eff} is the effective diameter of the PCM bed and the h_{pcm} to be 121 W/m^{2}K.

Substituting the values of h_{pcm} and h_{water} in the equation xxiv to get overall heat transfer U

Thus, U = 167.812 W/m^{2}K

From Equation, the required area of heat transfer 0.8 m^2.

The design of latent heat energy storage is done based on PCM paraffin Wax. The same equations can be used for other suitable PCM.

4. Theoretical Result Validation

The results obtained from a theoretical calculation using the set of equations presented above needs validation. To validate the calculated results, we have compared the calculation with the values presented by Florides et al.[10] and presented in

Table 5. The result is varied less than 10% as shown in Table 5 and hence within the acceptable range.

Table 5. Result comparison

Component	**Florides et al. [10]**	**Presented Work**	**Difference (%)**
Evaporator [kW]	1	1	-
Absorber [kW]	1.28	1.412	10.3
Generator [kW]	1.36	1.460	7.3
Condenser [kW]	-	0.965	

5. Result and Discussion

After the theoretical equations, assumptions and procedure used for calculation in the paper are validated. In this section, the effect of cooling capacity on the size of generator, absorber, and condenser is presented. The variation in the quantity of PCM required for the different cooling capacity has also been presented and discussed.

5.1 Effect of Generator Temperature on COP. The effect of generator temperature on COP is estimated with the following condition and assumptions and plotted in figure 4:

1. Condenser Temperature = 45˚C
2. Evaporator Capacity = 1 kW at 5˚C.
3. Solution Strength = 60%
4. Pressure in Evaporator and Absorber = 8.52 kPa

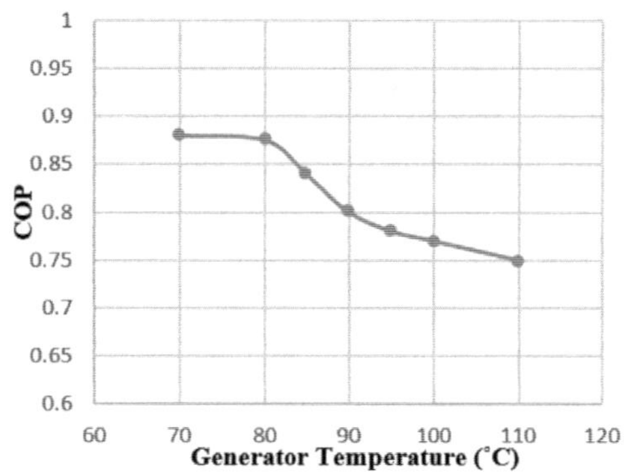

Figure 4. Effect of Generator Temperature on COP

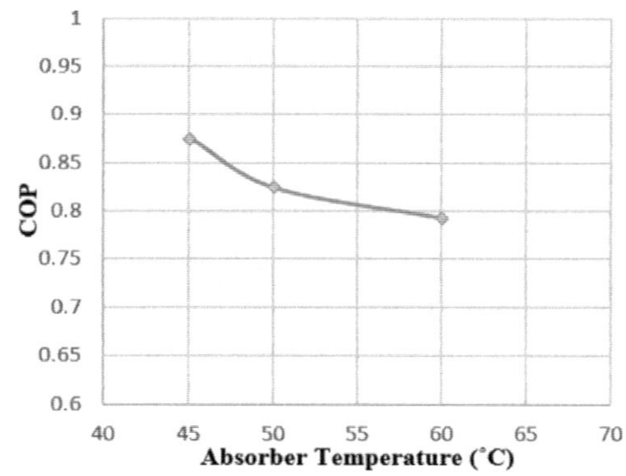

Figure 5. Effect of Absorber Temperature on COP

5.2 Effect of Absorber Temperature on COP

The effect of absorber temperature on COP is estimated with the following condition and assumptions:

1. Condenser Temperature = 45˚C.
2. Evaporator Capacity = 1 kW at 5˚C.
3. Solution Strength = 60%
4. Pressure in Evaporator and Absorber = 8.52 kPa

The variation of COP with absorber temperature is plotted in Figure 5.

5.3 Effect of Evaporator Temperature on COP

The effect of Evaporator temperature on COP is estimated for evaporator capacity of 1 kW, and other parameters are not changed.

S.No.	**Evaporator Temperature**	**COP**
1	5˚C	0.875
2.	10˚C	0.890
3.	15 ˚C	0.91

5.4 Effect of Component Capacity on COP

Figure 6 shows the generator, absorber, and condenser size for change in the machine's cooling capacity. The graph shows the different slope for generator, absorber and condenser for fixed change in cooling capacity. The actual sizes are presented in Figure.

5.5 Effect of Cooling Capacity on amount of PCM

Figure 7 shows the amount of required PCM for change in the cooling capacity of the VARS machine. The graphs show a linear variation in PCM quantity to cooling capacity. The higher the latent heat capacity of the PCM lower will be the mass required. The presented results are for the latent heat of fusion of 115 kJ/kg.

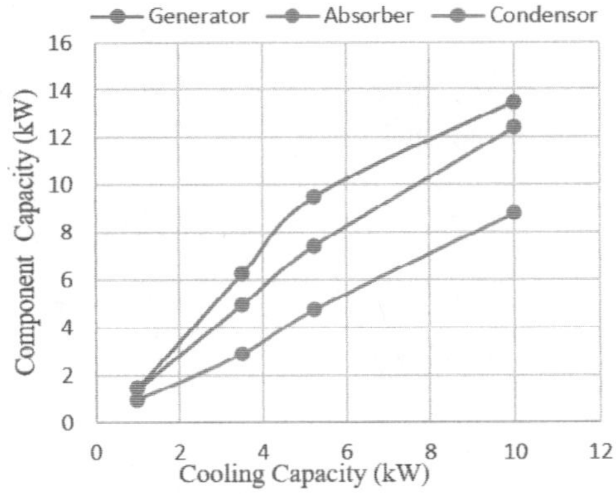

Figure 6. Required Component Capacity for different Cooling Capacity

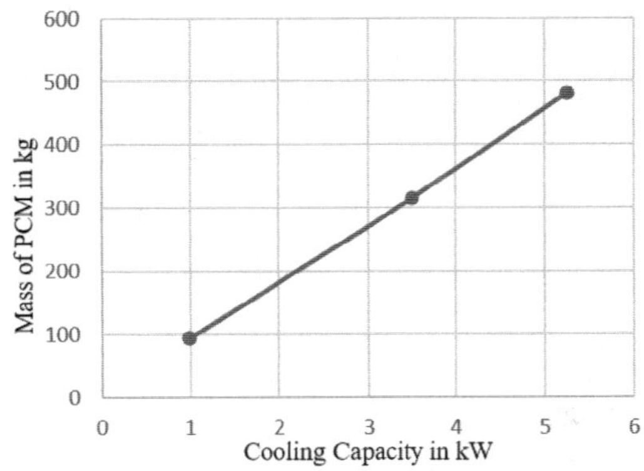

Figure 7. Required PCM for different Cooling Capacity

6. Conclusion

The presented theoretical design of solar-powered VARS coupled with PCM based latent heat energy storage is helpful to do initial sizing with some realistic assumptions and approximation. The obtained values can be used for the detailed design and sizing of the components.

The analytical design of each component is important and useful for detailed design. Component size varies linearly with cooling capacity and the amount of PCM required varies linearly with cooling capacity with different slopes.

References

[1] Tiwari A K, Sontake V C and Kalamkar V R 2020 Enhancing the Performance of Solar Photovoltaic Water Pumping System by Water Cooling Over and Below the Photovoltaic Array *J. Sol. Energy Eng.* **142** 1–9

[2] Liu M, Saman W and Bruno F 2012 Development of a novel refrigeration system for refrigerated trucks incorporating phase change material *Appl. Energy* **92** 336–42

[3] Montagnino F M 2017 Solar cooling technologies. Design, application and performance of existing projects *Sol. Energy* **154** 144–57

[4] Eames I W 2017 A LiBr-H2O absorption refrigerator incorporating a thermally activated solution pumping mechanism *Entropy* **19**

[5] Sarbu I and Sebarchievici C 2013 Review of solar refrigeration and cooling systems *Energy Build.* **67** 286–97

[6] Azzouz K, Leducq D and Gobin D 2008 Performance enhancement of a household refrigerator by addition of latent heat storage *Int. J. Refrig.* **31** 892–901

[7] Samanta S and Basu D N 2013 Modeling of a Solar Assisted Water-Libr Absorption Refrigeration System for Summer Airconditioning of an *Int. J. Emerg. Technol. Adv. Eng.* **3** 368–75

[8] Lee D and Kang C 2019 A study on development of the thermal storage type plate heat exchanger including PCM layer *J. Mech. Sci. Technol.* **33** 6085–93

[9] Kuppan Thulukkanam 2013 *Heat Exchangers: Introduction, Classification, and Selection*

[10] Florides G A, Kalogirou S A, Tassou S A and Wrobel L C 2003 Design and construction of a LiBr-water absorption machine *Energy Convers. Manag.* **44** 2483–508
[11] White M T and Sayma A I 2020 A new method to identify the optimal temperature of latent-heat thermal-energy storage systems for power generation from waste heat *Int. J. Heat Mass Transf.* **149**
[12] Jesumathy S P, Udayakumar M and Suresh S 2012 Heat transfer characteristics in latent heat storage system using paraffin wax *J. Mech. Sci. Technol.* **26** 959–65

IOP Conf. Series: Materials Science and Engineering 1146 (2021) 012027 doi:10.1088/1757-899X/1146/1/012027

Theoretical modelling of heat generation in batteries of electric vehicles for various operating environments

Jay D Patel[1], Rajesh S Patel[2]
[1]Assistant Professor, Ganpat University, Mehsana-384002, Gujarat, India ©
[2]Associate Professor, Pandit Deendayal Petroleum University, Gandhinagar-382421, Gujarat, India
Corsponding author E-Mail: jaydpatel1992@gmail.com;

Abstract. Due to high efficiency, cycling life, and specific energy, the Lithium-ion battery has been the best-favored selection for the Electric Vehicles (EV's) despite high cost. The performance of the battery pack is strongly dependent on the operation condition of the EV's and the environment. Under the extreme operation and environmental conditions of the battery, the rate of heat generation is very high which adversely affects the dissipation rate, decreases the battery life, and sometimes leads to the explosion and fire (i.e. thermal runaway). In electric vehicles, the battery thermal management technology plays a crucial role when the temperature of the cell rises exponentially during the operating environment. The temperature, therefore, between (15 to 35 ºC) should be controlled. Our work presents the modeling of heat generation in the battery pack under various operation conditions. The parametric studies have been conducted to study the effect of battery operation conditions on heat generation rate and battery pack temperature. This study would be very useful in the selection of battery thermal management technology to control the battery temperature to retain the optimum condition.

Keywords: Electric Vehicles (EV), Battery Cooling, C-rate, maximum battery cell temperature, maximum battery cell temperature difference

1. Introduction

Electrical Vehicles (EVs) have witnessed rapid evolution with ongoing developments in automobile sectors and favorable government policies [1]. Nowadays, there is an enormous demand for EVs in the automobile sector due to the growing sensitivity of various governments towards a cleaner environment. Battery cells are sources of electricity for EVs that produce electricity by electrochemical reactions during charging and discharging cycles. For battery packs in EVs, various battery kinds, dimensions, and capabilities are available. Batteries are an essential component of electric cars, which requires utmost attention at the time of design [2]. The output and life of the battery are susceptible to environmental factors, charging conditions, and vehicle operating conditions. [3]. For proper operation, the battery has to work in a strict atmosphere. Since the working of the batteries is quite temperature sensitive, running a battery within the range of temperature is very critical. [4].

Batteries produce vast quantities of heat as the operating conditions become extreme. If an appropriate Battery cooling system is not used in EV's to efficiently evacuate the heat, the temperature of the battery cells may raise. The increase in temperature will intensify the chemical reaction, which will further increase the production of heat. If this continues for a prolonged time, the battery pack may reach thermal runaway conditions [5]. For that, a Battery cooling system is employed to dissipate the generated heat during the operation.

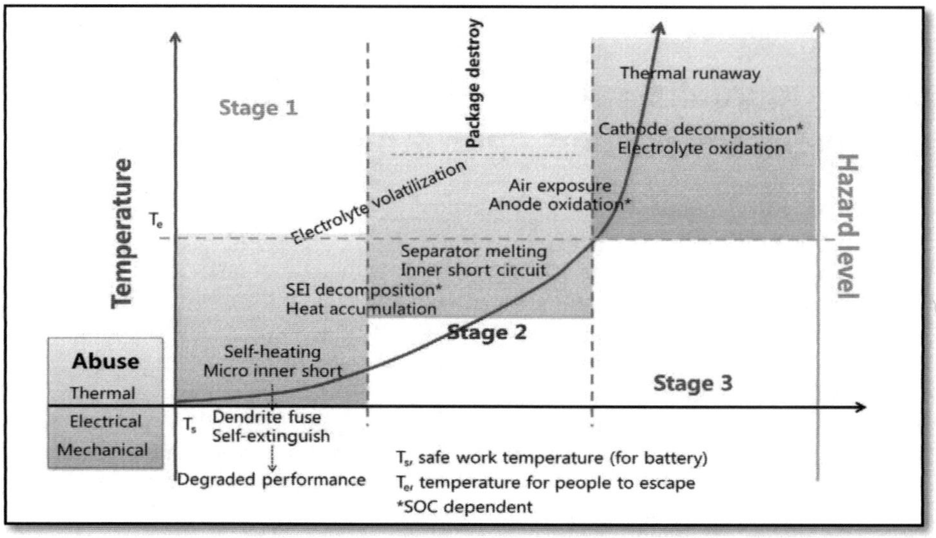

Figure 1.1 Phases of temperature rise in the Battery cell.

It is very essential to design and develop an appropriate Battery cooling system that increases the efficiency and effectiveness of EVs. The design of the Battery cooling system depends on the operation condition of EV's and batteries and its effect on battery pack temperature and battery life. Hence it is very essential to develop a theoretical model for parametric studies on the consequence of operation conditions on the rate of heat generation and temperature of battery pack.

2. Heat generation in Battery cell

Li-ion battery performance and lifespan depend on the thermal condition of the battery cell. Since the thermal behavior of the battery cell is closely related to the rate of heat generation within a battery cell, it's important to calculate the heat generation for different operating conditions. The reversible heat and irreversible heat is mainly responsible for heat generation in battery cells [6]. The heat generation rate in the battery cell depends on the electric current(I), the open-circuit voltage (U), and battery cell voltage (V) and is calculated by Eq.1.

$$Q = I(U - V) - I\left(T\frac{dU}{dT}\right) \qquad \text{Eq.1}$$

The factors which influence the heat generations in the battery cell are:

- The charging/discharging current. High solicitations result in an increase in battery temperature due to the dominating result of the Joule effect [7].
- The Depth of Discharge (DOD), which is related to the energy quantity which leaves from the battery.
- The battery temperature, the electrochemical reactions are accelerated and the internal resistance decreases when the temperature of the battery is increased [8].
- The state of Health (SOH), the amount of quality of a battery compared to its ideal condition [9].
- Available capacity [10].

The charging/discharging current plays a crucial role in heat generation compared to other factors which is evaluated by C-rate which is explained in the upcoming section 3.

3. C-rate

C-rate represents the rate at which a battery is discharged compared to its maximum capacity. If the whole battery is discharged by the discharge current in 1 hour, it's discharging at a 1C rate. If the same battery is discharged by the discharge current in 2 hours, it's discharging at 0.5C rate. If the same battery is discharged by the discharge current in 30 minutes, it's discharging at a 2C rate, and so on.

C-rate is dependent on the speed of the vehicle, gradient of slope, and number of seated persons. C-rate can be calculated from the driving conditions. Compared to the gradient of the slope, and the number of seated persons, the speed of the vehicle has a significant effect on C-rate.

The speed of the vehicle, gradient of slope, and number of seated persons influence the electric power consumption from the battery pack.

The total force required to drive a vehicle is calculated by Eq. 2

$$F = Fr + Fd + Fg + Fa \quad \text{.......... Eq.2}$$

Where Fr, Fd, Fg, Fa, and F represents the rolling force, drag force, gradient force, acceleration force, and total force required.

Electric power required to drive a vehicle is calculated by Eq. 3

$$P_{ele} = \frac{F*V}{\eta} \quad \text{.......... Eq.3}$$

Where P_{ele}, V, and η represents the electric power required, velocity of vehicle, and conversion efficiency of electric motor.

Effect of C-rate on electric power consumption is calculated by below Eq. 4 and 5.

$$I\_spc = \frac{P_{ele}/V_{BP}}{N_{CP}} \quad \text{.......... Eq.4}$$

$$C - rate = \frac{I_{spc}}{C_{bc}} \quad \text{.......... Eq.5}$$

Where I_{spc}, C_{bc}, V_{BP}, and N_{CP} represent the specific current, battery cell capacity, voltage of battery pack, and no. of battery cells in parallel.

3.1 Effect of speed of vehicle on C-rate

The effect of the speed of the vehicle on the C-rate is shown in Figure 3.1. As the speed of the vehicle increases, the C-rate increases rapidly. The C-rate is increased from 0.75C to 2.5C when the vehicle is accelerated from 60 kmph to 120 kmph.

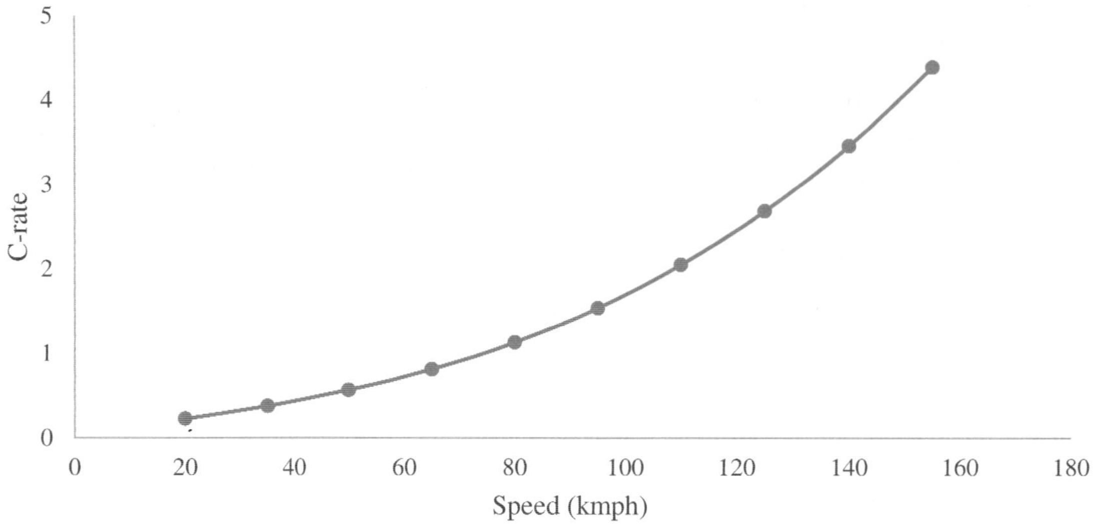

Figure 3.1. The effect of speed of vehicle on C-rate.

3.2 *Effect of road gradient on C-rate*

C-rate increases linearly with the slope of the road. C-rate increases from 1.2C to 1.82C when slope varies from 2^0 to 20^0 . The effect of the slope of road on C-rate is shown in Figure 3.2.

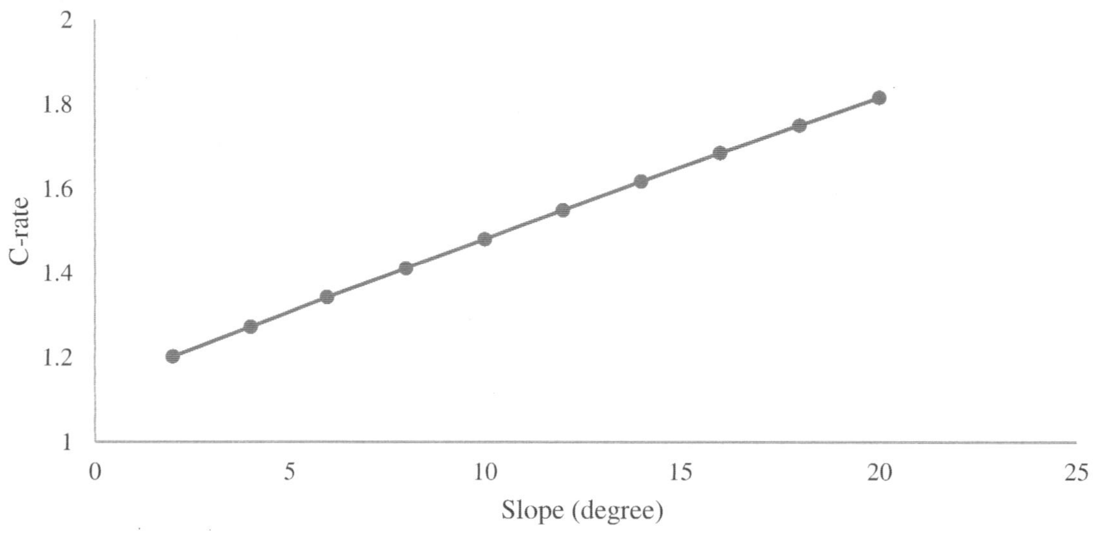

Figure 3.2. The effect of gradient on C-rate.

3.3 *Effect of seated person on C-rate*

According to the number of people seated in the vehicle, C-rate linearly increases. C-rate increases from 1.04C to 1.13C when the number of people varies from 1 to 5. The effect of a seated person on C-rate is shown in Figure 3.3.

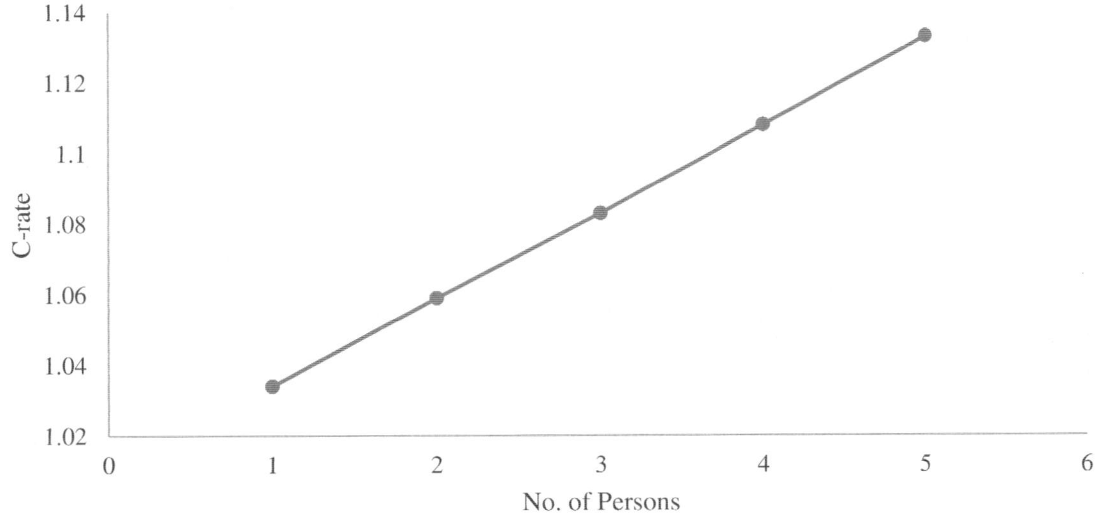

Figure 3.3. The effect of the seated person on C-rate.

3.4 *Effect of C-rate on rate of heat generation in battery cell*

As the C-rate increases, heat generation increases rapidly because of the large amount of current being drawn from the battery cell. The heat generation increases from 3.5 W to 10 W when the C-rate is increased from 0.75 to 5.5. The heat generation could be as high as 42.5 W which C-rate is 10 which can be seen from Figure 3.4.

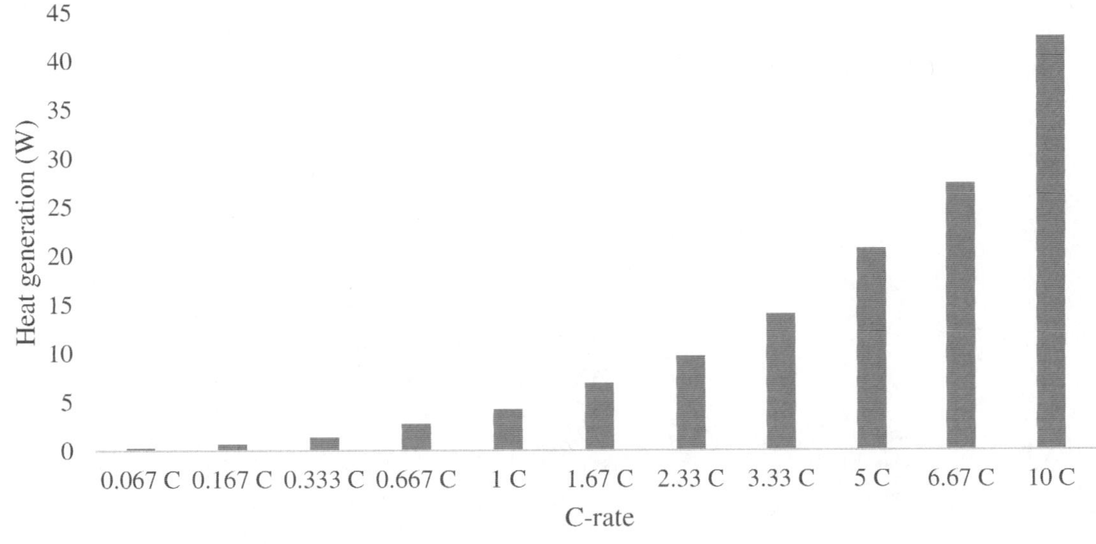

Figure 3.4. The effect of C-rate on heat generation in battery cells.

3.5 *Effect of C-rate on battery cell temperature.*

As the C-rate increases according to operating conditions, the battery cell temperature increases. It is seen from the result that the battery cell temperature could be as high as 80 °C for a time-lapse of 0.27 hours as the battery cell is discharged at 10 A current (5 C-rate). Battery cell temperature goes beyond the 55 °C for 5 C-rate conditions which necessitates the battery thermal management system which limits the battery cell temperature below the 55 °C for better performance and safety of battery pack.

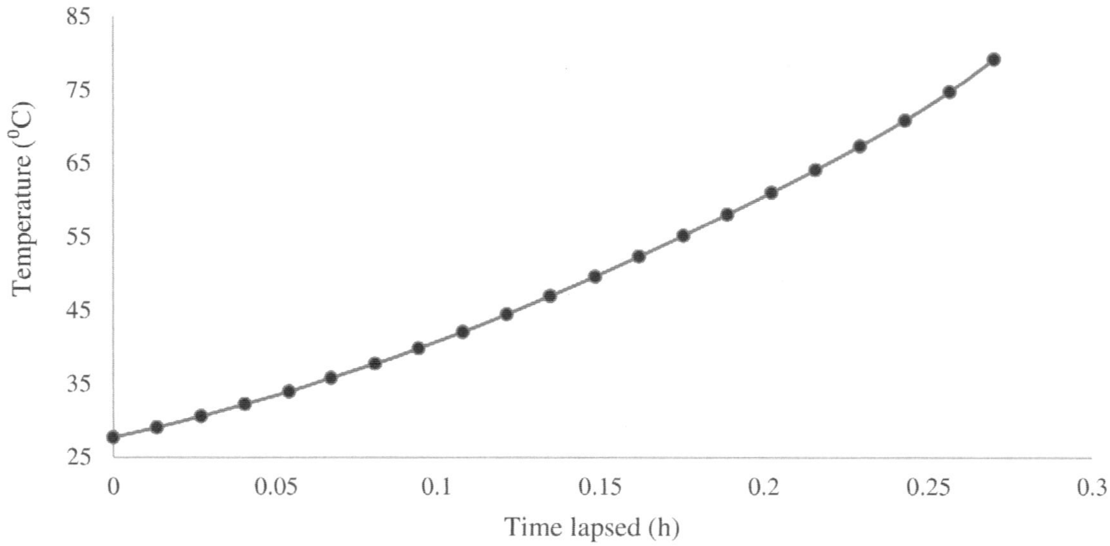

Figure 3.5. The effect of 10 A current drawing on battery cell temperature.

4. Effect of ambient temperature on battery cell terminal temperature

To understand the consequence of the environment on the rate of heat generation, ambient temperature is varied from 25 °C (winter) to 40 °C (summer) to evaluate the battery cell terminal temperature. The consequence of ambient temperature on the battery cell terminal temperature is shown in Figure 4 when a 5 A current is being drawn from the battery. It is seen from the result that the battery cell terminal temperature could be as high as 68 °C in winter and 84 °C in summer for a time-lapse of 0.55 hours.

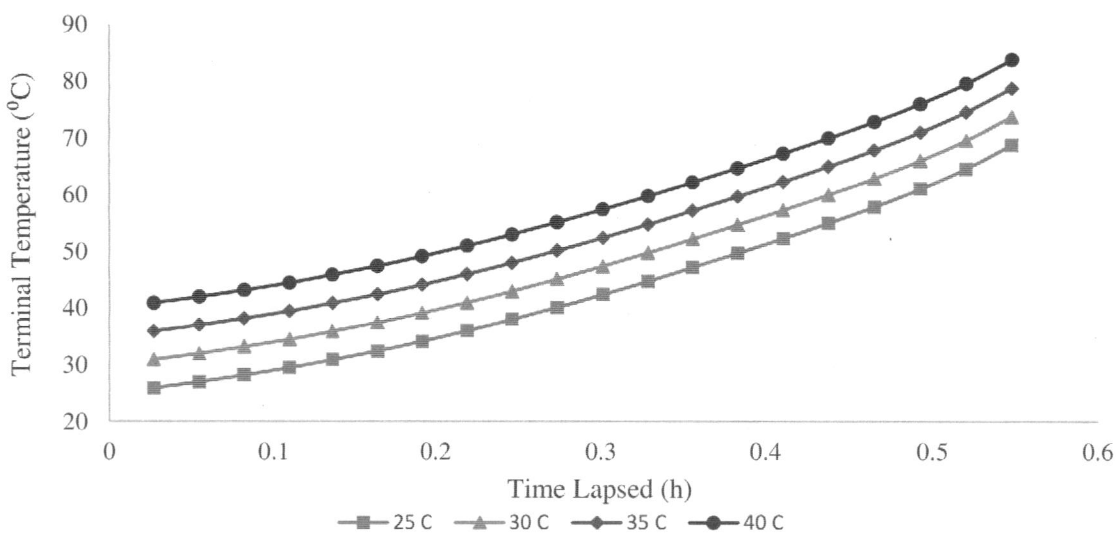

Figure 4. The effect of ambient temperature on terminal battery temperature.

Conclusion

It can be concluded from the above results that The efficiency and lifespan of the battery cells are susceptible to the vehicle's climatic conditions, battery conditions, and operational characteristics. For proper operation, the battery has to work in a strict state. It is vital to bound the battery cell temperature below 45-55 °C and the temperature difference between adjusting cells must not be increased over 5 °C. It is very essential to design and develop appropriate BTMS to increase efficiency and minimize the weight and size of EVs. BTMS is indeed not useful for extending the efficiency and life of a battery, but also from a protection perspective.

References

[1] "Global EV Outlook 2017," *Glob. EV Outlook 2017*, 2017, doi: 10.1787/9789264278882-en.
[2] S. M. A. S. Bukhari, J. Maqsood, M. Q. Baig, S. Ashraf, and T. A. Khan, "Comparison of Characteristics-Lead Acid, Nickel Based, Lead Crystal and Lithium Based Batteries," *Proc. - UKSim-AMSS 17th Int. Conf. Comput. Model. Simulation, UKSim 2015*, pp. 444–450, 2016, doi: 10.1109/UKSim.2015.69.
[3] T. M. Bandhauer, S. Garimella, and T. F. Fuller, "A Critical Review of Thermal Issues in Lithium-Ion Batteries," *J. Electrochem. Soc.*, vol. 158, no. 3, p. R1, 2011, doi: 10.1149/1.3515880.
[4] A. A. Pesaran, "Battery thermal models for hybrid vehicle simulations," *J. Power Sources*, vol. 110, no. 2, pp. 377–382, Aug. 2002, doi: 10.1016/S0378-7753(02)00200-8.
[5] X. Wu *et al.*, "Safety issues in lithium ion batteries: Materials and cell design," *Front. Energy Res.*, vol. 7, no. JUL, pp. 1–17, 2019, doi: 10.3389/fenrg.2019.00065.
[6] J. Kim, J. Oh, and H. Lee, "Review on battery thermal management system for electric vehicles," *Appl. Therm. Eng.*, vol. 149, no. September 2018, pp. 192–212, 2019, doi: 10.1016/j.applthermaleng.2018.12.020.
[7] C. Lin *et al.*, "Comparative study on the heat generation behavior of lithium-ion batteries with different cathode materials using accelerating rate calorimetry," *Energy Procedia*, vol. 142, pp. 3369–3374, 2017, doi: 10.1016/j.egypro.2017.12.472.
[8] J. F. Peters, M. Baumann, B. Zimmermann, J. Braun, and M. Weil, "The environmental impact of Li-Ion batteries and the role of key parameters – A review," *Renew. Sustain. Energy Rev.*, vol. 67, pp. 491–506, 2017, doi: 10.1016/j.rser.2016.08.039.
[9] C. Pastor-Fernández, K. Uddin, G. H. Chouchelamane, W. D. Widanage, and J. Marco, "A Comparison between Electrochemical Impedance Spectroscopy and Incremental Capacity-Differential Voltage as Li-ion Diagnostic Techniques to Identify and Quantify the Effects of Degradation Modes within Battery Management Systems," *J. Power Sources*, vol. 360, pp. 301–318, 2017, doi: 10.1016/j.jpowsour.2017.03.042.
[10] M. Li, "Li-ion dynamics and state of charge estimation," *Renew. Energy*, vol. 100, pp. 44–52, 2017, doi: 10.1016/j.renene.2016.06.009.

IOP Conf. Series: Materials Science and Engineering 1146 (2021) 012028 doi:10.1088/1757-899X/1146/1/012028

Tribological analysis of automotive material under wet lubrication condition using diesel, biodiesel and their blend

S K Chourasia [1, 2], A M Lakdawala [1] and R N Patel [1]

[1]Mechanical Engineering Department, SoE, Institute of Technology, Nirma University, Gujarat - 380015, India

[2] Mechanical Engineering Department, Gandhinagar Institute of Technology, Gujarat - 382721, India

E-mail: rajeshnpatel1969@yahoo.com; sajan1chourasia@gmail.com

Abstract. Biodiesel made from vegetable oil can be considered the best alternative fuel for a partial or complete diesel replacement. Earlier studies on biodiesel and its blends have proven that biofuels in the existing engine can be favorable for partial eradication of diesel from C.I. engines. Therefore, this study investigates the tribological impact of various biodiesels and diesel and their optimum blend at normal room conditions. The tribological analysis is performed on a pin-on-disc tribometer. During the tests, the coefficient of friction, friction force and wear of pins (Aluminium, Copper and Iron) were examined; apart from this, weight & volume loss of the metal pins were also investigated. The pin surfaces were characterized by surface microscopic analysis. Fuel constituents and fatty acid content were analyzed by GC-MS and FT-IR tests to understand the cause of tribological wear better. Castor and Rapeseed biodiesel show the lowest weight of pin in comparison to diesel. The wear rate of metals with diesel shows extreme rises in wear as the sliding distance increase. The coefficient of friction, friction force of pins was found to be highest in diesel, followed by biodiesel and their blend in almost every case.

1. Introduction

Although ample research has been done on biodiesel as a lubricant, very few researchers have compared and analysed the combined effect of two different biodiesel as a lubricant fuel on C.I. engine actual components materials. Keeping this in mind, an in-depth understanding of their tribological behaviour with engine component material is needed. Tribology can be defined as the branch of science & engineering that deals with the contacts of two surfaces in relative motion. The tribological study includes the deep study of wear, lubrication and friction at the surface of the material; therefore, this is chosen as the prime method for selecting a suitable lubricant for a particular application. For discovering a requirement of lubrication one has to understand the consequences of parameters related to tribology [1]. Therefore the lubricants are very essential for every moving component of the engine. The lubricants are those substances in solid or liquid state used to lubricate the surfaces that are in mutual contact with each other so that the material wear and friction between moving parts can be minimized. In most vehicles, the lubricating oil helps transfer power, clean, cool, and lubricate the engine components. The fuel used to run C.I. engine must have an excellent lubricating capability to minimise the wear and various tribological complications that are usually found in fuel lines and pumps. The good lubricating property of fuel is very important for fuel injection systems because fuel pumps are dependent on fuel for their rotational parts lubrication and cooling. Nowadays, the oil refineries have to produce a low sulphur content diesel because of stringent emission rules and regulations, which ultimately reduces its lubricating properties. The problems like high corrosion, extreme wear, erosion of fuel injector nozzle, difficulty in engine starting, high engine emission and reduction in engine power can be easily noticed in the engines using low sulphur fuel. The life of fuel distributor pumps is reduced by 94.8% with the continuous use of low sulphur diesel [2]. Another kind of lubricity like boundary lubricity, is formed because of mutual contact between two solid surfaces and diesel fuel impurities that stick on the metal surface. Surfactants (Polar), nitrogen, sulphur and heterocyclic aromatics are some of the most popular types of impurities found in diesel fuel. These impurities react or get themselves absorbed on rubbing metal surface, which minimizes the friction and wear by reducing the bond between mutual contact surfaces.

1.1. Tribology with hydrodynamic lubrication conditions

Baskar S & Sriram G conducted an experiment using a pin-on-disc tribometer for characteristics of the metal wear and friction of a bearing material using 20W40 synthetic oil and chemically improved Rapeseed methyl ester under hydrodynamic lubrication condition. The rotating disc is made from EN31 hardened steel and the pin is made from brass. They have kept 200N load constant for this tribology testing and a varying sliding distance from 2-10 m/s. It has been found that the friction coefficient of chemically improved Rapeseed methyl ester is lower and it also has wear-resistant properties [3]. Jain A K & Suhane A conducted a similar type of tribology testing by using pin-on-disc tribometer. They have chosen Castor, mahua oil and their blends for investigating its tribological behaviour as a bio-based lubricant. For this test mild steel rotating disc and carbon steel pin were chosen. Their results claim that with the use of bio lubricant, the mechanical operational efficiency and life of design application increases. The bio-based lubricant also reduces friction coefficient and lowers wear up to 63% [4]. The capability of mahua methyl ester towards lubricity for the applications like maintenance was investigated by Suhane et al. They created a blend of 90T gear oil with mahua methyl ester in various proportions. For this, they have also used pin-on-disc tribometer. This time, the test was conducted on hydrodynamic or wet lubrication conditions by using a blend of 90T gear oil with mahua methyl ester. Carbon steel and mild steel are used for the manufacturing of pin and rotating discs. The rotating disc's speed was kept in between 100 to 500 RPM with an interval of 100 RPM. Results show that the blend of 90T gear oil with mahua methyl ester has very good wear-reducing properties apart from this it also shows ecological benefits [5]. A similar kind of work was performed by Shanta et al. for studding the effect of mixing canola, soybean and peanut methyl ester with 15W40 lubricating oil on its tribological behaviour by using a pin-on-disc tribometer. The test was conducted on pin made from AISI 316stainless steel and disc made from AISI 1018 under hydrodynamic or wet lubricating conditions. They find that biodiesel's dilution in any proportion degrades the lubricating properties of 15W40 conventional oil [6]. Shahabuddin et al. conducted an experiment by using a blend of jatropha methyl ester with SEA 40 lubricating oil on a pin-on-disc tribometer under a constant load of 30N and 2000 RPM of a disc rotating speed for one hour to find its lubricating properties. For this experiment, the lubricating condition was kept as hydrodynamic and the pin and rotating disc were fabrication from aluminium and cast-iron. According to their claimed results, the accumulation of 10% of jatropha methyl ester in SEA 40 shows optimum results for an application like automobile. That blend shows best results with respect to viscosity, friction coefficient increment in temperature, wear, etc. [7].

2. Experimental methodology and materials

2.1. Selection of biodiesel

Although it has been recognized that biodiesel may be the superlative replacement of diesel and having the capability to fulfill the need of current diesel engines however deciding on superior biodiesel is a chunk of a tough challenge. For the current experiment, two biodiesels (Castor and

Rapeseed) and their optimum blend (15CA15R) are selected based on their availability, lubricating, performance, combustion, and emission properties are directly connected to the physical, chemical and fatty acid concentration of the biodiesel. The seeds of Castor and Rapeseed are found in abundant quantity in India and they are also non-edible. These plants does not require any additional care during their growth. Amongst the non-edible biodiesels, Castor and Rapeseed have been widely explored as useful sources of biodiesel. Various researchers have performed an experiment on single biodiesel and their biodiesel-diesel blend and found some pros and cons of every biodiesel after that, they suggested that combining two biodiesel together can improve their physical, chemical and lubricating properties. Retaining this as an initial objective for identifying the tribological behaviour of biodiesel two biodiesels and their blend were chosen. The Castor biodiesel has a high concentration of acid value, highest viscosity, density, lower cetane number and worst cold flow properties on the other hand, the Rapeseed biodiesel has lower acid concentration, lower viscosity, density, cetane number, flash point, improved cold flow properties. Because of the reasons mentioned above, they together can counter each other's properties. Therefore Rapeseed biodiesel is the suitable choice for making a dual biodiesel blend with Castor biodiesel.

This biodiesel was made via transesterification process by mixing raw Castor and Rapeseed oil with methanol separately at 60°C. Potassium hydroxide is used as a catalyst to enhance the reaction in both cases. The mixture of raw oil and methanol with potassium hydroxide is kept for 12 hours for a proper reaction; after that, the glycerine and biodiesel were separated. Fatty acid composition present in biodiesel, physical and chemical properties of biodiesels are shown in Table 1.

Table 1. Fatty acid composition and physical-chemical properties of Castor and Rapeseed biodiesel.

Fatty acid composition (%)	Biodiesel	
	Castor	Rapeseed
Caprylic (C8:0)	-	-
Capric (C10:0)	-	-
Lauric (C12:0)	-	-
Myristic (C14:0)	-	-
Palmitic (C16:0)	2.64	4.9
Palmitoleic (16:1)	-	-
Stearic (C18:0)	1.52	2
Oleic (C18:1)	4.69	62.8
Linoleic (C18:2)	8.4	19.4
Linolenic (C18:3)	-	8.9
Ricinoleic (C18:1:OH)	81.9	-
Arachidic (C20:0)	-	1.9
Gondoic (C20:1)	-	-
Behenic (C22:0)	-	-
Erucic (C22:1)	-	1.2
Lignoceric (C24:0)	-	-
Physical-Chemical properties		
Viscosity at 40 °C (mm^2/s)	280	34.7
Density ($\times 10^3 kg/m^3$)	0.961	0.92
Flash Point (°C)	240	321
Pour Point (°C)	2.7	-14.8
Cold filter plugging point(°C)	7.1	-12.8
Cetane number	39.48	54.21
Calorific value (MJ/kg)	38.6	36.8

2.2. *Functional group & chemical composition analysis*

The chemical functional group of diesel, biodiesels and their blend have been obtained by using G-FTIR. The spectra were obtained using a Fourier transform infrared spectrometer (Perkin-Elmer, model Frontier, USA), standard temperature LiTaO3 MIR detector, and an SNR value of 9,300:1, high-performance DTGS MIR detector with an SNR value of 14,500:1, temperature-stabilized. Optical system with windows of KBr for collecting data over a 8,300 – 350 cm^{-1} spectral range at a resolution of 0.5 cm^{-1} and the composition of biodiesel was analyzed by Gas chromatography-mass spectroscopy (GC-MS) in Shimadzu GC-2010 and using a Fused Silica capillary column having 30 m x 0.25 mm, 0.25 μm film thickness RTX-5. N-pentadecane (NC15) was used as an internal standard. A stock solution of acetone with a known amount of N-pentadecane was prepared a priori and used for analysis.

2.3. *Selection of metal for pin preparation*

A diesel fuel system's role is to spray a precise amount of atomized and pressurized fuel inside every engine's cylinder at the demarcated time. The diesel engine's essential parts are the fuel system, which comes in direct contact with diesel/biodiesel, i.e., injection nozzles, injection pump, fuel transfer pump, fuel tank, and fuel filters. Ferrous and non-ferrous materials, i.e., cast irons, steel, copper and aluminium alloys, come in direct contact with fuel. Rubber is the common non-metallic ingredients of the components used in the fuel systems [8-9]. Fuel gets transferred and gets in touch with these materials under different situations and these are the sources for deterioration and degrades of fuel supplying components diversely. The lift pump, fuel filter, plunger pump, injection pump, priming pump, diesel fuel tank, and injection nozzles are the fuel system's key components that directly contact diesel/biodiesel. Based on this concept, an elementary has been conducted on the existing engine parts (Make: Kirloskar, Model: TV1) and from that test, pins are prepared, pin composition is as shown in Table 2.

Table 2. Elemental composition of actual CI Engine components obtained by elementary analysis.

Elements (%)	Components of CI Engine					
	Connecting Rod Bearing	Piston	Piston Ring	Exhaust Valve	Inlet Valve	Cylinder Liner
S	0.023	-	0.03	0.074	0.074	0.028
Sn	-	0.016	-	-	-	-
Cr	16.94	0.016	19.11	0.043	0.043	17.02
Ni	10.95	0.66	9.06	0.046	0.046	11.05
Fe	68.99	0.93	68.77	93.651	93.651	68.74
Mg	0.013	0.49	0.0083	-	-	0.02
Pb	0.019	0.047	0.0025	-	-	0.012
Ti	-	0.027	-	-	-	-
Zn	0.014	0.63	0.014	-	-	0.015
Si	0.89	13.25	0.75	1.53	1.53	0.74
Mn	1.85	0.11	2	0.57	0.57	2.03
C	0.082	-	0.08	3.38	3.38	0.088
Al	0.02	82.23	0.0016	-	-	0.023
Co	0.02	-	0.0032	-	-	0.018
Cu	0.018	1.58	0.0119	0.46	0.46	0.013
P	0.03	-	0.045	0.14	0.14	0.032

Initially, before starting the tribo-wear test based on the elementary analysis of actual engine components, three metals each for pin has been selected for the pins preparation and disc EN31 is chosen as per ASTM G99-95a, the pins are having a size of 30 mm in length & 10 mm in diameter and disc having diameter and thickness of 175 mm & 10 mm. After fabrication and polishing, the grease was removed using a mixture of xylene–isopropanol in equal quantity v/v %. Before the test weight of each pin was recorded up to an accuracy of 0.1 mg [10].

2.4. Wear test parameters

Friction tests were carried out for the selected biodiesels to characterize their frictional properties under wet lubrication and pure sliding motion according to ASTM: G 99-95a standard using a pin-on-disc tribometer. Test fuels are diesel, two biodiesel (Castor and Rapeseed and their optimum blend 15CA15R (70% Diesel + 15% Castor + 15% Rapeseed). The test is performed using a tribometer configured for a pin-on-disc (Make: Ducom Instruments, Model: Micro POD). In this study, the metal disc is rotated against a stationary metal pin, the dimension and elemental composition have been selected as per ASTM: G 99-95a. To simulate the actual sliding condition the sliding speed and load were selected as 5.5 m/s and 70 N because as the engine used in our previous research was a constant speed CI engine (Make: Kirloskar oil engine ltd. Model: TV1) which runs at 1500 RPM and the full rated load of the engine was 12 kg as it is well known that the engine is not continuously running on its rated full load in real life. So as per the load factor provided by SCC-2270005055 of US EPA document titled as 'Median life, annual activity, and load factor values for non- road engine emissions modeling NR-005b' for the current engine load factor is provided as 0.59 or can be said as 59% [11]. That's why the sliding speed and load were selected as 5.5 m/s and 70 N others operating parameters are shown in Table 3.

2.5. Weight loss and volume loss analysis

To recognize weight reduction and volume loss because of wear, before and after the tribo-wear test the metal pins were appropriately cleaned so that all debris of metal, lubricant and dirt can be distant and after the cleaning, metal pins were investigated to get their weight with a precision of three digits after decimal so a minor variation in weight can likewise be noted down and this test is led with weighing machine-accessible at NABL licensed research laboratory. Using the weight loss and density of metal, one can quickly identify the volume loss. For this, the below-mentioned equation was used which is given by ASTM G 99-95a. Other details i.e., Density of metal, are mentioned in Table 3. The volume loss has been obtained by equation 1, provided by the ASTM standard.

$$Volume\ Loss\ (mm^3) = \frac{Before\ weight\ (g) - After\ weight\ (g)}{Alloy\ Density\ \left(\frac{g}{cm^3}\right)} \quad (1)$$

Table 3. Operating and working parameters configured during the wear test.

Track Diameter	RPM	Tip Velocity (m/s)	Lubricant Flow Rate Litre/Min	Pin Metal	Metal Density Kg/m^3	Pin (L&D)	Load (N)	Sliding Distance (m)
95 ± 0.01 mm	1105 ± 1	5.5 ± 0.1	0.333 ± 0.02 Max flow	Aluminium	2700 ± 1	30 ± 0.01 mm & 10 ± 0.01 mm	70 ± 0.01	10000 ±1
125 ± 0.01 mm	840 ± 1			Copper	8940 ± 1			
135 ± 0.01 mm	778 ± 1			Iron	7870 ± 1			

*Note: L: Pin length, D: Pin diameter

2.6. Microscopic Surface Analysis

After the tribo-wear test led by tribo-tester instrument, adequate data was found, but for the complete analysis, a detailed qualitative analysis has to perform for that an image was taken from scanning electron microscope (SEM), Make: Hitachi, Resolution SE: 3.0nm at 30kV (vacuum mode), Hitachi patented Quad Blass function instead of expensive LaB6-tips, Motorized: 5 axes in standard) at a magnification of 500X so that a detailed surface analysis can be done and along with this, the detection of wear type can also be recognized.

3. Result and Discussion

3.1. Weight loss analysis

Weight loss analysis can be defined as the reduction from the initial to the final weight of the pin due to rubbing on the disc. A very minute reduction in weight can be observed with the biodiesel, as shown in figure 1, their weight losses are gradually increasing as we move towards the 15CA15R and diesel. The probable reason might be due to the presence of fatty acid and also due to the high viscosity of biodiesel. It has good lubricating properties in comparison to diesel, which results in a reduction of pin weight. As with the government, new emission norms the diesel fuel has been more refined and the quantity of sulphur has to be reduced to 10ppm (BS-VI; nationwide), which ultimately leads to reducing the emission, but along with that the diesel fuel will lose its lubricating properties which can be seen from figure 1 there is 37, 369 and 150 mg reduction in weight by use of diesel with aluminium, copper and iron pin. With the use of other biodiesels, pins are showing variation marginally in their weight. The blend of diesel, Castor and Rapeseed biodiesel has shown relatively good results compared to pure diesel. This blend can fulfill the demand of the mordent automotive society for metal wear and regulatory emission norms.

3.2. Effect of biodiesel on the coefficient of friction

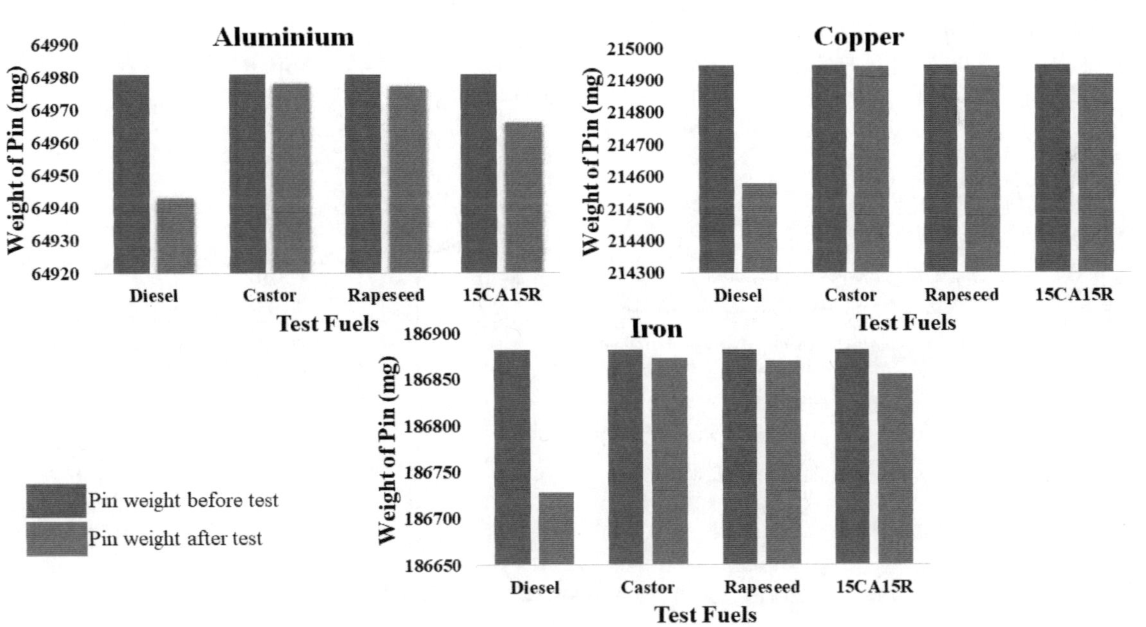

Figure 1. Weight analysis of metal pins (Aluminium, Copper & Iron) with fuels (Diesel, Castor, Rapeseed and 15CA15R).

Lubricated friction can be classified into three regimes: (BL – Boundary layer), (ML- Mixed layer) and (EHL- Elasto hydrodynamic). In contrast with the lubricated wet friction based on operating conditions such as the temperature, lubricant viscosity, sliding speed, etc., shown in Table 1 and 3. The trend of the coefficient of friction (CoF) follows the trend shown in figure 2. For the current case, all the operating parameters are kept constant and from the result, it has been identified that with every test fuel, the value of CoF till 1000 m is quite high. From figure 2 it has been identified that in comparison to aluminium and copper pins, the iron pins have more CoF the reason behind this is that the CoF depends on the materials used, i.e., the disc is also made from an iron alloy (EN31) and the pin is also having similar kind of composition due to that the CoF increases. Another finding is related to the drastic difference between the CoF value of pins with diesel in comparison to biodiesel and 15CA15R; the probable reason is due to the higher content of viscosity, which helps to reduce the friction in between the pin and disc, the lubricating film is found to be stable in biodiesel in compare to diesel which also leads to a reduction in CoF [12].

3.3. Effect of biodiesel on the friction force

The friction force is generated due to the opposite movement between the fixed pin and rotating disc surfaces. As no surface is perfectly smooth, which is the main cause of friction. In lubricated wet friction, a fuel separates two surfaces of pin and disc. The graph of friction force between pin and disc with each test fuel is shown in figure 3. Figure 3 shows that similar to CoF, the friction force also follows the same trend and the probable causes are already discussed above. During the experiment, it has been observed that scratching sound generated due to friction between pin and disc was found to be highest in the iron pins because of more friction force. The friction force trend remains more or less constant in the case of pins that are in contact with Castor, Rapeseed and 15CA15R after 2000 m till the last phase of the experiment. But pins with diesel have shown drastic fluctuation in their results from the beginning to end of the experiment. The probable reason for this is due to the low viscosity of diesel in comparison to other fuels. Oxygen bonds in biodiesel are superior lubricity providers in comparison to sulfur present in diesel.

3.4. Effect of biodiesel on wear

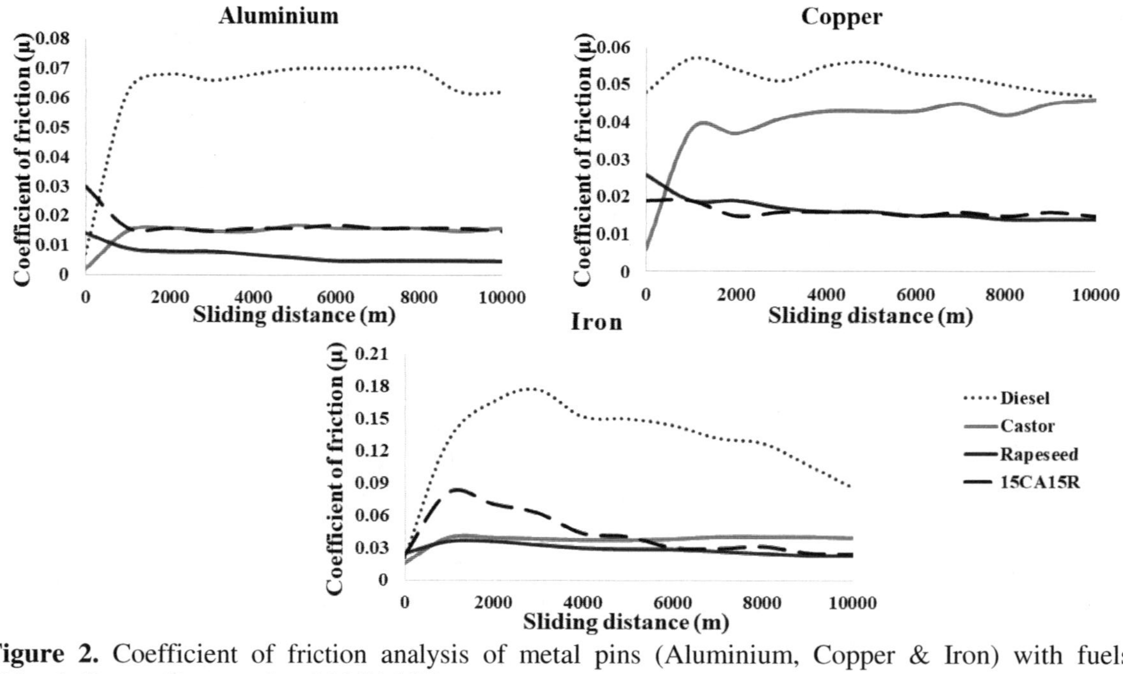

Figure 2. Coefficient of friction analysis of metal pins (Aluminium, Copper & Iron) with fuels (Diesel, Castor, Rapeseed and 15CA15R).

Wear can be defined as damaging, slow removal and distortion of material from surfaces. There are various causes of wear, but in the current case, due to sliding motion, fatigue, and creep, mechanical wear occurs between pin and disc. These types of study are also known as tribology as material wear has greater economic significance, as discussed by Jost Report [13]. Figure 4 shows that with diesel, the wear in aluminium, copper and iron pin is found to be approx. 150 µm, which is not even come closer to its blend 15CA15R.

Similarly, in almost all the cases, in the case of copper pins, after covering 2000 m of sliding distance, the wear has gradually reduced. An approx. of 290 µm of wear can be clearly seen with diesel fuel with iron; due to high CoF and FF, the fatigue, surface removal, or distortion occurred, which results in more wear in comparison to biodiesel and their blends. The absence of fatty acid composites which provide better lubricity against friction than hydrocarbons, due to the presence of polarity-imparting O radicals, free fatty acids, monoacylglycerols provides superior lubricity than pure esters and the presence of free OH groups, presence of double bonds and chain length are can be the probable reasons of the improved lubricating property of biodiesel in compare to diesel [14].

3.5. Microscopic Surface Analysis

Microscopic analysis can be defined as the in-depth study at the micro-level of the interface between two surfaces, from physics, chemistry and mechanical engineering, and material science. Figure 5 a detailed study of pin surface analysis, which shows the compound layer removal and deep grooved on the surface of aluminium and iron pin with diesel fuel. Secondly, with 15CA15R blend of Castor and Rapeseed, abrasive and plastic deformation can be seen at some point of pin surface. Due to the presence of free fatty acid and dissolved oxygen in biodiesel, the lubricity properties enhance, resulting in very fine surface removal and abrasion of metal from the surface. Compared to Castor and Rapeseed biodiesel, Castor biodiesel has better lubricity property, which is visible in figure 5. Due to the high presence of dissolved oxygen in Castor and Rapeseed biodiesel, the surface also gets some oxide film layer.

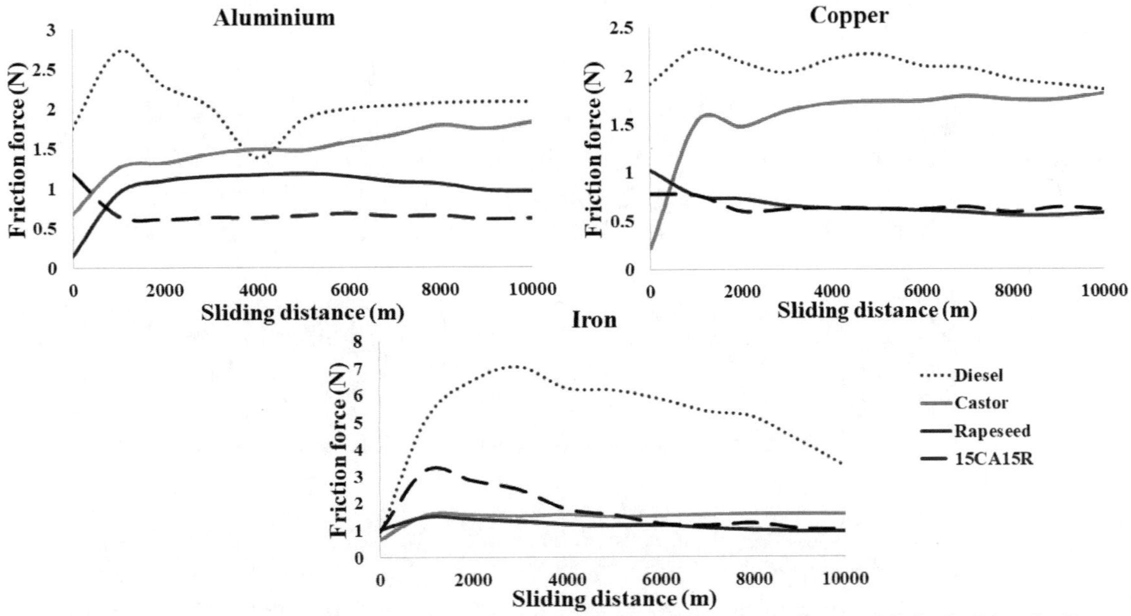

Figure 3. Friction force analysis of metal pins (Aluminium, Copper & Iron) with fuel (Diesel, Castor, Rapeseed and 15CA15R).

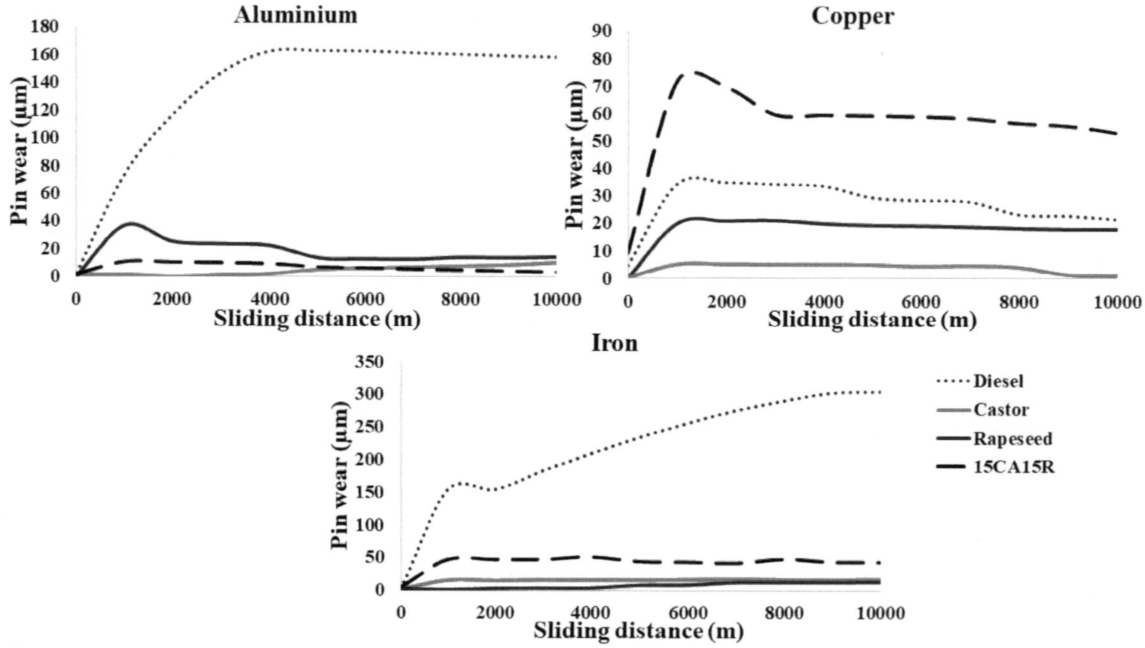

Figure 4. Wear analysis of metal pins (Aluminium, Copper & Iron) with fuel (Diesel, Castor, Rapeseed and 15CA15R).

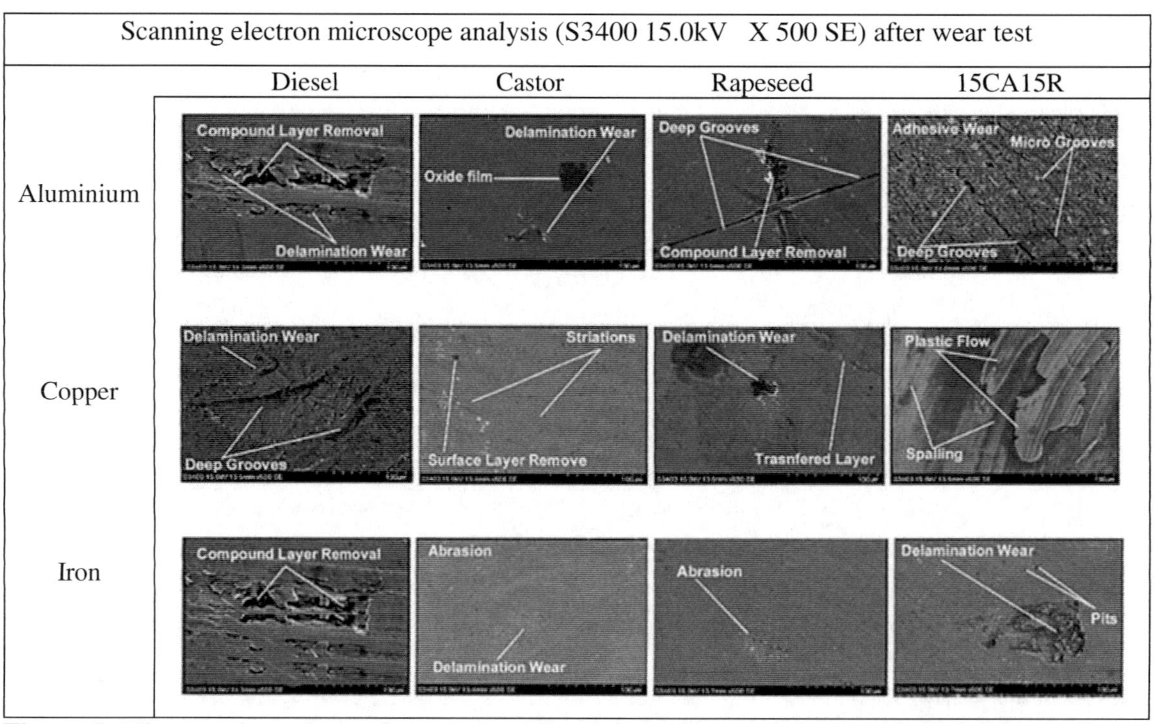

Figure 5. Microscopic analysis of metal pins (Aluminium, Copper & Iron) with fuel (Diesel, Castor, Rapeseed and 15CA15R).

4. Conclusion

The final consequence of the present experimental analysis has led to the following conclusion:

- The tribological behaviour includes friction force, coefficient of friction, wear and from the results, it has been proven that biodiesel is the best substitute for diesel.
- SEM results show lower surface wear and tear in biodiesel in comparison to diesel.
- 15CA15R has a nearly similar calorific value and better lubricating properties in comparison to pure diesel.
- B30 (15CA15R) blend can be considered as the best-compromised fuel for the existing diesel engines.

Acknowledgment

The authors would like to acknowledge the Nirma University financial support, Ahmedabad (Ph.D. Section), Fund by Grant No: NU/ Ph.D. /MRP/IT -ME /17-181 1269.

References

1. Odi-Owei S 1989 Tribological properties of some vegetable oils and fats. *Lubr. Eng.* **45** pp 685–90
2. Hadley J W, Owen G C and Mills B 2016 SAE TECHNICAL Evaluation of a high frequency reciprocating wear test for measuring diesel fuel lubricity 932692
3. Baskar S and Sriram G 2014 Tribological behavior of journal bearing material under different lubricants *Tribol. Ind.* **36** pp 127–33
4. Jain A K and Suhane A 2014 Investigation of Tribological characteristics of non edible castor and mahua oils as bio lubricant for maintenance applications *Proc. 5th Int. & 26th All India Manufacturing Technology Con. on Design and Research (IIT Guwahati)* (Assam: AIMTDR 2014) pp 1–6
5. Suhane A, Rehman A and Khaira H K 2013 Tribological investigation of mahua oil based lubricant for maintenance applications *Int. J. Eng.* **3** pp 2367–71
6. Shanta S M, Molina G J and Soloiu V 2011 Tribological effects of mineral-oil lubricant contamination with biofuels: A pin-on-disk tribometry and wear study *Adv. Tribol.* **2011** pp 1-7
7. Shahabuddin M, Masjuki H H and Kalam M A 2013 Experimental investigation into tribological characteristics of biolubricant formulated from Jatropha oil *Procedia Eng.* **56** pp 597–606
8. Stachowiak G W and Batchelor A W 2005 *Engineering Tribology*, 3rd edn (UK: Elsevier Butterworth-Heinemann) p 744
9. Eychenne V and Mouloungui Z 1998 Relationships between structure and lubricating properties of neopentylpolyol esters *Ind. Eng. Chem. Res.* **37** pp 4835–43
10. ASTM Standard 2006 G99, Standard test method for wear testing with a pin-on-disk apparatus *ASTM Int. West Conshohocken, PA* **v** pp 1–5
11. U.S. EPA 2010 Median life, Annual activity, and load factor values for nonroad engine emissions modeling *EPA-420-R-10-016. NR-005d. U.S. Environ. Prot. Agency* pp 1–47
12. Konishi T and Perez J M 1997 Properties of polyol esters—lubrication of an aluminum silicon alloy *Tribol. Trans.* **40** pp 500–6
13. Chattopadhyay R and Park M 2001 *Surface Wear: Analysis, Treatment, and Prevention*, ed C Terman and N Hrivnak (USA: ASM International)
14. Knothe G. S K R 2005 Lubricity of components of biodiesel and petrodiesel *Energy & Fuels* **19** pp 1192–200

Water desalination and wastewater reuse using integrated reverse osmosis and forward osmosis system

Dhaval Patel[1*], Anurag Mudgal[2*], Vivek Patel[3] and Jatin Patel[4]
[1]Junior Research fellow, India-H2O project, Department of Mechanical Engineering, School of Technology, Pandit Deendayal Energy University, Gandhinagar, India
[2,3,4] Department of Mechanical Engineering, School of Technology, Pandit Deendayal Energy University, Gandhinagar, India
Dhaval.20patel@gmail.com[1*], Anurag.mudgal@sot.pdpu.ac.in[2*]

Abstract. Rapid industrial growth and urbanization will be the major causes of environmental pollution, wastewater, and water quality deterioration. However, the technological advancement in water purification and the water treatment area will be the breakthrough. In recent ten years, membrane-based technology evolved for wastewater treatment, water desalination and power generation. Integration of technology like reverse osmosis(RO) and forward osmosis(FO) provides ambitious benefits like efficient recovery of osmotic energy from brine, improvement in the tendency of fouling and scaling and less quantity of chemical required for pre-treatment. This study reviews various technology that helps towards the wastewater treatment and water desalination with different integration and configuration. Majorly RO and FO integration gives promising outcome in both the situation and it could be price competitive with reverse osmosis systems when its recovery rate of the system is very high, the membrane flux and the membrane cost of the forward osmosis are like those of the reverse osmosis, and the electricity cost is expensive. The most important thing in commercializing the FO process is enhancing performance by improving flux and increasing recovery. An industrial-scale RO-FO hybrid system will be lucrative with membrane that provides greater flux, high selectivity, batter mechanical stability and robustness.

1. Introduction

In past decade, the water sacristy is becoming dangerous to some country and alarming notice to world. Rapid industrialisation and population growth are the major cause for the shortage of fresh water and became the environmental concern [1]. It also affecting the social and economic development. United Nations World Water Development Report (UN WWDR 2020) entitled 'Water and Climate Change' shows that in present scenario 4 billion people are directly or indirectly affected by the water scarcity and this number could be increase by 5.7 billion in 2050, which shows the significance of water as a natural and global resource [2]. Figure 1 show the water stress of different country in the world by the year of 2040.

Depletion of various fresh water source leads us towards the threatening level of scarcity, focus on various techniques that treat groundwater, wastewater and seawater leads to ways to meet global water demand. Absorption of such sources like brackish water and wastewater as well as seawater desalination gives abundant opportunity towards the water scarcity [3].

Membrane based technology is the promising future to tackle water issue [4]. From 1960, the reverse osmosis dominating technology in the field of purification and desalination, with highest percentage of market cover in recent years [5]. Around the globe, majority desalination and water purification plant are based on reverse osmosis technology because of simple nature of operation and cost effectiveness compare to other thermal based process. The global marketplace for desalination industries is based on reverse osmosis technology, which growing steadily and tends to reach its mature stage. Because of advancement in new technologies in terms of power consumption and membrane innovations, RO becomes the dominating technology in terms of water purification and water desalination [6]. Research and innovation in this type of technology mainly focuses on the reducing the energy consumption, use of renewable type of energy sources and advancement in the membrane. Researcher also focusing to address the various other types of problem like fouling and scaling effect on membrane and try to obtain membrane with suitable high flux [7].

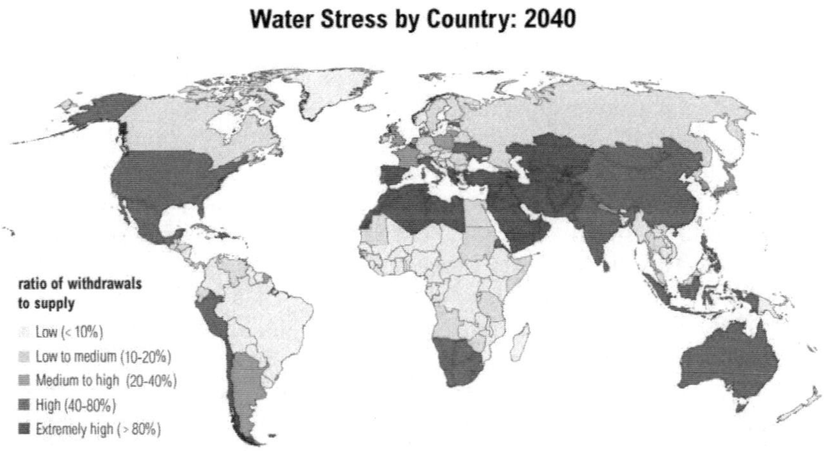

Figure 1. Projected water stress of different country by year 2040. [8]

Reliability of reverse osmosis for brackish/ground water and seawater is widely accepted and the technology is most suitable for the low cost operation of sweater desalination and water purification [9]. Feed water characteristics, membrane parameter and operating parameter are the parameter are the factors which affects the whole process efficiency of the reverse osmosis.

Forward osmosis is the process in which flow of two solution with different concentration are differentiated by semipermeable membrane, in which water particles cross from lesser concentration solution to higher concentration solution due to osmotic pressure dissimilarity. This process utilises very low amount of energy for only circulate the solution through the process. The aim of the forward osmosis process in integrated process is to weaken the seawater with incoming wastewater which reduce the pressure for next reverse osmosis process and makes the process energy efficient. Ultimately, the reutilisation of wastewater is advantageous for not only integrated process but also reduce the brine discharge also help to lower the final concentration of the brine.

Various literature has been caried out for FO-RO (Forward Osmosis – Reverse Osmosis) numerical models [10][11] and emphasis that implementation of forward osmosis may reduce the specific energy consumption (SEC) compared to traditional sweater reverse osmosis [12]. Integration of FO-RO can be advantageous in terms of energy and financial terms over the standalone systems.

2. Theoretical Background:

Osmosis is the phenomenon has been known to humankind since many years. It is a natural process, which occurs due to concentration difference (osmotic pressure difference), in which water molecules travel from lesser concentration solution to higher concentration solution through semipermeable membrane shown in figure 2A. Transfer of water molecules can be reversed or stopped using mechanically applied pressure or externally applied pressure. In such case like reversal flow due to externally applied pressure, the external applied pressure overcome the osmotic pressure and water particles are compelled to transfer in the opposite direction of the naturally occurred osmosis. This phenomenon is known as reverse osmosis which is seen in figure 2B. In osmotic equilibrium phenomenon, the water particles are passed through semipermeable membrane and reject the salt. The transfer of water molecules continues till the phase of osmotic equilibrium (figure 2C) where chemical potential becomes equal across the membrane.

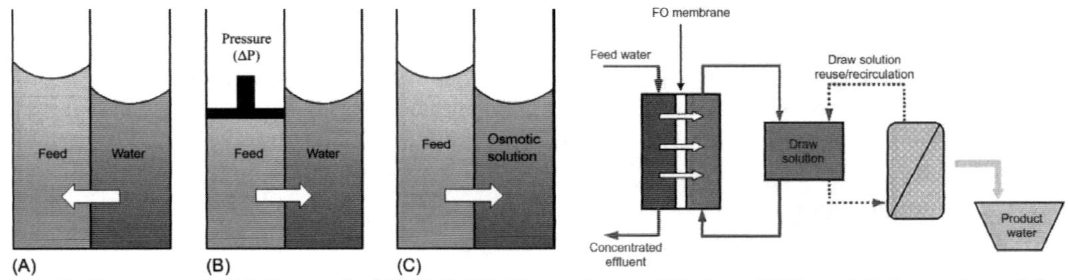

Figure 2. Process of (A) Osmosis (B) RO (C) Osmotic equilibrium [13] and Schematics of forward osmosis process [14].

Forward osmosis is the phenomenon based on the principle of the osmosis, in which two solution with different concentration is separated with selectively permeable membrane and passed side by side in parallel or in opposite direction. Where, membrane act as selective barrier which allows only water particles to passed and block the other large molecules. Here, on the completion of this process solution with lower concentration (feed solution) is become more concentrated and solution with higher concentration (draw solution) is become diluted. The difference between reverse and forward osmosis is the motive force. The forward osmosis process having numerous applications like wastewater treatment, in food processing industries, brackish/seawater desalination, in biotechnology, power generation using PAFO (pressure assisted forward osmosis) and also in medical applications. It is an alternative process for the treatment of wastewater and high quality water production [15]. The draw solution used in the forward osmosis process having distinct features, which can be reusable after extracting water from it by utilising draw solution regeneration process. Because of this reason forward osmosis is generally coupled with any draw solution regeneration techniques or post treatment step.

The use of draw solution is the distinctive feature of forward osmosis system. In forward osmosis system, the feed solution is becoming concentrated and draw solution become diluted, after that draw solution regeneration process is required to recover product water and/or reuse draw solution component. This is the reason why forward osmosis is generally combined with a post step to treat weaken draw solution. Draw solution recovery stage after FO becomes main power consuming system, which makes forward osmosis process less competitive [16]. Therefore, the implementation of forward osmosis technology in actual commercial level depends upon the efficiency of recovery system (draw solution recovery) and effectiveness of overall process.

3. Integrated RO-FO system

Hybridisation is the techniques where various combination is used in single treatment unit while integrated system is where assembly of different water and wastewater treatment process is in sequence. However, both the techniques have different capabilities. Following are some different purposes of hybrid/integrated water treatment system [17],

- Overall, it achieves better performance than any of the single components. Imperfection of any system can be minimised using integrated system.
- To get desired water quality (treated water quality).
- It overcome the trouble to extract water from nonconventional as well as polluted water source, and from untapped wastewater.
- To extend the life of system operation.
- It obeys with the rules and regulation.
- It makes sure that the efficiency of the treatment is good, and it removes various hazardous substance with good recovery of resource (like water as well as nutrients, energy etc.) in wastewater.
- To achieve zero liquid discharge with no liquid discharge to harm or pollute the environment.

3.1 Approach for sustainability and design of advanced integrated systems:

IOP Conf. Series: Materials Science and Engineering 1146 (2021) 012029 doi:10.1088/1757-899X/1146/1/012029

Implementation of integrated wastewater treatment and water desalination system has already in progress around the globe. Innovation in technology tied the research and industrial community together for further advancement in various design-based improvement and sustainable development. The latest is defined as the development that meets the needs of the present without compromising the ability of future generations to meet their own needs [18]. For the sustainable integrated wastewater treatment, following are serval approaches that can be focused,

- Benchmarking the technology for the perspective of sustainability (social, environmental, and economic) and global availability.
- Integration of these techniques with renewable energy sources like solar, wind etc. for better prospect of technology.
- Reduce, reuse, recover and conserve the resources that can be utilised in these techniques.
- Minimise the consumption of chemicals and reduce the secondary waste production.

The moto behind these types of sustainable hybrid/integrated wastewater treatment and water desalination technologies that to reduce/prevent the depletion of the resources, minimising the environmental impact and preserve the energy while achieving the target aims like provide sufficient and safe drinking water and retrieve and reuse water from the pollute source.

4. Integrated system for wastewater treatment:

Developing interest of drinking water and consciousness of the capability of wastewater (city wastewater) recuperation have draw out the advancement in wastewater treatment and asset recuperation methods. These arrangements can address the issue of water shortage and natural contamination. Membrane based technologies like reverse osmosis, ultra-filtration, nano filtration etc. can handled the most of dissolved impurities, organic chemical, constituents, natural organic matter and heavy metals. When wastewater is directly pass through the membrane, fouling is expected. For the irreversible type fouling in RO Membrane, Intensive chemical cleaning is required which lessen the lifespan of membrane and escalate the water treatment cost. As well as contrasting and customary water treatment, the activity at raised pressing factors likewise increments both capital and working expenses. Following are some challenges faced by the convention water treatment plants:

a) Emerging contaminants
b) Increasingly stringent environmental regulations
c) Pathogenic microorganisms
d) Disinfection by-products

Majority of integrated reverse osmosis – forward osmosis system for water reuse and wastewater treatment are generally used forward osmosis technology as a pre-treatment of a first stage of the integrated system. For example, seawater desalination generally used forward osmosis as a first stage of the integrated system. In which wastewater is taken as feed solution and oceanwater as a draw solution. Reverse osmosis utilised as posttreatment to extract water from draw solution and restore the draw solution. The utilization of wastewater as the feed solution for forward osmosis measure has given understanding into the idea of utilizing forward osmosis to dodge expensive treatment of effluents by regular treatment measures. Seawater as draw solution is one of the most cost-effective solution for integrated FO-RO system since higher osmotic pressure gives oceanwater advantage over the sewage water. The prime objective of forward osmosis in integrated system is to reduction of salinity or waste in stream.

Investigation on hybrid/integrated RO-FO system utilised for wastewater treatment also focuses on the secondary effluents. The schematic shows the integrated seawater desalination process, in which impaired quality water is applied as the feed solution in forward osmosis system while pre-treated seawater is utilised as draw solution in the system. The forward osmosis membrane with high rejection obstruct various suspended and dissolved solid [19]. Numerous examinations have announced that forward osmosis combined with reverse osmosis is found to function as a twofold boundary against chose demonstrating pollutants, boron, chemically dynamic mixes, chemicals, and other natural toxins [20]. Utilising municipal effluent (secondary effluent) as the feed of forward osmosis process and regular/common oceanwater as a draw solution will be resulting the rejection of majority micropollutant in forward osmosis system.

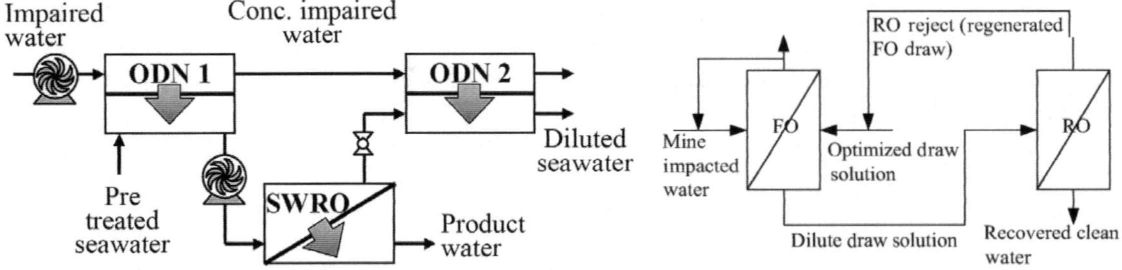

Figure 3. Schematic of integrated SWRO process [21] and FO-RO integration mode for matched volume. [22]

The system rejects majority of hydrophilic micropollutants about to 96 to 99 percentage. Hydrophobic unbiased and hydrophobic impartial mixes removal rate is low compared to the hydrophilic micropollutants [23]. Incorporation with low pressure reverse osmosis (LPRO), the FO-LPRO integrated system had compelling execution in dismissing low-sub-atomic weight micropollutants past 89.1% and other contaminants and mixes can be removed up to 99%. System shown in figure 3 is proposed and implemented by Cath et al. to dilute seawater using two step forward osmosis system with the integration of reverse osmosis system [24]. Which demonstrate the higher removal of natural synthetic compound. Other benefits of this system are the lower specific energy consumption and good recovery with less amount of fouling propensity.

Hybrid RO-FO system utilised for various reason in wastewater treatment and water desalination but majorly it utilised for the reduction of sweater osmotic pressure, so it can reduce the power consumption in next step which is reverse osmosis and other reason is the volume reduction via pre-treatment. The upside of decreasing quantity of wastewater can bring down the expense of energy utilization, transportation, and substance used for the treatment. Reutilisation of the water can increase the efficiency of the process [25].

5. Integrated RO-FO systems for water desalination

Seawater/Oceanwater desalination is an encouraging path for addressing the world's water shortage issue. Oceanwater desalination by reverse osmosis measure has been the biggest portion of worldwide desalination limit. Energy consumption in reverse osmosis is a burning issue because of it is operated on pressure driven mechanism [13]. Because of energy intensive nature of sea water reverse osmosis (SWRO), may be the problem turns form water scarcity to energy problem. Energy productivity of the system can address the issue and effort are made towards this direction to improve the desalination process.

Ge et al. established Forward osmosis combined with membrane distillation system which is polyelectrolyte based for the reuse of wastewater and recycle of acid dye. For this type of experimentation high performance polyvinylidene fluoride membrane were utilised to reuse of the draw solute and treat the wastewater [26].The study reveals that, systems still lacks in high performance FO membrane with high flux and high salt rejection and draw solution with high osmotic pressure, which is cost effective and have minimum reverse fluxes [27].

Forward osmosis is process which consume low energy utilising draw solution with sufficiently high osmotic pressure [28]. This is the motivation behind, why forward osmosis is utilized as energy saving cycle option in contrast to other regular reverse osmosis oceanwater desalination.

Various literature shows that the sweater desalination can be done through novel concept of combined forward osmosis and reverse osmosis system, where oceanwater is used as the feed solution and more concentrate non-volatile salt like NaCl, $MgCl_2$, etc. used as draw solution [29]. The water from the feed side is move alongside the membrane and dilute the draw solution, due to this process majority of the contaminants and salt are removed in forward osmosis process. The clean water is extracted form the draw solution in next step in reverse osmosis. This investigation carried out form bench scale to pilot scale test. Flux lost by fouling(irreversible) can be re-established by occasional actual cleaning and additionally osmotic discharging in forward osmosis [30].

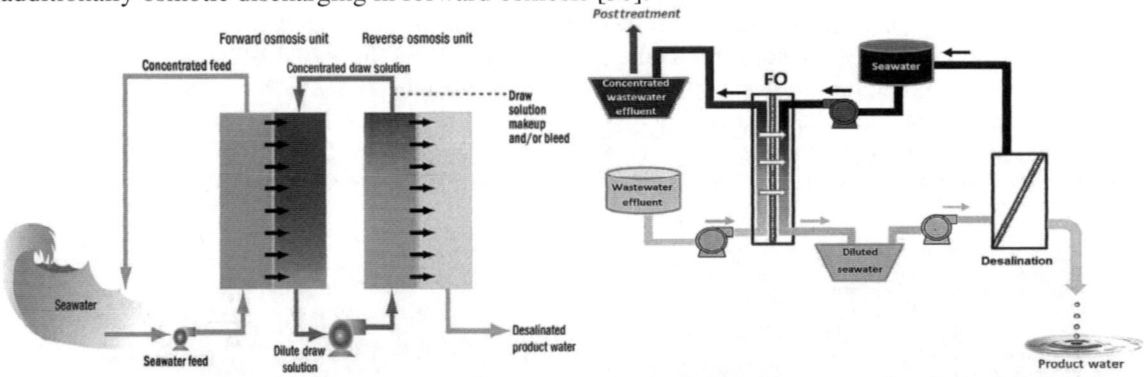

Figure 4. An integrated Forward osmosis –Reverse Osmosis hybrid systems for direct desalination [29] and Combined reverse osmosis forward osmosis system for indirect water desalination. [31]

Less energy requirement for pre-treatment, longer lifespan of the membrane and reduction in fouling are the major advantages of the periodic cleaning of the system. Theoretical analysis of the integrated forward osmosis – reverse osmosis suggests that high recovery up to 80% can be achieved in integrated FO-SWRO process [28]. A case study of Al Najdah of Oman suggest that combined FO-RO system provide numerous benefits like reduction in OPEX, reductio in ozone harming substance impression and lower consumption of the chemicals compared to conventional SWRO. It is challenge to extract water form draw solution using less energy irrespective to draw used, till the use of low cost energy using renewable energy source or utilisation of waste energy [4]. The integrated system may show the advantage of cost-saving by decreasing labour and upkeep interest for control of fouling and by changing membrane compared to conventional seawater RO system.

Challenges related to reverse draw solute can be overcome by improvement in design of membrane and proper draw solution selection. This is the process with two step including reverse osmosis and forward osmosis for indirect desalination shown in figure 4 [30]. This process used low salinity impaired water as a feed solution which is used to dilute the draws solution (seawater). Finally, this draw solution is supplied to posttreatment reverse osmosis process to extract the fresh water utilising low amount of energy input. The output of this forward osmosis process can be also treated by LPRO (low pressure reverse osmosis) with combination of forward osmosis system with LPRO system. Weakening of seawater by FO with free environmentally friendly energy, less working expenses, different boundaries for water sanitization, limited utilization of synthetic compounds, and decrease of reverse osmosis fouling and scaling is observed in many of investigation [32]. Indirect desalination via integrated reverse osmosis – forward osmosis system is the sore subject of investigation in research community and englaring this technology from lab scale to the large/industrial level sacle is the one of the main challenge.This is the reason to elobrate and provide good unterstanding towrds the process, economic feasibility and environmental impact of the system.

6. Batch Reverse Osmosis System:

Batch reverse osmosis process is a cyclic process which uses the minimum average feed pressure for the purpose of desalination, which minimise the specific energy consumption. The batch reverse osmosis process uses the variation of pressure along the cycle to get higher recovery ratio. In standard continuous RO process the pressure is remains constant during whole cycle.

A proposed batch reverse osmosis configuration by Sandra P. Córdoba et al. is shown in figure 5. The process occurs in following manner. At the initial stage of process, high-pressure pumps feed water to first side (side 1) of the piston in high pressurised tank. This is the process to feed the water in the system. After that, the high-pressure pump uses to feed the water in second side (side 2) and this water is used to displace the water on first side with piston arrangement which is directed to the RO modules with high pressure. Pure water from RO module is collected at particular location and brine/RO reject will be recirculated to the first side of the high-pressurised tank using recirculating pump.

The instant at which the piston arrives at the other end of the high-pressurised tank is the sign to complete stage 1 and a purge stage is started. A restriction shuts the recirculation loop, the recirculation pump discharges the salt water in the pipelines and replaces it using feedwater. Then, the high-pressure pump will be started to dispatch the feed water to the reverse osmosis modules and generate clean water in batch mode. At that point, the module purges the salt water/brine into the feed flow loop at high pressure on the other side of piston, where restriction again decrease the pressure to remove the salt water and reintroduce the feed. Later this process the batch RO system is again prepared to produce clean water. The figure 5 shows the different steps for feed, purge and refill the improved batch RO system.

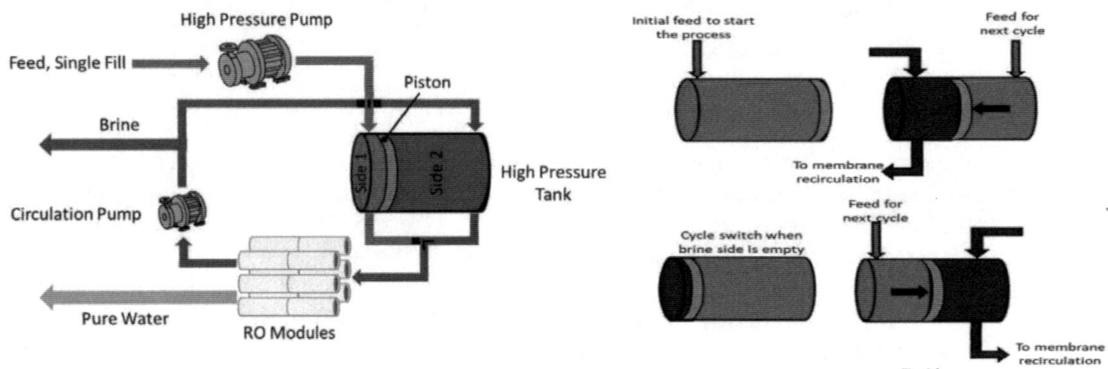

Figure 5. Batch reverse osmosis configuration and Steps in the batch RO configuration [33]

Integration of batch reverse osmosis (BRO) with various other technologies like forward osmosis, electrodialysis, electro coagulation etc. will give numerous amounts of benefit compare to stand alone system. This study focuses on the various application of integrated FO-RO system and lead this study towards the new approach of integrated batch reverse osmosis with other technology.

6..1 Comparison between batch reverse osmosis and convention continuous reverse osmosis

Tianyu Qiu et al. carried out study that compares the various type of reverse osmosis configuration [34]. The carried out this study for single stage, multistage, close circuit, and batch mode reverse osmosis. Figure 6 shows the variation of normalised energy consumption with recovery ratio for all above mentioned mode. Particularly, In Batch RO the specific energy consumption has lowest, which is reduced by 30% compared to close circuit desalination RO at 80 % recovery ratio. Ultimately, batch mode reverse osmosis operation gives higher recovery with less specific energy consumption and at less cost compared to conventional desalination system. During batch or close circuit operation (cyclic operation) pressurisation and depressurisation of fluid occurs on the outer area of the membrane, which affect the stability of the membrane. Fouling and scaling problem in batch reverse osmosis is less due to purging of brine is done using freshwater introduction and these becomes advantageous for the system.

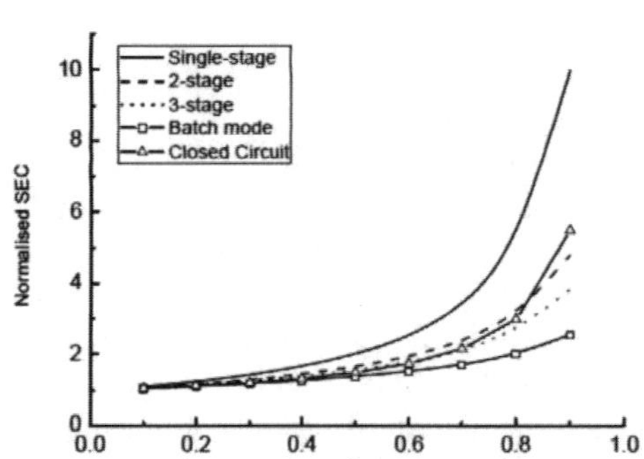

Figure 6. Normalised SEC vs Recovery ratio. [34]

Apart from energy consumption there are several other advantages that benefits the batch reverse osmosis configuration, which are listed below:

- System exhibits maximum recovery batch by batch with simple controls.
- Permeate water quality can be control by stopping the process and improved by total/partial second pass treatment.
- Batch reverse osmosis is flexible system when it changes the feed water quality.
- Cleaning is simple and easily implemented in this type of system.

This technology can potentially used in integration with forward osmosis system will provide benefits in terms of fouling and energy consumption.

7. Challenges and Future trend:

The integrated/hybrid system for water, wastewater treatment and water desalination has helps towards the various challenges related to clean water production, reject management, energy consumption issues as well as environmental pollution related issue. Although, apart from it there are various challenges in front of this technologies like technological maturity in term of membrane development, commercialisation of this technology, operation, and handling of these plants at large scale etc. Another constrain is that some of the integrated system requires pre- and post-treatment to eliminate some of the impurities and achieve zero liquid discharge which is advantage to system as well as the environment, but these technologies are more energy intensive and increase the capital and operation cost of the system. Various literature reported to recover the resources utilising during the process, but few of them were discussing the reutilisation of the resources and making the process more efficient.

Energy consumption is another issue for the integrated/hybrid technologies. As earlier mentioned, that some of the technologies are very energy intensive which increase the overall cost of the system. Integration of renewable energy source with the system leads to the clean energy and lower operating cost. These membrane-based technology needs frequent cleaning to eliminate impurities concentrated on the membrane surface. Generated effluent from this type of processes should be taken care of appropriately before released in to open climate or in channel. Indirectly, this waste management add up the operating cost. This is the issue that can be needs to address with better cleaning techniques or low-cost waste management techniques.

8. Conclusion

Reverse osmosis is most mature and established technology in wastewater treatment and seawater desalination. As well forward osmosis with no applied pressure and with less membrane fouling will be the favourable option for the water treatment processes. The advancement in wastewater treatment and water desalination technologies enables the adaptable combination of different treatment process to meet achievable target like reutilisation of the recovered resource and optimisation of process efficiency with less environmental impact. The integration of pressure driven, and osmotic driven membrane process (like integrated RO-FO) gives benefits of both system with synergetic effect, which makes this type of integration as a system with great potential. Like total energy consumption is reduced by 15% in integrated FO-RO system compare to seawater reverse osmosis system.

The limitation for this type of system is the availability of the membrane for the forward osmosis system, unavailability of ideal draw solution, performance of the system is bounded by the concentration polarisation, draw solute diffusion and membrane fouling. This will limit the economic feasibility of the system. By overcoming these challenges, potentially system can be achieved with great efficiency with minimum losses. Though, in depth studies on sustainable development of this type of technologies has been carried out, but sustainability of hybrid/integrated system have not been widely reported yet. Need to develop more understanding towards the filling the knowledge gap between sustainable development, system design and economic feasibility. The findings from this study will be utilised for the development and experimental investigation of integrated batch reverse osmosis – forward osmosis system.

Acknowledgments
This work has received funding from the Department of Biotechnology (Government of India, Project No. BT/IN/EU-WR/40/AM/2018) and the European Union's Horizon 2020 research and innovation programme under grant agreement No. 820906. We are also thankful to Pandit Deendayal Petroleum University, Gandhinagar to provide necessary facility and infrastructure for this research.

References:

[1] Wang J, Liu X. Forward osmosis technology for water treatment: Recent advances and future perspectives. J Clean Prod [Internet]. 2021;280:124354. Available from: https://doi.org/10.1016/j.jclepro.2020.124354
[2] WWAP (UNESCO World Water Assessment Programme), 2019, United Nations World Water Development Report 2020: Water and Climate Change. 2020. 12 p.
[3] Alvez A, Aitken D, Rivera D, Vergara M, McIntyre N, Concha F. At the crossroads: can desalination be a suitable public policy solution to address water scarcity in Chile's mining zones? J Environ Manage [Internet]. 2020;258(January):110039. Available from: https://doi.org/10.1016/j.jenvman.2019.110039
[4] Elimelech M, Phillip WA. The future of seawater desalination: Energy, technology, and the environment. Science (80-). 2011;333(6043):712–7.
[5] Rabiee H, Khalilpour KR, Betts JM, Tapper N. Energy-water nexus: Renewable-integrated hybridized desalination systems [Internet]. Polygeneration with Polystorage: For Chemical and Energy Hubs. Elsevier Inc.; 2018. 409–458 p. Available from: http://dx.doi.org/10.1016/B978-0-12-813306-4.00013-6
[6] Shenvi SS, Isloor AM, Ismail AF. A review on RO membrane technology: Developments and challenges. Desalination [Internet]. 2015;368:10–26. Available from: http://dx.doi.org/10.1016/j.desal.2014.12.042
[7] Malaeb L, Ayoub GM. Reverse osmosis technology for water treatment: State of the art review. Desalination [Internet]. 2011;267(1):1–8. Available from: http://dx.doi.org/10.1016/j.desal.2010.09.001
[8] Ahmed FE, Hashaikeh R, Hilal N. Solar powered desalination – Technology, energy and future outlook. Desalination [Internet]. 2019;453(November 2018):54–76. Available from: https://doi.org/10.1016/j.desal.2018.12.002
[9] Park PK, Lee S, Cho JS, Kim JH. Full-scale simulation of seawater reverse osmosis desalination processes for boron removal: Effect of membrane fouling. Water Res [Internet]. 2012;46(12):3796–804. Available from: http://dx.doi.org/10.1016/j.watres.2012.04.021
[10] Altaee A, Millar GJ, Zaragoza G, Sharif A. Energy efficiency of RO and FO–RO system for high-salinity seawater treatment. Clean Technol Environ Policy. 2017;19(1):77–91.
[11] Zaviska F, Zou L. Using modelling approach to validate a bench scale forward osmosis pre-treatment process for desalination. Desalination [Internet]. 2014;350:1–13. Available from: http://dx.doi.org/10.1016/j.desal.2014.07.005
[12] Blandin G, Verliefde ARD, Tang CY, Le-Clech P. Opportunities to reach economic sustainability in forward osmosis-reverse osmosis hybrids for seawater desalination. Desalination [Internet]. 2015;363:26–36. Available from: http://dx.doi.org/10.1016/j.desal.2014.12.011
[13] Rastogi NK. Opportunities and challenges in application of forward osmosis in food processing. Crit Rev Food Sci Nutr. 2014;(December):37–41.
[14] Li S, Li Z. Reverse osmosis and forward osmosis in integrated systems [Internet]. Current Trends and Future Developments on (Bio-) Membranes. Elsevier Inc.; 2020. 261–280 p. Available from: http://dx.doi.org/10.1016/B978-0-12-816777-9.00011-3
[15] Lutchmiah K, Verliefde ARD, Roest K, Rietveld LC, Cornelissen ER. Forward osmosis for application in wastewater treatment: A review. Water Res [Internet]. 2014;58(0):179–97. Available from: http://dx.doi.org/10.1016/j.watres.2014.03.045
[16] Linares RV, Li Z, Sarp S, Bucs SS, Amy G. ScienceDirect Forward osmosis niches in seawater desalination and wastewater reuse. 2014;6(4).

[17] Ang WL, Mohammad AW. Integrated and hybrid process technology [Internet]. Sustainable Water and Wastewater Processing. Elsevier Inc.; 2019. 279–328 p. Available from: http://dx.doi.org/10.1016/B978-0-12-816170-8.00009-0
[18] Keeble BR. The Brundtland Report: "Our Common Future." Med War. 1988;4(1):17–25.
[19] Bamaga OA, Yokochi A, Zabara B, Babaqi AS. Hybrid FO/RO desalination system: Preliminary assessment of osmotic energy recovery and designs of new FO membrane module configurations. Desalination [Internet]. 2011;268(1–3):163–9. Available from: http://dx.doi.org/10.1016/j.desal.2010.10.013
[20] Coday BD, Yaffe BGM, Xu P, Cath TY. Rejection of trace organic compounds by forward osmosis membranes: A literature review. Environ Sci Technol. 2014;48(7):3612–24.
[21] Cath TY, Hancock NT, Lundin CD, Hoppe-jones C, Drewes JE. A multi-barrier osmotic dilution process for simultaneous desalination and purification of impaired water. J Memb Sci [Internet]. 2010;362(1–2):417–26. Available from: http://dx.doi.org/10.1016/j.memsci.2010.06.056
[22] Thiruvenkatachari R, Francis M, Cunnington M, Su S. Application of integrated forward and reverse osmosis for coal mine wastewater desalination. Sep Purif Technol [Internet]. 2016;163:181–8. Available from: http://dx.doi.org/10.1016/j.seppur.2016.02.034
[23] Valladares Linares R, Yangali-Quintanilla V, Li Z, Amy G. Rejection of micropollutants by clean and fouled forward osmosis membrane. Water Res [Internet]. 2011;45(20):6737–44. Available from: http://dx.doi.org/10.1016/j.watres.2011.10.037
[24] Taylor P, Choi J, Kim H, Lee S, Hwang T, Oh H, et al. Desalination and Water Treatment Theoretical investigation of hybrid desalination system combining reverse osmosis and forward osmosis. 2012;(December 2014):37–41.
[25] R. Valladares Linares a, *, Z. Li a, V. Yangali-Quintanilla b, N. Ghaffour a GA a, T. Leiknes a JSV. Life cycle cost of a hybrid forward osmosis e low pressure reverse osmosis system for seawater desalination and wastewater recovery. Water Res. 2016;88:225–34.
[26] Ge Q, Wang P, Wan C, Chung TS. Polyelectrolyte-promoted Forward Osmosis-Membrane Distillation (FO-MD) hybrid process for dye wastewater treatment. Environ Sci Technol. 2012;46(11):6236–43.
[27] Chung TS, Li X, Ong RC, Ge Q, Wang H, Han G. Emerging forward osmosis (FO) technologies and challenges ahead for clean water and clean energy applications. Curr Opin Chem Eng [Internet]. 2012;1(3):246–57. Available from: http://dx.doi.org/10.1016/j.coche.2012.07.004
[28] Chung TS, Zhang S, Wang KY, Su J, Ling MM. Forward osmosis processes: Yesterday, today and tomorrow. Desalination [Internet]. 2012;287:78–81. Available from: http://dx.doi.org/10.1016/j.desal.2010.12.019
[29] Shaffer DL, Yin N, Gilron J, Elimelech M. Seawater desalination for agriculture by integrated forward and reverse osmosis : Improved product water quality for potentially less energy. J Memb Sci [Internet]. 2012;415–416:1–8. Available from: http://dx.doi.org/10.1016/j.memsci.2012.05.016
[30] Bennett A. Desalination and water reuse: What's the future for forward osmosis? Filtr Sep [Internet]. 2013;50(5):28–30. Available from: http://dx.doi.org/10.1016/S0015-1882(13)70202-7
[31] Li Z, Siddiqi A, Anadon LD, Narayanamurti V. Towards sustainability in water-energy nexus: Ocean energy for seawater desalination. Renew Sustain Energy Rev [Internet]. 2018;82(November 2017):3833–47. Available from: http://dx.doi.org/10.1016/j.rser.2017.10.087
[32] Bitaw TN, Park K, Yang DR. Optimization on a new hybrid Forward osmosis-Electrodialysis-Reverse osmosis seawater desalination process. Desalination [Internet]. 2016;398:265–81. Available from: http://dx.doi.org/10.1016/j.desal.2016.07.032
[33] Sandra P. Córdoba, Abhimanyu Das DMW. IMPROVED BATCH REVERSE OSMOSIS CONFIGURATION FOR BETTER ENERGY EFFIENCY. IDA 2019 World Congr World Congr Water Reuse Desalination, Dubai, UAE, Oct 20-24, 2019. 2019;
[34] Qiu T, Davies PA. Comparison of configurations for high-recovery inland desalination systems. Water (Switzerland). 2012;4(3):690–706.

AUTHOR INDEX

Institute of Physics
Dirac House, Temple Back
Bristol BS1 6BE UK

ISSN: 1757-8981
ISBN 978-1-7138-3416-8

9 781713 834168

CURRAN ASSOCIATES INC.
proceedings.com